Elements of Murder

Elements
of Murder

JOHN EMSLEY

OXFORD
UNIVERSITY PRESS

OXFORD

UNIVERSITY PRESS

Great Clarendon Street, Oxford OX2 6DP

Oxford University Press is a department of the University of Oxford.
It furthers the University's objective of excellence in research, scholarship,
and education by publishing worldwide in

Oxford New York

Auckland Cape Town Dar es Salaam Hong Kong Karachi
Kuala Lumpur Madrid Melbourne Mexico City Nairobi
New Delhi Shanghai Taipei Toronto

With offices in

Argentina Austria Brazil Chile Czech Republic France Greece
Guatemala Hungary Italy Japan South Korea Poland Portugal
Singapore Switzerland Thailand Turkey Ukraine Vietnam

Oxford is a registered trade mark of Oxford University Press
in the UK and in certain other countries

Published in the United States
by Oxford University Press Inc., New York

British Library Cataloguing in Publication Data

Data available

Library of Congress Cataloging in Publication Data

Data available

ISBN 0–19–280599–1

1

Typeset by RefineCatch Limited, Bungay, Suffolk
Printed in Great Britain by
Clays Ltd., St Ives plc

Contents

—◆+◆—

Acknowledgements

—◄◆►—

It is with heartfelt thanks that I acknowledge the support of the following friends and acquaintances for their help in writing this book. Some provided me with information I would not otherwise have had, some were prevailed upon to check that the contents of chapters were scientifically correct, and some were even prevailed upon to read the completed manuscript. In alphabetical order they are as follows:

Dr John Ashby, of Leek, Staffordshire, who works for the Central Toxicological Laboratories, Cheshire, checked the arsenic chapters and provided extra information about this element.

Dr Alan Bailey, of the Analytical Services Centre of the Forensic Science Service, London, veted the chapters about thallium and checked the glossary.

Thomas Bittinger, of the Marketing Division of Reckitt Benckiser, translated the paper concerning the poisoning of Pope Clement II.

Paul Board, of Fugro Robertson Ltd, Llandudno, provided material relating to the death of Mozart, and the murderess Zoora Shah, plus several other items. He also read the completed manuscript.

David Dickson, director of the Science and Development Network *www.scidev.net* brought to my notice the research that is being done to find an answer to the arsenic contamination of water supplies in Bangladesh and Bengal.

My wife, **Joan Emsley**, read the complete text and helped me clarify subjects where it appeared I was assuming the average reader would have a degree in chemistry.

Raymond Holland, of Bristol and Chairman of the Bristol & South West Section of the Society of Chemical Industry, reviewed the chapters on mercury and provided data on the use of mercury(II) chloride in timber preservation.

Steve Humphrey, of the Toxicology Department of the Forensic Science Service, London, read the chapters on arsenic.

Dr Michael Krachler, of the University of Heidelberg, Germany, gave up-to-date advice on antimony analysis and checked the chapters on this element.

Professor Steve Ley and **Rose Ley**, of the Department of Chemistry, University of Cambridge, read the complete text and made invaluable suggestions for improving it.

Sylvia Countess of Limerick CBE gave me a copy of the *Expert Group to Investigate Cot Death Theories: Toxic Gas Hypothesis*, which she chaired in 1988, and she also read the antimony sections of this book.

C. Harrison Townsend, of Vancouver, Canada, told me about the fur-trappers who were poisoned to death by snow.

Professor William Shotyk, of Heidelberg University, provided information about antimony, and also checked the chapters on that element.

Dr Michael Utidjian, of Wayne, New Jersey, provided lots of interesting items about arsenic and mercury.

Dr Trevor Watts, Head of Department of King's College Dental School, checked the section on dental amalgams.

Introduction

An earlier book of mine *The Shocking History of Phosphorus* (published as *The Thirteenth Element* in the USA) dealt with the way that dangerous element had impinged on human beings over the centuries. There were chapters on its environmental impact, together with those on its uses, misuses, and abuses in everyday life – not least by murderers. These chapters provided most of the human interest stories and it was this that led me to think about other dark elements and *Elements of Murder* is the result. Just as with phosphorus, the story of the dangerously toxic elements began in the days of alchemy when for hundreds of years they were used in vain attempts to discover a way of creating unlimited wealth via the Philosopher's Stone, or securing health and longevity via the Elixir of Life. Of course both searches failed but along the way these alchemical experiments poisoned famous scientists and even killed a king, as we shall discover.

There are currently 116 chemical elements in the periodic table. Thirty of them are unstable and dangerously radioactive and are rarely encountered outside nuclear facilities or research laboratories. Thankfully most of the remainder are harmless, but some are moderately toxic and a few are highly toxic. There are about 80 elements that comprise the Earth's crust and each of us has detectable traces of all of them in our body including gold, platinum, and even uranium. We also have measurable amounts of the poisonous elements such as arsenic, mercury, and lead and these are the ones which most of this book is about. Before you begin your journey into the darker side of the periodic table it may be helpful to know a little about the chemical composition of the human body. This requires 25 chemical elements for its growth and maintenance, and these are called the *essential* elements. They are listed in the Appendix on page 386.

As you might expect, the toxic elements antimony, lead, mercury, and thallium are not essential, although arsenic might be – the jury is still out on that one – but there are elements that are both essential and highly toxic, such as fluorine, selenium, and chromium. Even elements such as sodium and potassium can be deadly under certain circumstances. Elements like these will be dealt with in the final chapter of the book.

Murder by poison may be a dying art, thanks in no small part to advances in forensic analysis that make it almost certain that a toxic agent will be identified if poisoning is suspected as the cause of death. In *Elements of Murder* we will see that all kinds of food, drink, and medicines were employed to create a fatal brew, and in one of the murders described the poison was even administered as an enema. What is fascinating about the classical poisoning cases is that we are now able to reassess them in a way that enables us to understand them in ways that earlier generations could not. In former times it was always difficult to prove that someone had been murdered by poison and good legal counsel could play upon the lack of scientific knowledge to ensure that a murderous client walked free.

Things to bear in mind

Elements of Murder is a popular science book and as such will be using terms that you may not be familiar with; these are to do with names, money, and units of measurement.

Names: The histories of some of the elements will take us back to the times when the names used to describe chemicals were very different to those of today. In the Glossary there are tables giving these historical and medical names, the correct chemical name, and the chemical formulae.

Money: I have endeavoured to relate currencies of the past to those of today but such conversions can never be exact even in the case of the pound sterling which has been around for more than a thousand years. (A pound was equal to 20 shillings, and a shilling was made up of 12 pence.) A pound in Saxon times [1000s] was wealth indeed; a pound in Elizabethan times [1600s] was still many times the average weekly wage; a pound in Victorian times [1900s] was what an ordinary man could earn in a week and support a family on; a pound today will not even buy a Sunday newspaper. By the time you are reading this, the pound sterling may have disappeared into history and the euro taken its place. Throughout the book I have tried to give some guidance as to the current value of the amounts of money that are being quoted in terms of pounds sterling or US dollars. If I have been somewhat inconsistent then the fault is mine.

Units of measurement: Very little of a poison may be present in a victim's body after death and small units are needed to discuss the tiny amounts detected by forensic analysis, such as milligrams (mg), which are a thousandth part of a gram, and micrograms (μg), which are a millionth part of a gram. In Imperial units these correspond roughly to a 28,000th part of an ounce and 28 millionth part of an ounce respectively. Earlier generations who used pounds and ounces, had the *grain* as their smallest measure but even this is relatively large in modern terms being around 65 mg. Alternatively, in discussing the amount of a toxin which is present in a sample that has been analysed, it is more informative to talk in terms of parts per million (ppm) which is equivalent to a milligram of a substance in a kilogram (or a litre), and parts per billion (ppb) which is equivalent to a microgram in a kilogram (or a litre).

You are about to enter a world that was once a closed book to the human mind. Today we can unravel the mysteries that early generations struggled to understand, and appreciate all that has been done to remove toxic elements from our lives. While we have made the world a safer place, we can still learn from tales of the days when chemical elements poisoned millions – and sent a few inconvenient individuals to an early grave.

The poisonous
elements of alchemy

THE South Sea Bubble of 1720 saw prices on the London stock market rise to unsustainable heights and new companies were launched to take advantage of the public's willingness to invest. As in the dot.com bubble of the late 1990s, many of the companies were little more than hope and hype and among them was one for 'transmuting of quicksilver into a malleable fine metal'. Back in the early 1700s the idea that it was possible to convert mercury into gold was still widely accepted, even by eminent scientists such as Isaac Newton. He had spent much of his earlier life carrying out alchemical experiments, as we shall see. Nor was he alone. Indeed alchemy was actively encouraged by dukes, emperors, monarchs, and popes.

The company that sought to make gold was banned, along with a hundred others, in July 1720 as the Government tried to control the South Sea Bubble (which finally burst in September of that year). Just how insane things had become was demonstrated by another company whose prospectus declared that it was 'for carrying on an undertaking of great advantage, but nobody to know what it is'. The fraudster who set that one up didn't issue shares, only the right to buy its shares at some future date, and these rights he sold at £2 per right per share. He took £2,000 within 5 hours of opening an office in Cornhill (now equivalent to something like £1 million) and at the close of the day he simply pocketed the money and disappeared.

That was sometimes the way that alchemists behaved once they had found a gullible patron, but not all alchemists were con men. Some of them genuinely tried to make gold, and such alchemists were driven towards three, ultimately unattainable, goals: the Philosopher's Stone that could turn base metals into gold; the Elixir of Life that would confer longevity on those who drank it; and the Alkahest, the universal solvent that could dissolve anything and everything. Because these aims were incapable of ever being realized it is little wonder that most of what they did was scientifically meaningless. Nevertheless, down the centuries the alchemists developed basic types of laboratory apparatus and discovered a few important chemical compounds.

The risks to their health were great because their research invariably involved particularly poisonous of elements, especially mercury. They were very fond of this metal because they believed that all other metals, including gold, were composed of mercury, sulfur, and salt, with mercury being the most important. Much of this chapter will be concerned with the effects that this highly poisonous metal had on some of them.

So when did alchemy begin? Who were the alchemists? And did they really poison themselves?

The alchemists

Alchemy flourished in China, India, the Middle East, and Europe, wherever gold was valued and the desire for more was ever present. In the West, alchemy can trace its roots to ancient Egypt where one of the earliest identifiable alchemists lived. He was Democritus and he dwelt in the Nile Delta about 200 BC. He wrote *Physica et Mystica* [*Natural and Mystical Things*], which included not only useful recipes for dyes and pigments, but also some for making gold, although his instructions were written in obscure language making them difficult to understand. This might have been done because they were really recipes for making *fake* gold.

A later Egyptian alchemist was Zosimos who lived around 300 AD, and he described such chemical processes as distillation and sublimation, crediting an earlier alchemist, Maria the Jewess, as the inventor of these. She too had lived in Egypt about 100 AD and she had experimented with mercury and sulfur, although her best known invention was the *bain-Marie* which is still used in cooking when gentle heating is required. Zosimos also left obscure writings on how to turn base metals into gold, and he wrote of 'the tincture' and 'the powder' which later generations of alchemists took to be the Elixir of Life and the Philosopher's Stone respectively. Another alchemist who lived about this time was Agathodiamon who wrote of a mineral that, when fused with natron (naturally occurring sodium carbonate), gave a product that was a 'fiery poison' and which dissolved in water to give a clear solution. It seems certain that he had made arsenic trioxide and that the mineral he had used was either realgar or orpiment which are arsenic ores. Of this we can be certain because when he put a piece of copper into the solution it turned a beautiful green colour, which is what would happen if copper arsenite were formed. This pigment was to turn up again 1,500 years later, and to lead to massive contamination of the domestic environment and to many deaths – as we shall see.

By the time of Zosimos, alchemy was starting to decline, along with the Roman Empire, but a lot of its writings were saved by a sect of dissident Christians, the Nestorians, who fled to Persia around 400 AD. This information eventually passed to the Arabs, in whose hands alchemy again flourished and the word alchemy comes from Arabic. Early Muslim rulers encouraged all branches of learning and as their empire expanded into Spain around 700 AD so it brought the new alchemy to the attention of those in Western Europe. The two great Arab alchemists were Abu Musa Jabir ibn Hayyan (721–815 AD), known in Europe as Geber, and Abu Bakr Muhammad ibn Zakariya Ar-Razi (865–925 AD), known in Europe as Rhazes. Their writings were translated into Latin and became widely known throughout Europe, and influenced all who followed.

More than 2,000 works are attributed to Geber. He said that everything was composed of the four elements, fire, earth, water, and air, and that these combined to form mercury and sulfur, from which all metals were made, varying only in the proportions of these basic components. Geber knew that when mercury and sulfur were combined, the product was the red compound cinnabar (mercuric sulfide) yet he believed that if the perfect proportion could be found, then gold would be the result.

Rhazes wrote the influential *Secret of Secrets*, which contained a long list of chemicals, minerals, and apparatus, including several kinds of glassware. He was the first to distil alcohol and use it as an antiseptic, and he also recommended mercury as a strong laxative. Another product he knew about was a mercury chloride called corrosive sublimate. An ointment made from this was used to cure 'the itch', which we know as scabies and which thankfully is now rare. It is caused by a mite which burrows below the surface of the skin and produces almost unbearable itching especially in the genital region, and it is transmitted during sexual intercourse. The poisonous nature of mercury and its ability to penetrate the skin made it an effective treatment of the disease.

Indian alchemists were also active by 700 AD and their ancient lore is encapsulated in a text written around 800 AD and called *Rasaratnakara*. This dealt mainly with mercury and its reactions with other compounds, again claiming that mercury was endowed with the power to make gold. It was also capable of making humans immortal, once it had been transformed into a 'nectar'. Traditional Indian medicines, and those of China, still use mercury and its compounds as ingredients.

In the early Middle Ages in Europe there were several noted alchemists, such as Avicenna (985–1037), Albert Magnus (1193–1280), Roger Bacon (1220–1292), and Thomas Aquinas (1225–1274), some of whom became better known for their theological writings. The most famous alchemist of this period was a Spaniard who also called himself Geber, hoping thereby to give his writings more credence by association with the earlier Geber. As a result his works were widely read and he was in fact the first person to report how to make nitric acid, silver nitrate, and red mercury oxide. His books were best known for the clear descriptions he gave of the apparatus he designed and how these were to be used, and this made them influential beyond the field of mere alchemy.* In effect Geber made alchemy respectable.

European alchemists slowly added to the body of chemical knowledge and one of their most notable discoveries was *aqua regia*, a mixture of concentrated nitric and hydrochloric acids that was capable of dissolving gold itself. This discovery fuelled the belief that gold was transmutable. When this solution was diluted with oil of rosemary the gold stayed soluble and this potion, called *aurum potabile*, was even prescribed as a cure-all. Unfortunately most alchemists were wedded to arcane language and it is almost impossible to understand the manuscripts they wrote, often because they used several different names to describe the same substance. Mercury, for example, was known as the doorkeeper, May-dew, mother egg, green lion, and bird of Hermes, to name but a few.

Nicholas Flamel (1330–1418) was one of the most famous French alchemists and it was widely believed that he had found both the Philosopher's

---

* His books were: *Summa perfectionis magisterii* [The Sum of Perfection], *Liber fornacum* [Book of Furnaces], *De investigatione perfectionis* [The Investigation of Perfection], and *De inventione veritatis* [The Invention of Truth].

Stone and the Elixir of Life because of the great age to which he lived and the wealth he accumulated, which he used to endow churches and build hospitals. There were reports that in January 1382 he had converted mercury to silver and three months later he was reported to have converted a large amount of mercury to gold. It is more than likely that his wealth, and old age, stemmed from his miserly and abstemious lifestyle. He probably became rich through money-lending and debt collecting, but there is no doubt that in his early years he was an alchemist, and used his alchemy in later life to disguise the real source of his wealth.

England too had its famous alchemists, such as George Ripley (born in the early 1400s) who came from Bridlington, Yorkshire. He studied in Italy for 20 years, where he eventually became a domestic chaplain of Pope Innocent VIII. He returned to England in 1477 and published *The Compound of Alchymy; or the Twelve Gates Leading to the Discovery of the Philosopher's Stone*. The 12 gates were the various chemical techniques such as distillation and sublimation. Because he was very rich it led contemporaries to believe that he too had discovered how to make gold, but on his deathbed he confessed to wasting his life on futile ventures and urged those who came across his writings to burn them, saying they were based not on actual experiment but on mere speculation.

Bernard of Treves (1406–1491) began searching for the Philosopher's Stone in his early teens and was still looking when he died aged 85. He was lucky in that he was born into a wealthy family and so could afford to spend his whole life as an alchemist, although there is plenty of evidence that some of those who joined him in his search were simply con men. One of these was a man known as Master Henry, whom he met in Vienna in 1464, and with his help Bernard performed an experiment that failed miserably. He gave Master Henry 42 gold marks, which he sealed in a vessel with mercury and olive oil, and heated them for 21 days. Surprisingly, when the vessel was opened there were only 16 gold marks to be found.

Fraudulent alchemists like Master Henry had several tricks for conning the gullible, such as using double-bottomed crucibles in which gold could be hidden, or inserting gold leaf inside pieces of charcoal that were added to the crucible, or, simplest of all, pre-dissolving some gold in mercury and then distilling this off leaving gold behind. No one doubted that transmutation was possible and in 1404 a law was passed in England in the reign of King Henry IV that forbade the making of gold or silver by alchemical methods.

This law, known as the Act of Multipliers, remained on the statute books until the 1660s when it was repealed thanks to the efforts of scientist Robert Boyle who was convinced that it was discouraging research that might well make the nation wealthy.

Alchemy flourished in the 1500s and 1600s and its practitioners became noted figures of their day: Georg Agricola (1494–1555), Paracelsus (1493–1541), John Dee (1527–1608) and his close associate and con man Edward Kelley (1555–1595), Michael Sendivogius (1566–1636), Jan Baptista van Helmont (1577–1644), and Joseph Francis Borri (1616–1695). The last of these came from Milan and he spent much of his life searching for the elusive Philosopher's Stone under the patronage of various dukes and monarchs, including his most protective patrons the ex-Queen Christina of Sweden and King Frederick III of Denmark, although he spent the last 20 years of his life a prisoner of the Pope in the Castle of St Angelo. Paracelsus became famous for his *medical* use of alchemical materials such as mercury, and we will hear more of him in later chapters. Sendivogius probably discovered oxygen, which he made by heating nitre (potassium nitrate).

Scams were often perpetrated on alchemists, and they appear to have been easily fooled. The Swiss scientist Johann Helvetius lived in The Hague, and in December 1666 he was visited by a man who said had discovered the Philosopher's Stone. He sold a small piece to the scientist to investigate, with the promise that he would return the next morning to show him how to make more. Helvetius' wife urged her husband to try it out that evening, and indeed they used it to convert half an ounce of lead into the finest gold. A local goldsmith pronounced it genuine, and Helvetius became famous when the news got out. Sadly the mysterious visitor never returned to reveal how the Stone had been made.

Alchemical fraudsters have continued in business right up to modern times. Long after alchemy had given way to chemistry there were those who still claimed transmutation was possible. The Emperor Franz Joseph was conned out of the equivalent of $10,000 in 1867 by three supposed alchemists, and as late as 1929 a German plumber Franz Tausend was still conning people. He persuaded a group of financiers to allow him to demonstrate his method. At the State Mint, and before an audience which also included a judge, the state attorney, and a police detective, he produced a tenth of a gram of gold from a gram of lead. All his equipment had been thoroughly tested beforehand and it appeared he had achieved a genuine transmutation.

In fact the gold had been smuggled into the room inside one of his cigarettes.

The 1600s saw the gradual emergence of chemistry from alchemy and in this period we find several men we now recognize as true scientists who were in their time secret alchemists, such as Robert Boyle (1626–1691), John Mayow (1641–1679), and Isaac Newton (1642–1727). By the end of the 1700s, however, alchemy was no longer respectable, at least in scientific circles, although even in the late 19th century some alchemists were still at work, including August Strindberg (1849–1912) the great Swedish writer. He devoted a considerable amount of effort to the project and believed he had succeeded in 1894 when he sent samples of his 'gold' to the University of Berlin and published his method in a fringe journal, *L'Hyperchimie*. Like all before him he was deluded, and later analysis of his samples showed them to be iron compounds, which can sometimes appear a deep gold colour.

The chemistry of the alchemists was really quite superficial in that it consisted of heating mercury with sulfur and any other ingredient that the alchemist had to hand. Mercury was known to dissolve all metals except iron and the amalgams so formed were then heated with sulfur. The resulting material could take on a variety of hues, especially if arsenic oxide was also added to the vessel, so much so that they would lead the alchemist to think a different process had occurred each time. Alchemical elixirs can still be purchased on the Internet, where there are recipes for making gold, and the subject can still be studied at the Paracelsus College Australia, which is based in Adelaide and has its own website at elevity.com/alchemy/parcoll.hmtl. This gives useful access to translations of many of the writings of the alchemists of the Middle Ages.

Mercury vapour is known to be highly dangerous. What is somewhat surprising is that many alchemists lived to old age, suggesting that either this toxic metal did them little harm or, more likely, that they spend more time theorizing about transmutation than attempting to carry it out. It seriously affected some of them, as we know from the experiences of those who were practizing alchemists in England in the late 1600s.

The first chemist

Today Robert Boyle (1627–1691) is regarded as one of the founding fathers of chemistry. He was the brother of Lady Ranelagh and he lived at her

London home, Ranelagh House, which was situated in the fashionable St James's district. Robert Boyle was a complex character. He was a life-long bachelor, a staunch Christian, a giver to charity on a large scale, a scholar, a world-renowned scientist – and an alchemist. Despite his ground-breaking work on the study of gases that led to Boyle's Law, which relates volume to pressure, he spent a great deal of his life searching for the Philosopher's Stone. He too was conned out of a great deal of money by a Frenchman George Pierre des Clozets who promised to reveal the recipe for making gold and admit him to a secret society of true alchemists. Boyle fell for the scam and paid dearly for it.

The fact that Boyle had been an alchemist for most of his life was to prove an embarrassment to the scientific establishment in later years because of the need to present him as the first true chemist. His book *The Sceptical Chymist* is today regarded as the seminal work that severed the link between chemistry and alchemy but is not just an attack on alchemy. Indeed among Boyle's papers when he died there was one he had partly written called *Dialogue on Transmutation and Melioration of Metals* in which he described a well-documented transformation of base metal into gold performed by a French alchemist, and which he said had been witnessed by several eminent people. Boyle believed his search for the Philosopher's Stone was justified because it would not only transmute metals but would be an 'incomparable' medicine.

Boyle himself published a paper in the Royal Society's *Philosophical Transactions* of 21 February 1676 entitled 'On the Incalescence of Quicksilver with Gold'. This reports a 'mercury' which, when mixed with gold, causes it to react and evolve heat. Lord Brouncker, President of the Royal Society, attested to the efficacy of Boyle's new 'mercury' in that when it was mixed with gold powder on the palm of his hand, he felt the heat it generated.

In another of his publications, *Producibleness of Chymical Principles*, Boyle reports on a 'mercury' that could dissolve gold instantly but refuses to reveal its nature because he feels it would 'disorder the affairs of mankind, favour tyranny and bring a general confusion, turning the world topsy-turvy'. We can only guess what this 'mercury' was but it was probably an antimony-copper-mercury amalgam. Boyle's instructions for making it, though, were written in alchemical language:

> Take pure Negerus, Dakilla, imbrionated banasis ana, mix them very well together & drive off all that you can in a retort with a strong fire of sand. It dissolves gold readily and that with sensible heat.

Negerus was mercury, Dakilla was copper, and imbrionated banasis was antimony. The danger inherent in carrying out such an experiment was breathing the mercury vapour that would be given off during the experiment, and indeed Boyle's regular exposure to mercury might well explain his chronic sickness. Undoubtedly Boyle was exposed to mercury fumes but there is no evidence that he was disabled by them and indeed most of his experiments were performed by an apprentice. He had taken up residence at Ranelagh House in 1671 and lived there until his death in 1691. In 1676 he persuaded his sister to allow him to build a laboratory in the garden and then to enlarge it in 1677. It was equipped with a furnace, retorts, flasks, and other alchemical apparatus together with a range of simple chemicals. It was there that he carried out a series of experiments that revealed him to be a true chemist rather than an alchemist.

In 1669 Hennig Brandt, an alchemist of Hamburg, discovered phosphorus, which he believed would lead him to the Philosopher's Stone on account of its almost miraculous ability to shine in the dark and spontaneously burst into flames. He sold some of this to a Daniel Kraft who demonstrated it around the courts of Europe, eventually arriving in London in 1671, where he even put on a private demonstration for Boyle at Ranelagh House, to which other members of the Royal Society were invited. Boyle was duly impressed and asked how it was made, only to be told that it was derived from 'something that came from the body of man'. Boyle deduced, rightly, that this was urine, but he could not extract phosphorus from it no matter how he tried until his apprentice, Ambrose Godfrey, went to Hamburg and met Brandt who told him that it required high temperatures. In fact phosphorus was obtained by heating to red heat the residues from boiled-down urine, and in this way Boyle obtained what he desired. What he did next distinguished him as a true chemist: he researched the properties of phosphorus and its reactions with other materials and published his findings not in the secret language of the alchemists but in plain English, and in a manner that would allow even a modern chemist to repeat what he had done. Whether they would want to repeat his observation that 'if the privy parts be rubb'd [with phosphorus] they will be inflamed for a good while after', is doubtful.

Phosphorus came too late in the age of alchemy to have much impact. It was neither the Philosopher's Stone nor the Elixir of Life although others assumed it might well be and experimented accordingly. It was not recognized to be a *chemical* element for another century. Indeed there were only a few elements as we know them which they used in alchemical recipes, and these were mercury, arsenic, and antimony. Of these, mercury was the material that forever tantalized, promising so much and yet delivering so little, and all the while it may have been affecting health and mental stability. It is worth examining this remarkable liquid a little more closely before looking at two men whom it seriously affected.

Mercury

Mercury was known to the earliest civilizations in China, India, and Egypt. The oldest known sample of mercury metal was found by the German archaeologist Heinrich Schliemann (1822–1890) in an ancient Egyptian tomb at Kurna which dated from around 1600 BC. The name mercury, by which we know this element, comes from the name of the planet and its first recorded use was by the Greek philosopher Theophrastus around 300 BC. The Romans called it *hydrargyrum* and it is from this word that today's chemical symbol for mercury, Hg, is derived. The early English name of quicksilver derived from the old English word *cwic*, meaning living, as in the phrase: 'the quick and the dead'. The Romans knew that on heating cinnabar it was reduced to globules of metallic liquid mercury. At the other side of the world, the Chinese were also observing the same phenomenon and the alchemist Ko Hung (281–361 AD) wrote of the wonder of turning bright red cinnabar into silver mercury simply by heating.*

Mercury has a strong affinity for sulfur atoms, and the two combine to form insoluble mercury sulfide, HgS, which is how it occurs as the main mercury ore, bright red cinnabar. When used as a pigment, cinnabar is known as vermilion and it was even used by cave painters 20,000 years ago in Spain and France. Vermilion was especially popular with the Romans, who decorated whole rooms in their villas with it. The Roman writers Vitruvius and Pliny refer to mercury metal but were of the opinion that the mercury which was found naturally in the mines of Spain was somehow superior to that which was obtained by roasting cinnabar; the former they referred to as *argentum vivum* (living silver) and the latter as *hydrargyrum*

---

* The sulfur is oxidized by the oxygen of the air, forming sulfur dioxide gas, and mercury metal is left behind.

(silver water). Pliny was clearly familiar with mercury and wrote of it as follows:

> It acts as a poison upon everything, and pierces vessels, even making its way though them by the agency of its malignant properties. All substances float upon the surface of quicksilver, with the exception of gold, this being the only substance it attracts to itself. Hence it is an excellent refiner of gold, for on being briskly shaken in an earthen vessel with gold it rejects all the impurities that are mixed with it. When once it has expelled these superfluities, there is nothing to do but to separate it from the gold.

Pliny reported that more than 4 tonnes of mercury metal were imported into Rome every year. He also said that men who worked with the ore protected themselves against the dust by covering their heads with bladders.

Down the centuries mercury continued to fascinate all those who were attracted to alchemy. There was nothing quite like it and it seemed to have almost magical properties. Mercury chloride still has its uses in magic even today, witness the 'psychic' Uri Geller who used it to demonstrate his supposed mind-over-matter mental powers in night clubs in Israel in the 1970s. According to Joe Schwarcz in his book *The Genie in the Bottle*, Geller would demonstrate his remarkable ability to heat metal by thought alone. A member of the audience would be invited on stage to hold a piece of aluminium foil which would mysteriously get hotter and hotter until it was too hot to hold, during which time Geller closed his eyes and supposedly focused his mind on the metal, supposedly willing it to heat up. The trick was to put a tiny amount of mercury chloride* powder on the aluminium and fold it over. A chemical reaction between the aluminium and the powder slowly begins to take place and eventually it gives off a lot of heat.

Mercury was important to the Scientific Revolution for barometers and thermometers, and while these uses could coexist with alchemy there was a discovery about mercury which fatally undermined belief that this metal was somehow forever different from all other metals. For alchemical theory it was *the* element, a component of all metals, and so held the key to the transmutation of base metals into gold. It uniquely represented the quintessential property of *fluidity*. Reports from Siberia, that mercury could freeze solid and become like any other metal were dismissed as little more than travellers' tales.

---

* This is the higher chloride, mercury(II) chloride, formula $HgCl_2$.

What could not be discounted was a report from two Russian scientists, A. Braun and M.V. Lomonosov, of St Petersburg. In December 1759 they had experimented with snow to see how low a temperature they could achieve. Mixing snow with salt causes its temperature to fall by several degrees and they thought that mixing acids with snow might produce even lower temperatures – and so it did. Suddenly the mercury in their therm-ometer stopped moving, and indeed it appeared to be solid. When they broke away the glass they found it had become a solid metal ball with a protruding piece of wire, which they could bend, just like other metals. Mercury was just a metal with a low freezing point of $-39°C$.

What was still not truly appreciated was the toxicity of mercury vapour and it is this which could have insidious effects on alchemists, and even on amateur dabblers including a famous king and his most intelligent subject.

The madness of Isaac Newton

Isaac Newton was one of the greatest scientists of all time. His achievements were impressive: he explained the nature of light and colour; he established the theory of gravity and deduced how the solar system works; he devised the laws of motion; and he invented an early form of differential calculus. What is less well known is that he spent most of his time when he was Professor of Mathematics at Trinity College, Cambridge, as an alchemist. When, in 1940, the economist John Maynard Keynes opened a box of Newton's papers that had lain undisturbed for 250 years, he was amazed to discover a collection of notebooks in which Newton had recorded his numerous attempts to make gold. In the years when he was writing his great works on physics and mathematics, he was actually spending much of his time carrying out alchemical experiments and copying out ancient alchemical texts.

Newton believed that the ancient alchemists knew how to make gold but that the secret had been lost. Nor was he alone in this belief. As we saw above, the great Robert Boyle thought it was possible, and John Locke the philosopher believed likewise. Indeed, Newton even cautioned Boyle about the need to remain silent about their alchemical interests.

Newton first experimented with mercury by dissolving it in nitric acid and then adding other things to the solution. When such experiments pro-duced nothing worthwhile he turned to heating mercury with various metals in a furnace, and his assistant and room mate John Wickins tells how he would sometimes work through the night. In one of his experiments he produced a kind of 'living' mercury that made gold swell. When nothing came of this, he turned his attention to antimony and by 1670 he had made the so-called Star Regulus, a dramatic form of antimony – see below.

In 1675 Newton wrote up his findings in a 1200-word manuscript known as the *Clavis* [Key]. By now he was 41-years-old and had gone grey, which he jokingly said was due to quicksilver. Although there is no connection between the two, there is a link between the body burden of several metals and their level in hair. Mercury, lead, arsenic, and antimony, are particularly attracted to the sulfur atoms in the keratin of hair and so it is possible by the analysis of a strand of hair to show whether that person had been exposed to a large dose of these toxic metals.

Newton's alchemical experiments appear to have reached a climax in the summer of 1693 when he wrote an account that is a combination of bizarre alchemical symbols and comments and is known as the *Praxis* [Doings] and this showed how unbalanced he had become. Isaac Newton was well known for being temperamental. Criticism of his work aroused in him an abnormal hatred of a rival and his feuds with other eminent scientists of the day such as Robert Hooke and Gottfried Leibniz were more emotional than rational. At times, Newton withdrew into virtual isolation and in 1693, when he was 50-years-old, his behaviour became so abnormal that his sanity was even questioned.

The published correspondence of Newton contains a noticeable gap from 30 May to 13 September 1693, when he wrote a letter to Samuel Pepys in which he said that he had been suffering from poor digestion and insomnia for the past year and admitted that he had not been 'of my former consistency of mind'. In the same letter he displayed evidence of this by rebuking Pepys for suggesting that he had ever asked favours from him or from King James, and ended the letter by saying that he never wanted to see Pepys or any of his friends again. He later wrote to the philosopher John Locke, among others, to apologize for the things that he said to them earlier. He asked Locke to forgive him for saying that Locke had been trying to 'embroil me with women'. In another letter, written to a friend of Pepys, he asked him to explain to Pepys his odd behaviour and said that he had suffered 'a distemper that seized his head, and that kept him awake for about five nights together'.

From these and other letters, Newton's physical symptoms are revealed to be severe insomnia and loss of appetite, while his mental symptoms were delusions of persecution, extreme sensitivity to remarks that he saw as implied criticisms, and loss of memory, all typical symptoms of mercury poisoning. In 1979, two articles appeared together in the *Notes and Records of the Royal Society of London* which confirmed that Newton had indeed suffered this way. The first was by L.W. Johnson and M.L. Wolbarsht, the second by P.E. Spargo and C.A. Pounds. According to Johnson and Wolbarsht, Newton's symptoms were consistent with mercury poisoning. Proof that this might well have been the cause comes from the paper by Spargo and Pounds. They analysed samples of Newton's hair by neutron activation and atomic absorption analysis and found high levels of toxic elements – see Table 1.1 – which shows that he had about four times more lead, arsenic, and antimony than normal, and 15 times more mercury. Two authentic samples of Newton's hair had been preserved in the Earl of Portsmouth's family along with other relics of Newton's that went to his niece, whose daughter married the first Earl of Portsmouth. Samples of Newton's hair are also kept at Trinity College Cambridge, and a single hair was found in one of his original note books which was assumed to be from his head. One of Newton's hairs had a mercury level of 197 ppm and another had a lead level of 191 ppm, both of which would be a strong indication of chronic mercury and lead poisoning at some stage in his life.

These findings are not surprising because we read in his alchemical note-books that he experimented with lead, arsenic, and antimony, and some of these he tried to volatilize by heating to high temperatures. He also admits to evaporating mercury over a fire, which was a particularly dangerous thing to do. Although there is no date for when these samples of hair were collected, most would probably have been cut from his head when he died in 1727. This being so then the level of mercury to which he was exposed in the critical period of 1693 would certainly have been much higher. In all cases

Table 1.1 Analysis of toxic elements in Newton's hair

	Mercury	Lead	Arsenic	Antimony
Normal level / ppm	5	24[*]	0.7	0.7
Level in Newton's hair / ppm	73	93	3	4

[*] This refers to the average level in hair in 1979; today it would be much less.

they reveal a remarkably high level of exposure, suggesting that he was in fact exposed to other sources of mercury. One of those might well have been from the decorations in his rooms. Newton had a desire to be surrounded by the colour red and to this end he had the walls of this rooms painted red, in which case it is more than likely that vermilion was the pigment used.

Yet if Newton's strange behaviour in 1693 represented the effects of mercury, it did him little permanent harm because he lived to the ripe old age of 84. It is not easy to say to what extent Newton's paranoid behaviour was due to mercury poisoning. He had such a sad childhood that his behaviour throughout life can be explained as due to his upbringing. His father died before he was born, his mother married again when he was 2, and his stepfather, a parson, wanted nothing to do with him, so it was left to his grandmother to raise him. All his life he had pronounced psychotic tendencies but his exposure to mercury may well have contributed to his mental instability. Newton was never insane and indeed he was entrusted with overseeing the operations of the Royal Mint, was elected President of the Royal Society in 1703, and was knighted in 1705.

The strange death of King Charles II

King Charles II was not an alchemist as such, but he was very interested in science and especially 'chymistry'. He had a laboratory built in the basement of his palace at Westminster and there, with the aid of one or two assistants, he spent time smelting and refining mercury, and indeed he became accomplished in the experimental techniques of the alchemists. Charles had his laboratory staff extract mercury from cinnabar and even distil it. No doubt his aim was to transmute base metals into gold, and thereby solve his financial difficulties. He was at odds with parliament, who had the power to vote him money, and had to rely on massive financial subsidies from his old friend King Louis XIV of France, for whom he was in effect a client king.

In fact Charles's interest in 'chymistry' had started in 1669 when he had established the office of Chemical Physician to the King, appointing Dr Thomas Williams to it, and providing him with a salary of 20 marks per year and research facilities where he could 'compound and invent medicines', some of which he did with the help of the King himself. Charles's laboratory was even visited by the diarist Samuel Pepys who, on the morning of Friday 15 January 1669, was walking to Whitehall when he met the King who invited him to come and inspect his new laboratory. This he did and he described it as 'the King's little laboratory under his closet, a pretty place, and there saw a great many Chymicall glasses and things, but understood none of them'.

In 1684, Charles began to exhibit some of the symptoms we now recognize as due to chronic mercury poisoning: he became irritable and easily depressed which was quite out of character for the man who was renowned for his cordiality, his many mistresses, and his love of the good life. Something clearly happened when he was in his laboratory during the last week of January 1685, something that exposed him to a lot of mercury vapour.

On Sunday, 1 February, he spent the evening with three of his courtesans listening to love songs and enjoying a meal with them, but going to bed alone. He awoke the following morning feeling quite ill. *The Calendar of State Papers – Domestic* reported what happened:

> When his Majesty arose yesterday morning, he complained that he was not well and it was perceived by those in his chamber that he faltered somewhat in his speech, notwithstanding which he went into his closet, where he stayed a considerable time. When he came out he called for Follier, his barber, but, before he got to the chair, he was taken with a fit of apoplexy and convulsions which drew his mouth to one side (this was about 10 minutes past eight), and he remained in the chair while he had three fits, which lasted nearly an hour and a quarter, during which time he was senseless. His physicians blooded him and he bled 12 oz freely. Then they cupped him on the head, at which he started a little, then they gave him a vomit [an emetic] and a glyster [an enema] and got him to bed by ten. He spoke before one. He called for a China orange and some warm sherry, in which time both the vomit and the glister wrought very kindly, which his physicians say are very good symptoms. He mended from one to ten last night, when they were laying him to rest, his physicians having great hopes that the danger of the fit is over, since that hand they feared was dead, he of his own accord moved and drank with it and complained of soreness, which they say is an extraordinary good symptom. Last night they sat up with him three Privy Councillors, three doctors, three chirurgeons [surgeons], and three apothecaries, and this morning, Dr Lower, one of the physicians that sat up, says that he rested very well, and that naturally, and not forced. This morning he spoke very heartily, so that they hope the danger of this fit is over.

But Charles had been fatally poisoned and the remission of his symptoms that Tuesday was not to last. On the Wednesday, the King took a turn for the worse, suffering more convulsions, and his skin became cold and clammy. He was given a strong laxative, which 'had good operation', and two doses of quinine ('the Jesuits' powder'). On the Thursday he had more convulsions and his life was now clearly in danger, so much so that he was visited by his brother James, heir to the throne, who brought along a Roman Catholic priest who received him into the Church of Rome and he took the Eucharist. (Charles, who had ruled as a Protestant king throughout his entire reign, was secretly a Catholic.) The following morning, Friday, 6th February, he was propped up in bed to watch the sun rise and even ordered that his eight-day clock be wound up, but by seven o'clock he was having difficulty breathing, by eight-thirty he was clearly failing, and by ten o'clock he was unconscious and obviously dying. Extreme remedies were administered including King's Drops, which was an extract of human skull and had been invented by a Dr Jonathan Goddard (and even prepared in Charles's own laboratory), and Oriental Bezoar Stone, which was made from the stones sometimes found in the stomachs of animals. These were remedies of last resort and clearly useless as antidotes to mercury poisoning, which of course had not been diagnosed as such by his physicians. Charles died just after noon that Friday.

Frederick Holmes, in his book *The Sickly Stewarts*, has considered the various accounts of Charles's death, including the report of his autopsy, which was carried out the day after he died and observed by a group of physicians. While the original report was lost in a fire at Whitehall in 1697, a copy has survived and is now in the archives of the College of Physicians of Philadelphia, USA. Holmes, who was the Edward Hashinger Distinguished Professor Medicine at the University of Kansas Medical Center, and a Fellow both of the American College of Physicians and of the Royal Society of Medicine, says there is only one conclusion that fits all the facts: Charles's death was due to mercury poisoning.

Charles died of an acute insult to his brain, which caused the epileptic seizures he exhibited during his last few days of life. The hand paralysis following the first of these is a common complication of epileptic seizures known as the post-ictal state. However, it is the autopsy which explains why the 54-year-old King, who up to then was in remarkably good health for a man of his age, was suddenly taken ill and died. The autopsy showed the outer parts of the brain to be engorged with blood while the ventricles of the brain contained much more water than normal. The rest of his organs were sound.

Until the 20th century, it had been assumed that Charles had suffered a stroke but, says Holmes, this was not the case and the onset of epileptic seizures suggests a serious disease of the brain. It has been suggested that he died of malaria and that this was the cause of the brain disease, but that does not fit with the facts either. Holmes concludes that the King died of mercury poisoning, caused by his exposure to it in his laboratory. This theory was first put forward in the 1950s by the romantic novelist Barbara Cartland in her book *The Private Life of Charles II: The Women He Loved*, but it was more seriously argued by two American scientists, M.L. Wolbarsht and D.S. Sax, in 1961 in a paper published in the *Notes and Records of the Royal Society of London*. They noted that Charles often spent his mornings in his laboratory where he was obsessed with the idea of 'fixing' mercury, in other words combining it with other materials, a process that included distilling large quantities of it. The air of that room must have been heavily polluted with mercury vapour and he would be totally unaware of if because it has no smell. Other great scientists were to suffer some degree of mercury poisoning due to poor laboratory conditions in the centuries to follow, including Michael Faraday (1791–1867). They were exposed to enough vapour to cause the symptoms of mild mercury poisoning although it was not recognized as such.

Breathing mercury vapour causes no respiratory symptoms, unless the dose is very high. The metal is absorbed by the lungs and passes into the blood stream and thence to all parts of the body but the nervous system is particularly affected. The brain is most vulnerable because mercury can move across the blood-brain barrier which is there specifically to protect this vital organ against toxins, and once inside the brain it causes all kinds of symptoms such as lack of energy, unsteady gait, insomnia, etc. In his last year of life, Charles showed some of the signs of mild mercury poisoning, and he became less physically active. We know he was exposed to mercury because the analysis of a strand of his hair showed ten times the expected level. John Lenihan and Hamilton Smith, of the University of Glasgow, used nuclear activation analysis techniques to measure this in 1967. The sample of hair had been obtained through a radio broadcast the previous year, when a listener in Wales sent them a lock of Charles's hair attached to a card which bore the words:

This lock of hair was taken from the head of King Charles the 2nd, by the mother of Sir John Jennings Kt, and given to Miss Steele of Bromley by Phillip Jennings Esq. nephew to the Admiral Sir John Jennings above said 1705.

The analysis showed the hair to contain 54 ppm of mercury, which is about ten times greater than normal, and while there is no record of when the hair was cut from the King's head, it was probably after he died and it certainly provides evidence that he was exposed to the metal during his final months of life.

While such analysis reveals exposure, it does not prove that he was necessarily putting his life at risk with his experiments. What killed him was *acute* mercury poisoning. In other words, in the days before he became ill he had done something in his laboratory which filled the air with mercury fumes and these he breathed in, maybe for an hour or more. The other possible explanation of mercury in the King's body could be mercury-based medication taken to treat syphilis, but neither his medical history, nor his autopsy, nor state records, indicated that any of his several mistresses had infected him with venereal disease.

Holmes uses his expert knowledge to show how the King's deathbed symptoms are consistent with mercury poisoning caused by breathing a lot of the vapour. This was the only route by which it could have entered his body without affecting other organs and yet kill him so quickly. When the blood-brain barrier is breached by mercury, the protein-containing part of the blood, the serum, leaks into the crystal-clear fluid surrounding the brain, the cerebrospinal fluid. This is exactly what the post-mortem revealed, all the cerebral ventricles were filled with a kind of serous matter, and the substances of the brain itself were quite soaked with similar fluid. The mercury that found its way into his brain then damaged the brain cells themselves causing the seizures that were observed. These seizures were not due to the other likely causes such as an abscess, tumour, meningitis, or internal bleeding because these would have been noted at autopsy. It was quicksilver that killed the King.

Arsenic

Humans appear to have been exposed to arsenic for more than 5,000 years and we know this because hair from the Iceman, who was preserved in a glacier in the mountains of the Italian Alps for this length of time, contained high levels of the element. His exposure to arsenic is thought to indicate that he was a coppersmith by trade since the smelting of this metal is often from ores that are rich in arsenic. The arsenic is volatilized as arsenic trioxide and it deposits in the flue of the furnace or on nearby surfaces.

Theophrastus, Aristotle's pupil and successor and who lived around 300 BC, recognized two forms of what he referred to as 'arsenic' although these were not the pure element, but the arsenic sulfide minerals orpiment (As_2S_3) and realgar (As_4S_4). The ancient Chinese also knew of them and the encyclopaedic work of Pen Ts'ao Kan-Mu mentions them, noting their toxicity and use as pesticides in rice fields. The mineral realgar was recommended as a treatment for many diseases as well as for banishing grey hair. Arsenic compounds are also referred to in Democritus's *Physica et Mystica*, and the Roman writer Pliny wrote that the Emperor Caligula (12–41 AD) financed a project for making gold from orpiment and while some was produced it was so little that the project was abandoned.

The link between arsenic and gold was not forgotten and arsenic really came into its own in the Middle Ages. Realgar was found to yield so-called white arsenic by fusing it with natron (natural sodium carbonate). Petrus Oponus (1250–1303) showed that both orpiment and realgar could be converted to white arsenic, which we now know as the dangerously toxic arsenic trioxide, and which in the hands of the unscrupulous was to wreak such havoc down the ages. If white arsenic was mixed with vegetable oil and heated it yielded another sublimate, arsenic metal itself, and this may be how the discoverer of the element, Albertus Magnus (1193–1280), first made it. What was also noted in the Middle Ages was that when arsenic was applied to copper metal it turned it silver, and this too appeared to be a kind of transmutation.

Antimony

The origin of the word antimony is uncertain. One theory is that it came from the Greek *anti-monos* meaning not-alone. Another theory is that it is derived from its use as mascara in preference to the mineral minium (black lead), in other words it was anti-minium and so became anti-mony. A more likely derivation of the word is from the Greek *anthemonion* meaning 'flower-like' because of the beautiful flower-like crystals of the antimony ore stibnite.

Constantine of Africa, who died in 1078 AD, first used the word antimony and he is believed to have coined the name, although he was not referring to the element itself. He was born a Muslim, and was educated in Baghdad, but eventually embraced Christianity and became a monk. The chemical symbol, Sb, comes from the Latin word *stibium* which was the name of the mineral by which antimony sulfide was known in ancient times.

One of the first to write about antimony was Roger Bacon in the 1200s, and he was well aware of the metal and several of its compounds, and wrote about them openly. His interests were purely scientific. In the more secret world of alchemy, antimony played a key role. Like gold, it could only be dissolved by the king of acids, *aqua regia*, which suggested some affinity between the two metals, but of course no matter what the alchemists did to antimony it stubbornly refused to be transmuted into the much more desirable element. Others viewed antimony as a possible route to the Elixir of Life, but again they were to be disappointed, although John of Rupescissa, writing around 1340, suggested that the medical men might use some of its compounds to treat their patients.

The compounds of antimony that were known in the Middle Ages probably derived their names from those used by the alchemists, so we have regulus of antimony for the metal itself, golden sulphuret for antimony sulfide, butter of antimony for antimony chloride, and powder of Algaroth for antimony oxide chloride. But antimony had a much more ancient pedigree than the alchemists of the Middle Ages realized.

The Chaldean civilization, which flourished in the sixth and seventh centuries BC, in what is now Iraq, had craftsmen who were capable of working with antimony metal, as shown by a vase of that period which was analysed in 1887 by a French chemist Pierre Berthelot (1827–1907) who showed it to be almost pure antimony. It is now in the Louvre museum. Whether the ancient metallurgists were capable of extracting the metal from stibnite (antimony sulfide, Sb_2S_3), or whether they used samples of native antimony, which are sometimes to be found, is not known. Egyptian women of the oldest profession certainly had a fondness for stibnite powder which they applied as a form of mascara known as *kohl*. One of the most infamous practitioners of her craft, and user of *kohl*, was the temptress Jezebel whose exploits are recorded in the *Bible*, which twice warns of women who painted their eyes: *Second Book of Kings* (chapter 9, verse 30) and *Ezekiel* (chapter 23, verse 40).

The Chaldean craftsmen were able to make yellow lead antimonate and this was used in the glaze of the ornamental bricks which adorned the walls of Babylon during the reign of Nebuchadnezzar (604–561 BC). This pigment was still being made in the 1900s and became known as Naples yellow. How the Chaldeans made it can only be guessed at but they probably heated together stibnite and red lead (lead oxide, Pb_3O_4) and produced it from the chemical reaction of these.

The Ancient Greeks and Romans regarded antimony metal as a type of lead but made little use of it. Their successors in the Byzantine navy, however, found a new use for antimony sulfide as an ingredient of the famous weapon they employed against enemy ships. So-called Greek fire was a burning liquid that was squirted from their warships, rather like a flame-thrower, and which brought terror to those exposed to it because it was impossible to extinguish and it even burned on the surface of water. How it was made has remained a secret to this day; indeed it was a capital offence to reveal it. It was last used in the defence of the capital, Constantinople, in 1453. The most likely composition of Greek fire was crude oil, stibnite and saltpetre (potassium nitrate), a combination that would be highly flammable and almost impossible to extinguish with water. Once it is ignited, antimony sulfide generates a lot of heat, and its flammability was put to use in the early forms of household matches, whose red tips were due to the colour of this compound.*

The alchemists were always fascinated by antimony and one of their names for the metal, which they obtained by heating stibnite with iron powder, was *regulus* [king] *of antimony* or *martial regulus* [king of Mars] implying an impure form of regal gold. Needless to say, they never achieved their objective of converting one into the other, but the spin-off from their researches greatly added to the store of knowledge about antimony. The alchemists had another name for antimony, *lupus metallorum* [wolf of metals], based on its remarkable ability to alloy with other metals and change their character. The alchemists were probably the first to discover butter of antimony, which they got by heating the metal with corrosive sublimate. They purified the product and then heated it in a sealed vessel for several

---

* Antimony trisulfide has a different role to play in modern warfare. It is used in camouflage paints because it reflects infrared radiation in the same way as green vegetation.

months until it had become a red powder, which they called the 'powder of projection'. It was said that if this was sprinkled on other metals, along with the addition of mercury, it would transform the metal into gold. Of course they were wrong.

Antimony sprang to prominence quite openly in the 1400s when it became an essential part of the new craft of printing. Molten antimony has a unique chemical property of expanding as it solidifies, and by adding it to molten lead it produced a cleaner type face. This property of antimony had also been appreciated in the ancient world, where it was used to produce finely cast objects. Not only that but the lead alloy it produces is much harder than lead itself, again something that was appreciated by printers because it made stronger type. The preferred alloy for type consisted of 60% lead, 30% antimony, and 10% tin and this was used for more than 400 years.

The accumulated knowledge about antimony appeared in a very influential book published in 1604 and called *The Triumphal Chariot of Antimony*. It opens with an introduction about the author, a mysterious monk called Basil Valentine who apparently lived in the 1400s, and belonged to the Order of St Benedict. We are told that Valentine hid his manuscript inside a pillar of the church in Erfurt where the monastery was located, and there it rested until one day a bolt of lightning split open the pillar and it was revealed. In fact there never was a Benedictine monastery at Erfurt and no monk of this name has ever been traced, although that did not prevent other writers mentioning him and his book and even putting his date of birth at around 1400. The book popularized the use of antimony and its compounds in the treatment of disease, and thus started the widespread use of antimony which continued for 300 years.

The Triumphal Chariot of Antimony begins with a mixture of alchemy, pious utterances, and abuse directed at the physicians and apothecaries of the day. When the author eventually gets down to the origin and nature of antimony the tone becomes alchemical. The section on the compounds of antimony reveals a knowledge of these which suggests he had practical experience in dealing with them. He mentions antimony metal, the oxide-sulfide glass, an alcoholic solution of the glass, an oil, an elixir, the flowers, the liver, the white calx (oxide), a balsam, and others. In its pages are described antimony trichloride, prepared by heating a mixture of antimony sulfide and mercury dichloride in equal proportions.

The Triumphal Chariot of Antimony was in fact written by its publisher

Johann Thölde, who was also a pharmacist and part owner of a salt works at Frankenhausen in Thuringia. Two pieces of literary evidence prove it was a later work and that Thölde was the author. The first is the reference to syphilis as the new malady of soldiers, which only appeared during the French invasion of Naples in 1495, and, more convincingly, parts of the text reproduce parts of Thölde's book *Haliographia*, particularly the section on how salts can be obtained from metals. *The Triumphal Chariot of Antimony* shows that not all the trials and tribulations of the alchemists had been in vain, but the compounds of antimony which they had discovered were quite poisonous, as we shall see in Chapters 7 and 8, and yet they were widely used by doctors.

And what of alchemy today? Despite its tenets being based on a completely false set of beliefs, which puts it quite outside the boundaries of scientific investigation, it continues to flourish. Peter Marshall, the author of *The Philosopher's Stone*, says that alchemy is alive and well and living in places like China, India, Egypt, Spain, Italy, France, and Prague, where he went to search out its secrets and interviewed several latter-day alchemists. While some still cling to the idea that transmutation is possible, others are more concerned with the alchemy of the human mind. The theory is that we must search within ourselves for the Philosopher's Stone which can transform our inner being from dross to noble metal. Alchemy continues to attract other adherents who still believe in the transmutation of metals and who search the ancient literature looking for clues to the supposed secrets of the Philosopher's Stone and the Elixir of Life, secrets that they believe to have been discovered on more than one occasion but which have then been lost. They will forever seek in vain.

Mercury

Mercury poisons us all!

For more technical information about the element mercury, consult the Glossary.

ERCURY is everywhere and we cannot avoid it. The average adult contains around 6 milligrams of mercury – assuming they have no mercury amalgam fillings in their teeth – and this is something we have to live with because we can do almost nothing to reduce it. Our average intake of mercury is about 3 micrograms per day for adults, and about 1 microgram for babies and young children. At these levels the amount we consume in a lifetime is less than a tenth of a gram, although in previous centuries people would consume more than this in a day in the form of medication, generally for embarrassing diseases, such as the unspeakable syphilis or, even worse, the unmentionable constipation. We shed mercury from our body through our urine, faeces, and even our hair. We could excrete mercury via our saliva glands, which are greatly stimulated by mercury, but the mercury in saliva tends to return to the stomach.

So where does it all come from? The answer is mainly from the food we eat, although a little comes from the air we breathe and the water we drink, and some may even come from our own body if we have mercury amalgam fillings in our teeth. Agricultural soils may hold as much as 0.2 ppm of mercury and this finds its way into plants and food crops. Grass contains relatively little mercury, around 0.004 ppm, which explains why grazing animals are not really contaminated, and meat and dairy products have low levels. Seawater contains even less mercury than the cleanest soil and has only 0.00004 ppm, yet some fish absorb mercury to the extent of concentrating it in excess of 1 ppm.

Are we harmed by this amount of mercury? Probably not. In December

1997, the US Environmental Protection Agency (EPA) published a seven-volume report on mercury and announced a safe daily dose of 0.1 microgram per kilogram bodyweight, which for an ordinary adult would be 7 micrograms. Were this limit to be acted upon then it would outlaw the sale of all swordfish, shark, and most tuna, whereas the Food and Drugs Administration (FDA), which has a more pragmatic view of mercury, bans their sale only if their mercury content exceeds 1 ppm. As we shall see, the EPA guideline is somewhat unrealistically low in that it is probably exceeded by all those in the population with amalgam fillings in their teeth, and yet the EPA claims that more than 600,000 children are born each year in the USA with learning deficits due to exposure to mercury while in the womb.

A person's reaction to mercury is unpredictable. Some can tolerate it in large amounts without showing signs of poisoning, while others were so sensitive that when mercury-based drugs were injected into them they were dead within seconds of the injection. One boy aged 4 actually died as the hypodermic needle was being withdrawn from his arm!

In this chapter we will look at how mercury could be affecting us. However, before we look at the effects on humans we should first spare a thought for the environment.

Mercury in the environment

Every plant and living creature contains some mercury, and this has been true for millions of years. Mercury stirs restlessly through the environment, and through the biosphere, and it can do this because it exists in different forms. It can be present as elemental mercury and as such is volatile, which means it can circulate via the atmosphere. Mercury can exist as methyl mercury compounds produced by bacteria and thereby become more soluble. It can exist in one of two oxidation states of which mercury(I) is less common and less soluble, while mercury(II) is more common and more soluble, unless it meets a sulfur atom and precipitates as the totally insoluble mercury(II) sulfide (formula HgS).

In the past 500 years the amount of mercury released to the environment has increased dramatically due to the activities of humans. William Shotyk of the University of Heidelberg has studied the mercury levels in peat bog in remote areas of Canada, Greenland, and the Faroe Islands where it is possible to measure the amount of mercury being deposited from the atmosphere stretching back more than 14,000 years. Shotyk has shown that soil accumulated around 1 microgram of mercury per square metre per year although this was sometimes as high as 8 micrograms following a major volcanic eruption. Then from around 1500 AD onwards the amount of mercury being deposited began to increase slowly so that it had doubled by the 1700s, thereafter rising with the onset of industrialization until it reached over 100 micrograms in the mid-1950s. It has since declined and is now around 10 micrograms per square metre of soil per year.

Mercury liberated naturally into the environment every year has been estimated to be about 1000 tonnes, but this is far exceeded by the human contribution. Cathy Banic of the Meteorological Service of Canada, has been tracking airborne mercury, and reported in 2003 in the *Journal of Geophysical Research* that the Earth's atmosphere contains about 2,500 tonnes of the metal of which one-third comes from natural sources.

The level of mercury in the air over the Atlantic Ocean still continues to increase by around 1% per year. Ninety per cent of this is elemental mercury which comes mainly from coal-fired power stations (65%) and waste incineration (25%). Coal-burning in the USA adds 48 tonnes of mercury per year to the atmosphere although set against the global total of several thousand tonnes it is relatively modest.

Mercury's behaviour in the atmosphere can be quite puzzling. Sometimes it disappears for no apparent reason. For instance, during the weeks of the Arctic winter, when the sun never rises above the horizon, the level of mercury in the atmosphere builds up, but as soon as the first ray of sunshine appears the mercury vanishes for about three months. This mystery remained unsolved until 1998 when it was discovered that the airborne mercury suddenly deposited itself on the surface of the snow. Nor does this happen only over the North Pole; a team of German researchers also showed it occurs in Antarctica.

The explanation is that the mercury builds up in the atmosphere as long as it is not oxidized and these are the conditions during the sunless winter days. Once the sun shines, however, it triggers a sequence of chemical changes that speedily bring about oxidation of the mercury. Some is oxidized by ozone (O_3) that is being formed and some by the chlorine and bromine radicals that are generated within the aerosols formed from sea spray. Together these produce mercury oxide, mercury chloride, and mercury bromide which fall to the snow-covered land. Then as the polar summer unfolds, they may be reduced back to elemental mercury by the action of ultraviolet rays and this returns it to the atmosphere.

Bacteria in an environment that is heavily contaminated with mercury can adjust to it by getting rid of a lot of it, and this they do by converting the mercury to soluble methyl mercury or even to volatile dimethyl mercury. Other bacteria have had to adapt to living with methyl mercury by deactivating it, which they do by breaking the methyl-mercury chemical bond, converting the methyl to methane gas and the mercury to mercury metal, which then is volatilized.*

Research carried out in the 1990s by Steve Lindberg of the US Oak Ridge National Laboratory's Environmental Sciences Division revealed that trees and soils emit *elemental* mercury. He reported that the level of emission over a forest canopy was 100 nanograms per square metre per hour, and 7.5 nanograms per square metre per hour over soil – nor was this from soil near industrial sites, but from areas where there had been no possible contamination by human activity. Using specially grown saplings in laboratory chambers, it was shown that plants take up mercury from the soil in which they grow and release it through their leaves to the air. However, when the concentration of mercury in the air reaches a certain level, the process reverses and the plants re-absorb mercury.

In 1996 a team at the University of Georgia in Athens, USA, led by Richard Meagher, reported that they had genetically engineered plants that could detoxify land polluted with mercury compounds by absorbing them and reducing them to elemental mercury, which was then released from the plants' leaves. They had taken a gene from bacteria that can carry out this process and inserted it into a weed called arabidopsis, which was then shown to thrive on a growth medium that contained twice the level of mercury chloride that would kill normal plants.

The United Nations Environment Programme agreed at a meeting in Nairobi, Kenya, in 2003, that there should be a *global* effort to reduce mercury emissions. The report they issued identified the main sources as coal-fired power stations and waste incinerators, which together accounted

---

* These bacteria have two enzymes that do the job; the first is *lyase* that cleaves the bond to form methane (CH_4) and the mercury ion Hg^{2+}, while the second enzyme is a *reductase* that reduces this ion to mercury metal.

for 70% of emissions, and that 60% of the mercury from these sources was being emitted by Asian countries. The global emission of mercury by human activity is estimated to be around 2200 tonnes per year of which 1,500 tonnes come from power plants, 200 tonnes from metal production, 100 tonnes from cement production, 100 tonnes from waste disposal and 300 tonnes from small-scale gold mining. All of this contaminates the planet but it does little harm to us as individuals. The threats to our welfare come generally from more direct contact with this element.

Dental amalgam

It is impossible to estimate how many dental fillings there are in the world but in the USA alone more than 100 million cavities are filled every year, in the UK it is around 25 million, and in Germany around 40 million. This kind of treatment had been performed by dental surgeons for about 200 years with no apparent ill effects, but in the early 1990s there was suddenly a lot of debate about the safety of dental amalgams on account of their high mercury content. Could these release sufficient mercury to be affecting the health of those who had them? Was the air in dental surgeries dangerous to those who worked there? And what was happening to the mercury when bodies were cremated? More than a decade later, mercury amalgams are still a major use of mercury, mainly because they do the job so well and last a long time.

The dangers of mercury amalgam fillings have been discussed in scientific and medical journals from the time they were first introduced almost 200 years ago, and their advantages seemed obvious and far outweighed any risks. Their opponents accused dental fillings of causing mercury poisoning. In the 1800s the American Society of Dental Surgeons required its members to sign an oath that they would not use them, but it had to disband in 1856 because it had so few members. Its successor was the American Dental Association which supported the use of mercury-silver amalgams.

Concern about mercury amalgam fillings surfaced again in the 1980s and 1990s when they were aired in the media. On 23 December 1990 the influential programme '60 Minutes' broadcast on CBS-TV in the USA was devoted to the subject, claiming that mercury amalgam fillings were dangerous to health. (In the UK in 1994 a similar programme was broadcast by the BBC-TV called 'Poison in the Mouth'.) Dr Hal Huggins of Colorado

Springs was the chief opponent of mercury fillings which he saw as the cause of many ailments ranging from depression to cancer, and including such diverse diseases as multiple sclerosis, arthritis, ulcers, and allergies. He advised those with such fillings to be tested for mercury exposure and have their fillings removed and replaced, which many people did.

Professor Murray Vimy of the Faculty of Medicine, University of Calgary, Alberta, Canada, was another leading adversary of dental amalgam. He summarized his opposition in an article in *Chemistry & Industry* in January 1995 accusing such fillings of poisoning millions of people. Vimy estimated that eight such fillings – and this is a not uncommon number (I have ten) – would release about 10 micrograms per day. He said that fillings constituted the largest non-occupational source of mercury in the general population, and speculated that mercury might be involved in Alzheimer's disease, the brains of whose victims were reputed to have more mercury than normal.

Research had been carried out on sheep whose teeth were drilled and filled with mercury amalgam. They were then monitored and some disturbing findings came to light. For example, a sheep that was pregnant was found to pass mercury to its developing foetus, and there it tended to concentrate in the foetus's liver. Other sheep had their own liver function impaired, with their filtering ability reduced by half within a month. Other effects on the health of the sheep were also noted although they, like humans with filled teeth, did not display outward symptoms of mercury poisoning.

Not everyone agreed with all these alarms about the dangers of mercury amalgam fillings; indeed few now believe that they pose any threat to health at all. Vimy's remarks and finding were criticized by members of the British Dental Association. Huggins even had his licence to practise dentistry revoked in 1996 on the grounds that his treatments were a sham. (Some of those who came for advice were diagnosed as suffering from what he detected to be 'mercury toxicity' even though they had no mercury amalgam fillings!) The research on sheep was criticized on the grounds that the amalgam used had been obsolete for almost 50 years and the way the sheep had been treated would have allowed them to swallow scraps of amalgam when the dental work was being carried out. Moreover the filled teeth were placed opposite to one another so they would grind together all the time.

The dental amalgam story began in 1812 when a British chemist Joseph Bell found that mixing powdered silver with mercury produced a curious paste that was malleable like putty for a short time but then became hard like metal. It seemed ideal for filling cavities in teeth because it would mould itself to the shape of the cavity and then set firm with enough strength to withstand the pressures that teeth normally experience when biting and chewing. Unfortunately the silver he used was obtained by filing down silver coins and being impure it produced an amalgam that tended to expand slightly with heat, eventually cracking the tooth in which it was placed.

What was needed was a better alloy and this was invented by a US dentist C.V. Black in 1895, and consisted of 70% silver and 30% tin, which was added as a powder to the mercury. This formed a plastic mass that could be pressed into a tooth cavity with a reassuring 'squeak' and which would harden in about 5 to 10 minutes. It expanded slightly as it solidified thereby exactly filling the hole, but thereafter it was unaffected by changes in temperature. The modern version of the alloy powder consists of 60% silver, 27% tin, and 13% copper. Mixing the amalgam used to be done by the dentist or dentist's assistant, with the result that all dental surgeries soon became heavily contaminated with mercury.* Today the amalgam comes in a sealed plastic capsule and the mixing can be done in the capsule using a special agitator.

What is the ultimate end of all the mercury that people are carrying around in their mouths? A lot of it in Europe exits up the flues of crematoria. When bodies are cremated, the temperature is around 700°C, high enough to decompose any amalgam in the teeth and volatilize the mercury. It has been estimated that as much as 3 grams of mercury is released by the average corpse, adding up to around 10 kg of mercury per year being emitted from a crematorium chimney. However, these figures have been disputed because many who are cremated are elderly and have lost most or all of their teeth (80% of those aged over 80 come into this category) and indeed this is true for two-thirds of cadavers. It is possible to prevent the loss of mercury during cremation by adding a capsule of a selenium compound to the coffin and this reacts with the mercury vapour to form involatile mercury selenide. Alternatively the flue gases can be filtered through a charcoal filter before being emitted to the atmosphere and that too will remove the mercury.

---

* Raymond Holland told me that surplus mercury was squeezed out of the amalgam through a cloth!

Mercury and metabolism

The human body can tolerate quite large amounts of mercury, although if the total exceeds 4 grams there is serious risk of death. (A fatal dose of methyl mercury is probably around 200 mg.) About one person in ten has a level of mercury in their body that would make them unsuitable as food for any cannibals who followed the nutritional guidelines regarding excess mercury levels in meat, that it should not exceed 0.05 ppm. Today most mercury comes from our food, but in former times the use of mercury in various household products such as floor polishes, laundry aids, and paint served to increase absorption of the metal especially as this metal can easily penetrate the skin. Absorption this way has led to severe poisoning, and sometimes children with ringworm died when they were treated with antifungal cream containing mercury. A 9-year-old girl died five days after a solution of mercury(II) chloride in alcohol had been applied to a patch of ringworm on her scalp.

Mercury can also enter the body through the stomach wall, the lungs, and in certain cases has even been absorbed through vaginal douching and via the rectum as an enema – more of this in Chapter 4. The result of excessive exposure to mercury is violent vomiting and diarrhoea, and these may begin within 15 minutes and persists for hours. This reaction happens by whichever route the mercury has gained access to the body. Every microgram of mercury that enters our blood stream will interfere with some part of our body until we excrete it, and that may be a slow process. In 1960 a study on rats, using the radioactive isotope mercury-203, showed that mercury collected first in the liver from where it passed to the kidneys before being slowly excreted. Meanwhile it can disperse to other organs and will accumulate in the muscles, liver, kidneys, and bone. It is rapidly lost from the muscles and kidneys, but only slowly from the liver and bone. Mercury can collect in the skeleton and there are reports of drops of mercury being found in exhumed bones of people who have been treated with medicines containing metallic mercury during their lifetime.

The toxicity of mercury varies according to the form in which it comes. Methyl mercury is particularly dangerous and this will be dealt with in more detail below. Liquid mercury metal is the least toxic; mercury vapour is more dangerous, while the poisonous nature of mercury salts depends on their oxidation state. Compounds of the lower state, mercury(I), are generally much less toxic than compounds of the higher state, mercury(II), partly because the latter tend to be more soluble and can pass through the gut wall within minutes of being swallowed. The upper limit for a medical dose of mercury(I) chloride, calomel, was 30 times higher than that of mercury(II) chloride, or corrosive sublimate, and consequently calomel became the ingredient that doctors preferred to prescribe.

Acute mercury poisoning from a large dose of a soluble mercury salt causes damage to the kidneys, intestines and mouth, and the symptoms are vomiting, stomach pains, weak pulse, and difficulty in breathing. If a fatal dose has been taken, death may intervene before any but the most obvious symptoms manifest themselves, but usually death takes about a week, although some people have been known to last 3 weeks before dying.

Mercury has a particular attraction to sulfur and it will attach itself to the sulfur atoms of certain amino acids. When these amino acids are part of an enzyme's protein it may result in that enzyme being rendered inactive. The enzyme *Na/K-ATPase*, which is essential to the working of the central nervous system, is particularly sensitive to mercury and this leads to the most noticeable symptom of mercury poisoning, the 'shakes'.

The mouth is also affected by mercury which stimulates the flow of saliva, a reaction that can appear within hours of taking the poison, although in some individuals it may never occur. If the poisoning continues, the odour from the mouth gradually worsens and the lips, gums, and teeth become inflamed and later are covered with a grey film. The condition may become so bad that the teeth become loose and drop out and parts of the jawbone may eventually be exposed.

Mercury targets the kidneys, and the first effect is to increase the rate of urine flow, thereby acting as a diuretic. In 1886 calomel was first prescribed for this purpose and was replaced in 1919 by another mercury-based drug called novasurol, which had been designed as a treatment for syphilis but which was noted to have a powerful diuretic effect. Later, novasurol gave way to other mercury compounds. Although a little mercury may have a stimulating effect on the kidneys, a lot has a disastrous effect, and eventually they cease to function and urine flow stops. The body then becomes poisoned by its own waste products because these cannot be removed, and even if the kidneys begin to work again, it may be too late to save that individual.

The most distressing action of mercury is on the nervous system and the brain. The mental deterioration seen in cases of chronic poisoning by mercury is known as erethism and is characterized by symptoms such as timidity alternating with anger and aggression, lack of concentration, loss of memory, depression, insomnia, listlessness, and irritability. All forms of mercury can penetrate to the brain, methyl mercury particularly so, but the metal and mercury salts have the least ability to cross the blood/brain barrier.

Mercury metal and mercury vapour

Little Willie from his mirror
Licked the mercury right off,
Thinking in his childish error
It would cure the whooping cough.
At the funeral his mother
Brightly said to Mrs Brown:
''Twas a chilly day for Willie
When the mercury went down.'

[Harry Graham, *Ruthless Rhymes
for Heartless Homes*, 1899]

Liquid mercury is the form of the element that most people are familiar with and this is comparatively harmless. Little Willie was not likely to have met his end by this means. If you swallow liquid mercury it is not likely to kill you, because it is not absorbed as it passes through the gut. Patients who have bitten the end off a thermometer, and swallowed the mercury that escaped from it were never at risk of dying from mercury poisoning. People have been known to drink a cupful of mercury, weighing about a kilogram, as a laxative, and one man drank an ounce of mercury every day for 5 years for this very reason. A famous case of drinking liquid mercury was that of the Margrave of Brandenburg who accidentally drank a large draught of it on his wedding night, in 1515, but was reputed to have suffered no ill effects.

When mercury has been injected as an emulsion it has led to fatalities although one patient was given 27 g of mercury by injection and he survived. A nurse even tried to kill herself by injecting liquid mercury into her veins and while she was ill for a time, she eventually recovered. When she died 10 years later, of tuberculosis, the post-mortem revealed globules of mercury still in her body. On the other hand, others who have wanted to commit suicide have been successful, although the longed-for release has often taken several weeks to arrive, and some would-be suicides were saved. A 21-year-old woman who injected 30 g of metallic mercury into her thighs survived as a result of being given the antidote Dimercaprol (see Glossary).

Liquid mercury may be a metal but it is also a *volatile* liquid and in the form of a vapour it can be much more dangerous. Admittedly mercury does not evaporate like water or other common solvents, but enough mercury passes into the air to make it dangerous to breathe. The great scientist Michael Faraday (1791–1867) first demonstrated the volatility of mercury, and we now know that the evaporation rate of mercury from the surface of the liquid is 800 mg per square metre per hour. (This can be prevented by covering the surface with water or oil.) Yet for centuries the dangers of breathing its vapour were not realized and many worked in an atmosphere of it. The popular image of the mad scientist may have had some basis in fact and could have been due to the effects of mercury vapour because every laboratory was polluted by it, and even when spilled mercury was cleared up, it was almost impossible to track down every last drop.*

In the 1920s a German chemistry professor, Alfred Stock, discovered how ubiquitous mercury was, having been diagnosed as suffering from chronic mercury poisoning although his researches did not directly involve it. In fact it came from the mercury manometers he was using to measure the pressure inside the glass apparatus which he needed in order to study air-sensitive gases. At the base of each manometer was a cup of mercury and it was the evaporation from these many pools of mercury which caused him to become seriously ill. Stock found ways to analyse for mercury in tiny amounts and as a result he was able to prove it was present in the air of every laboratory and workplace – and that included schools. The air in poorly ventilated laboratories can register 9 mg of mercury per cubic metre, which is greatly above the recommended safe upper limit of 0.1 mg per cubic metre. Stock also discovered that there were traces of mercury in almost everything we eat.

Stock's predictions about the dangers of mercury contamination in laboratories were to prove embarrassing for those who were unaware of his warnings. The physicists in the world-famous Cavendish Laboratory of the Physics Department of Cambridge University moved to a new building in the 1970s and their old accommodation was refurbished for use by social scientists, 43 of whom were eventually housed there. Soon they were complaining

---

* Powdered sulfur was spread on floors to remove traces of mercury in the belief that this would react to form mercury sulfide.

of feeling ill and indeed they were being poisoned by the mercury vapour in the atmosphere from the many spills that had taken placed during earlier years. When they were tested in the 1990s, half were judged to be affected and six were even found to have mercury levels in the blood and urine that were more akin to those suffering from industrial exposure. As much as 80% of inhaled mercury is retained by the lungs and absorbed into the blood stream. There it remains for a long time, and following heavy exposure it takes about 60 days for the level to drop by a half, and even after a year there will still be a few per cent of the inhaled mercury circulating in the blood.

Mercury vapour may be dangerous, but under normal conditions so little of it evaporates from a container as not to cause problems unless of course it is spilled, and even then it is not life threatening although it can cause severe illness. This may be difficult for a doctor to diagnose if there is no reason to suspect mercury poisoning. It is a different story if mercury metal is deliberately heated sufficient to cause it to evaporate. The results of breathing that vapour can indeed be fatal, as proved to be the case for the merry monarch, King Charles II.

What killed kings has also killed commoners. Mercury vapour was responsible for the death of a child whose father demonstrated the volatility of mercury to his family by putting it on a hotplate in the kitchen and watching it evaporate. He and his wife were very ill and their child died a few days later. It could also affect people on a grand scale as happened in 1810 when a now-forgotten mass-poisoning of sailors occurred. At the time the dangers of breathing the vapour of mercury was unrecognized although the medical man who had to deal with its results deduced that this was the cause of the outbreak of a curious illness.

The Royal Navy 74-gun man-of-war HMS *Triumph* arrived at Cadiz in February 1810. A month later a Spanish vessel laden with mercury and destined for the mines of South America was driven ashore nearby in a gale. The *Triumph* sent its long boat to her assistance even though the wreck was in range of the guns of a fort held by the French, then at war with Britain. The ship was a write-off but its cargo was worth salvaging. The sailors from the *Triumph* were able to recover 130 tonnes of mercury from the wreck by working secretly at night. The mercury was taken back to Cadiz and stowed in various parts of the *Triumph* and aboard a smaller ship, the sloop *Phipps*.

To begin with, the mercury was placed in the hold where the crew's spirit rations were kept, but there was so much of it that soon the bags were being stowed in sleeping quarters as well, such as those of the petty officers, pursers, and surgeons, all of whom became badly affected. They found their tongues swelling and their mouths were salivating to an alarming degree. The salvaged mercury had been held in leather bags in wooden boxes, but it was only the bags that were salvaged. Many of these now split and spilled their contents.

Soon large amounts of the metal were sloshing about below decks and indeed some of the officers had it rolling about on the floor beneath their bunks. *The Edinburgh Medical and Surgical Journal* of 1810 published a short letter from a reader in Lisbon which told of the incident and in which the writer speculated that it was an 'effluvia' caused by the interaction of the mercury and the leather of the bags that was to blame for the illnesses that afflicted the men. As more and more sailors became affected it was clear that the mercury was to blame and indeed the *Phipps* was beached and holes were bored in its bottom to allow the mercury to run out.

William Burnett MD was the Inspector of Hospitals to the Mediterranean Fleet at the time and later reported what had happened in the *Transactions of the Royal Society* in 1823. In this he noted that Michael Faraday had recently carried out research into mercury and shown that it did indeed emit vapour and it was probably his reading of Faraday's researches that caused Burnett to put pen to paper, because at last he could explain for the strange happening aboard the two ships. By 10 April 1810 around 200 men on board the *Triumph* were suffering from mercury poisoning which caused excess salivation in some, while others were semi-paralysed and many suffered 'bowel complaints'.

The sick were taken to other ships where they soon recovered, while the *Triumph* itself was sent to Gibraltar to be decontaminated. Not that this was effective because a new crew also started to suffer in the same way. The ship was despatched back to England on 13 June and then things did begin to improve somewhat, thanks to the movement of the vessel and the ventilating of the lower decks. Even so, 44 sailors and marines had to be transferred to other ships in the fleet and they had recovered by the time they reached Plymouth on 5 July. All the sheep, pigs, goats, and poultry on the *Triumph* died, as did the ships cat, a dog, the mice, and rats – and a canary. Five men eventually died, two of gangrene of the cheeks and tongue. A woman passenger, who had a fractured leg and was confined to bed during the voyage, lost all her teeth and the skin on the inside of the mouth all peeled away.

Doctor Burnett prescribed sulfur for those who were sick but reported that taking this did not relieve their symptoms. The only effective remedy was to be removed from the ship. He also noted that various metal objects on board were found to be covered with black stains, and this included a gold watch and silver and gold coins. The ship carried 7,940 pounds of ships biscuit and this was all condemned as unfit to eat, and some was even found to contain globules of metallic mercury.

The air of all rooms in which mercury is spilled soon become contaminated with the vapour. Unless this is known to have happened then it may result in illnesses that are wrongly diagnosed because the true cause is unrecognized. Sometimes the doctor treating those affected may be sufficiently knowledgeable to realize the cause of his or her patient's condition. One 10-year-old boy who took a flask containing 250 ml home from school, played with it and eventually spilled most of it on furniture and carpets. The family then began to suffer from mercury poisoning which most affected his mother and 14-year-old sister, who became very ill. It was the girl's symptoms that enabled the cause of their illness to be diagnosed because the doctor recognized that she was suffering from pink disease – see later – and knew that this was caused by mercury.

Inorganic mercury salts

Mercury(II) chloride, once better known as corrosive sublimate, is a strong disinfectant and was use both in hospitals and homes for this purpose during Victorian times. It came in the form of Van Swieten's solution which consisted of 0.1% mercury chloride dissolved in a water/alcohol mixture. Naturally accidents occurred, but egg white was found to be a suitable antidote and prompt first aid treatment with this followed by an emetic or stomach pump was capable of preventing the mercury chloride from being absorbed into the body tissues and doing its deadly work. A large dose of mercury(II) chloride will cause an immediate reaction typical of a heavy metal poison, whereas similar dose of mercury(I) chloride (calomel) would not produce such a violent reaction. Nevertheless, even this would still cause some noticeable effects such as copious salivation.

Mercury(II) chloride is soluble in water to the extent of 7 grams in 100 mls at 20°C. A lethal dose is thought to be about 500 mg, although death has been caused in children by as little as 100 mg. A lot depends on how much of the poison is absorbed before the body's defences, such as vomiting, come into play. People have been known to survive doses of 5,000 mg even before the use of modern antidotes, provided first aid treatment was promptly given. Modern methods have saved the lives of people who have taken a teaspoonful of the chloride, which amounts to 20,000 mg, provided antidotes were administered without delay.

Quite simple treatment could sometimes be effective. In 1581 at Baden in Germany a condemned criminal who was due to be hanged struck a bargain with the judge: he would consume a deadly poison instead, provided he could be given some clay to eat at the same time. The judge agreed and the man was given one and a half dram (about 6,000 mg) of corrosive sublimate which was known to be more than double the fatal dose. Five minutes after taking the poison, the prisoner swallowed a similar amount of an earth known as *terra sigillata* stirred into wine. The prisoner was then closely observed for many hours and it was reported 'the poison did extremely torment and vex him; yet in the end the medicine overcame it, whereby the poor wretch was delivered and being restored to his health, was committed to his parents.'

The prisoner had proved what had long been suspected: that certain clays were capable of acting as antidotes, and especially to metallic poisons which by their very nature have the metal positively charged. This means they are attracted to and held by the negative silicate particles of finely divided clay and so carried out of the body. *Terra sigillata* was ideal, being a clay that originally came from the Greek islands of Lemnos and Samos, and it was used to make the red pottery that was especially popular in Roman times and was known as samian ware.

Mercury(I) chloride was widely used, and as calomel was on sale in over-the-counter medicaments, especially suitable for dogs (as a de-wormer), humans (as a laxative), and babies (as teething powder). Even so it was poisonous. In babies, mercury poisoning would manifest itself as pink disease, so named because of the pink coloration of the fingers, toes, cheeks, nose, and buttocks; medically it was known as acrodynia or erthroedema. In adults, the pink coloration of the skin was seen as a beauty aid and until recently there was a Mexican beauty cream, Crema de Belleza Manning, which relied on calomel to achieve the desired effect. Even as late as 1996 the US Centres for Disease Control and Prevention issued a warning about it saying that it could contain up to 10% by weight of calomel. More than 230 people in Arizona, California, Texas, and New Mexico responded to a warning about the cream and of the 119 who underwent tests, 87 were found to have elevated levels of mercury in their urine. Three people who used the cream in Texas actually needed clinical treatment for mercury poisoning.

The effect of mercury on the gums was at one time thought to be beneficial for babies and young children cutting their first teeth. The sleepless nights for both child and parents which accompanies this phase of their development led many to seek relief with so-called teething powders. In the north of England, where they were particularly popular, the sale of such powders had reached a figure of 7 million annually by 1953. The powders, which were first introduced in 1812, relied on the calomel they contained to cause excessive salivation. Some babies that were especially sensitive to mercury died, although for most babies the occasional powder was unlikely to be fatal. One 10-month-old girl who died had been given 64 such powders over a period of six months, and before her death she had been excreting mercury at the rate of 3.9 mg per litre of urine, over ten times the normal rate.

Pink disease had a mortality rate of 10% and accounted for one in 25 hospital admissions in the 1940s. It was thanks to two American doctors, D. Warkany and J. Hubbard, that the cause was traced to teething powders, and in particular to Steedman's Teething Powders which contained 26% mercury(I) chloride. The reason it took so long to make the link between mercury and pink disease was that only one child in 500 who was given the teething powder developed the pink condition.

Medicinal mercury

Until the advent of modern drugs, there were certain conditions where mercury compounds were beneficial, and so they found widespread use. Solutions of mercury(II) salts made very effective antibacterial agents because the mercury(II) ion, Hg^{2+}, forms insoluble salts with their proteins and kills them. As late as the 1970s mercury compounds were still to be found in the pharmacopoeias of most countries although they had fallen into disuse as new and better products took their place. Nevertheless, Table 2.1 shows ten such mercury-based preparations that were available until recently.

The use of mercury in medicine has a long history. Paracelsus (1493–1541) was a strong advocate of its benefits. Lord Chamberlain Hunsdon in Queen

Table 2.1 Mercury-based medicaments available until recently

Medicament	Ingredients	Specified treatment
Blue pill*	Mercury with sugar	Strong laxative
Calomel	Mercury(I) chloride, Hg_2Cl_2	Laxative; dusting powder for eczema and anal itch
Channing's solution	Mercury(II) iodide, HgI_2, solution in water	Antiseptic; disinfectant
Golden eye ointment	Mercury(II) oxide, HgO	Conjunctivitis
Grey powder	Mercury with chalk (calcium carbonate)	Laxative; syphilis
Harrington's solution	Mercury(II) chloride, $HgCl_2$, solution in alcohol	Antiseptic; disinfecting skin prior to surgery
Mercuric cyanide solution	Mercury(II) cyanide, $Hg(CN)_2$	Eye lotion
Mercuric nitrate solutions	Mercury(II) nitrate, $Hg(NO_3)_2$	Concentrated solution used to remove warts. Dilute solution used as nasal drops and ear drops, and to bathe eczema and psoriasis
Mercury ointment	Mercury metal dispersed in beeswax	Boils
Red mercuric iodide ointment	Mercury(II) iodide, HgI_2, in animal fat such as lard	Ringworm

* President Abraham Lincoln took these, which may well have accounted for his mood swings.

Elizabeth I's reign was a great believer in mercury therapy. The Chamberlain Restorative Pill was concocted from cinnabar, and was prescribed for scurvy and other diseases. Personal letters of his still survive and his handwriting was very shaky indeed, which is just what is to be expected of a man who dosed himself on pills containing mercury.

Metallic mercury dispersed in fat was used as an ointment for skin complaints. The grinding of metallic mercury with oils and fats was referred to as 'killing' the mercury and the ointments so produced were a stock in trade of pharmacists up to the 20th century, going under such names as blue ointment (25% mercury, 24% lard, 50% vaseline), and grey oil (mercury, lanolin, and olive oil). The mercury was finely dispersed in these mixtures and this accounted for their action. Such salves first appeared in the 13th century and their application was known to induce excess salivation, which we now realize meant that the mercury was being absorbed through the skin and into the blood stream. Giovanni de Vigo (1450–1525), who became the personal surgeon to Pope Julius II, was very aware of the new disease syphilis that had appeared in Europe in the 1490s and which was characterized by skin eruptions in the genital regions. He invented a plaster to be applied to these open sores. Its active ingredient was grey oil, and it proved effective, so much so that Vigo plasters remained in use for 300 years.

Syphilis spread throughout Europe in the 1500s and the older medicines were powerless to deal with it. It was the ability of mercury remedies to alleviate the symptoms of this that led to the rise of the empiricists, of whom the best known was Paracelsus. He concocted pills made from the red oxide of mercury and cherry juice, and used dilute solutions of corrosive sublimate in lime water as a lotion for bathing venereal ulcers. Mercury compounds were effective in treating syphilis because they killed the organism *Treponema pallidum* that caused the disease, but the 'cure' as it was known, was risky and almost as feared as the disease itself.

It is still possible to discover those who were given this treatment by examining a lock of their hair, if it has been preserved. Hair contains a lot of sulfur-containing amino acids and these attract mercury and provide a permanent record of the level of exposure to the element. It was often the custom in earlier times to cut locks of hair from those who had died, and especially of famous people, and when these have been analysed by modern methods, unusually high levels of mercury have been found, leading some to suspect that the person has been medically treated with mercury medicaments and by implication they might well have been suffering from syphilis for which mercury was the only effective treatment.

It has been tempting to assume that when famous people were treated with mercury the reason was syphilis and King Henry VIII of England and Ivan the Terrible of Russia were both dosed this way. Samples of their hair are not available for analysis, but that of eminent individuals in later generations has been analysed. Hair from the Scottish poet Robert Burns (1759–1796) contained a lot of mercury, suggesting that this famous lecher had been treated for venereal disease at sometime in his life. Napoleon's hair also shows an above average level of mercury, probably due to his being given calomel when he was taken ill on St Helena, but this was not the cause of his death; arsenic probably was the poison that killed him – see Chapter 7.

Not all treatments need imply syphilis or alchemy. One of the saddest deaths of a scientist was that of the Danish astronomer Tycho Brahe (1546–1601) who suffered with prostate trouble. At a royal banquet in Prague he dare not leave the table to relieve himself, with the result that his bladder split and he died of urinary poisoning a few days later. Analysis of a strand of his hair, which had the root intact, showed that the day before he died he was given a mercurial medicine in an effort to save his life.

Mercury is still part of traditional Chinese medicine and two preparations, Antidotal pills and Cinnabar sedative pills contain around 4% of mercury compounds. The former is given at the rate of four pills three times a day for poisonous insect stings, delivering a dose of 0.2 g of mercury, the latter at five pills three times a day as a calmative, delivering as much as a gram.

Theodore Turquet de Mayerne was expelled from the College of Physicians in Paris in 1603, even though he was the favourite doctor of King Henry IV. This was because of his religious and medical beliefs. He was both a Protestant and an empiricist, and critical of the doctrines of the great physician of the Roman Empire, Galen, whose teaching still held sway. Mayerne's crime was to have written a defence of mercurial and antimonal medicines. Mayerne was judged to be 'stuffed with lying reproaches and impudent calumnies, which could not have proceeded from any but an unlearned, impudent, drunken, mad fellow' and that he was 'unworthy to practise physick in any place because of his rashness, impudence, and ignorance of true physick'. His contemporaries wanted rid of him, and once expelled from the College of Physicians he could no longer practise his profession. He arrived in London in 1611 and became James I's physician. He believed in mercury and is attributed with the discovery of calomel. Actually, Mayerne was not its discoverer, and there is evidence that it was known not only to the Arabs but even to Democritus, the Greek philosopher.

Several mercury compounds were in use by Mayerne's time such as *turpeth* (basic mercury sulfate, $HgSO_4.HgO$) and *sal alembroth* (mercury chloro amide, $HgNH_2Cl$), both of which were discovered during the 1500s. Throughout the following centuries, mercury compounds like these continued to be popular in medicine, especially in the treatment of syphilis. Turpeth mineral tablets were sometimes inserted into the vagina as a contraceptive, although this could result in ulceration of the vagina, and occasionally death. Baron von Swieten gave his name to a solution of mercury(II) chloride in whisky which was to be taken morning and night and involved an

intake of 30 mg of mercury(II) chloride a day, which is about a twentieth of
the dose that puts life in danger. In general practice a first dose of mercury(II)
chloride of 15 mg (a fifth of a grain) was deemed safe, but only to be followed
by much smaller doses of around 3 mg (a twentieth of a grain).

Although the sale of over-the-counter mercury medicines was ended by
law in the UK in 1955, those affected by their use before this time continued
to present themselves for treatment for the following 40 years, and at one
London hospital more than 120 such patients were seen in the years 1975–
1993. In other countries mercury products took longer to disappear from
commercial use. In 1981 in Argentina more than 1500 infants had to be
treated for mercury poisoning, the cause being a mercury-based disinfectant
that was used in the laundering of nappies (diapers). And as late as the
1990s, mercury iodide (1–2%) was added to soap as a disinfectant, and
Africans who used it found their skin turning a shade lighter. Such soaps
were being widely sold in countries like Nigeria, mainly for their cosmetic
effects.

Less dangerous was the supposed medical benefit of having a small sample
of metallic mercury next to the body. This tradition started in the 1600s
when pharmacists in London sold tubes of mercury that were to be carried
on one's person and which they claimed would ward off rheumatism. In the
1990s they re-emerged in Canada as the Tenex Elbow Shock which contained 30 grams of mercury in a plastic container. This was to be strapped
to the wrist and which it was claimed would prevent tennis elbow. More
than 100,000 were sold but were completely useless and are no longer
manufactured.

Mad cats and mad hatters:
accidental mercury poisoning

THERE are two kinds of mercury poisoning: chronic poisoning in which the body is subjected to regular small doses of mercury which exceed the amount it can excrete every day, and acute poisoning, in which a person is exposed to a life-threatening dose. It is the former type of poisoning that this chapter is about. Large doses deliberately given will be the topic of the next chapter.

Chronic mercury poisoning used to be an occupational hazard for many employees. Those affected suffered from the physical symptoms of fatigue, general weakness and a tremor of the hands, to the extent that their handwriting became spidery, and these symptoms were due to the effects on the central nervous system. More serious were the psychological symptoms such as irritability, depression, and a paranoid belief that other people were persecuting them, all of which came as a result of mercury seeping into the brain. The groups of workers most at risk from chronic mercury poisoning were gilders, hat makers, dentists, those in the electrical industries – and detectives. Most of these occupations no longer use mercury, and in those that do it is strictly controlled so that the risks are now negligible.

Monitoring those exposed to mercury in their employment can be done via their urine or blood. Yet it was a long struggle to make people aware of the dangers this metal posed, and along the way there were some major examples of exposure involving hundreds of thousands of individuals, many of whom had their lives made wretched by mercury. Indeed the campaign against mercury really started 300 years ago when an Italian physician was the first to become interested in the link between occupation and illness.

That physician was the surgeon Bernardino Ramazzini (1633–1714), who is today regarded as the founder of occupational and industrial medicine. In 1700 he wrote the first book on the subject: *De Morbis Artificum Diatriba* [The Diseases of Workers]. In it he outlined the health hazards associated with various chemicals, dust, and metals encountered by those working in 52 different occupations, including the miners who worked in the mercury mines. He was even aware that the mercury treatment for syphilis could affect doctors as well as patients even though the former wore leather gloves when applying the mercury ointments to syphilitic sores. Ramazzini was aware that the mercury could penetrate the leather and poison the doctor, especially if there were many patients to treat. The answer, said Ramazzini, was to give the ointment to the patient and tell him to apply it himself!

Mercury mining, and mining with mercury

Native mercury occurs naturally as tiny droplets in mineral deposits of the ore cinnabar, which is mercury sulfide (HgS).* Mercury itself is abundant in some cinnabar mines and in the Rattlesnake mine, in Sonoma County, California, it would sometimes come spurting out of the ore when a pick was sunk in. Generally it was released by heating the ore which caused the sulfur component to react with the oxygen of the air and be lost as sulfur dioxide gas, leaving metallic liquid mercury behind.

World production of mercury is currently around 1,500 tonnes per year, and it is still traded traditionally in so-called 'flasks', a quantity that has been used since Roman times, and these contain 76 imperial pounds (just over 35 kg) of mercury. Mineable reserves of mercury are said to amount to around 600,000 tonnes, mainly located in Spain, Russia, and China, with small deposits in Slovenia and Italy. World production of mercury had been declining in recent years but it used to be around 10,000 tonnes annually, of which 6,000 tonnes would eventually be lost to the environment. The demand for mercury in the USA has declined markedly during the past

---

* Cinnabar is one of three common mercury-containing minerals. The others are calomel which is mercury(I) chloride (formula Hg_2Cl_2), and metacinnabar, which is an alternative form of mercury(II) sulfide.

25 years, from a high of 2,200 tonnes per year in 1975 to less than 200 tonnes now. Mining for mercury in the USA ended in 1990.

The cinnabar deposit at Almaden in central southern Spain has been mined for more than 2,500 years and deposits in China have been mined for even longer. Who discovered these deposits is not known, but the discovery of one economically important deposit is on record, and that was the one at Idrija, Slovenia. In 1490 a barrel-maker found mercury droplets at the bottom of one of his barrels, which he had filled from a local well, and digging soon revealed the bright red ore. The mine at Idrija was the largest after that at Almaden, and from its cinnabar ore 107,000 tonnes of mercury were extracted during the 500 years of its operation, amounting to 13% of all the mercury ever produced. More than a million tonnes of rock were extracted from its shafts, which eventually extended for more than 700 km. Elemental mercury occurs in the shale of these galleries and it was known to be a particularly dangerous mine down which to work. Indeed when the level of mercury vapour was measured it was found to be as high as 6 mg per cubic metre.

The mines at Idrija were important for another reason, and that was that the miners came to the attention of the great physician Paracelsus, who described the various symptoms that they had to live with: most notably profuse salivation, mouth ulcers, and tremor of the hands. Not that much was done to alleviate the miners' plight, although in 1665 the working day was reduced from 8 hours to 6 hours in an effort to improve their health. Not only the miners suffered from Idrija's mercury. In 1803 the mine caught fire on account of the oil-bearing shale in which the cinnabar was to be found. As a result, mercury vapour escaped from its shafts, polluting the surrounding countryside and affecting 900 people and their livestock.

In historical times almost all the mercury that was mined was destined to be blown to the winds, and literally so. It was used to extract and purify gold, which was the basis of all currencies as well as being used for religious objects, personal adornment, and tableware of the wealthy. Mercury has the remarkable ability to dissolve gold and it was used to extract gold from alluvial sediments. The amalgam so formed would then be heated to drive off the mercury, leaving pure gold behind. Although it was possible to condense, collect, and re-use some of the mercury, the process generally resulted in about half of it being lost to the atmosphere.

Because of their access to the mercury mines of Almaden, the Spanish were ideally resourced to extract the gold of the New World from the 1500s onwards. For every tonne of gold and silver that was shipped from the Americas to Europe, about a tonne and a half of mercury made the journey in the opposite direction. And mercury really was shipped in large quantities. For example, the inventory of the *El Nuevo Constante*, which left Cadiz on 7 December 1765, had on board 85 tonnes of mercury in 1,300 boxes, each holding two flasks, which it delivered to Veracruz, Mexico, on 27 February 1766. (The ship was lost in a storm on the return journey in May that year, loaded with bars of silver.)

Eventually a source of mercury was discovered in Peru and there the Huancavelica mines provided much of the mercury used to extract gold and silver in South America. In North America, where gold mining developed much later, the mercury came from the New Almaden mercury deposit in Santa Clara County, California, which was discovered by a Mexican settler in 1820 although mining it did not begin until 1845. This mercury found a ready market in the California Gold Rush of 1849, and its supply was more than adequate to meet the needs of the gold diggers. In the 400 years of mining in the Americas the cumulative total of mercury lost to the environment has been estimated to be about a quarter of a million tonnes.

Pollution of the Amazon by mercury from small-scale gold mining occurred extensively in the 1980s and 1990s and by 1994 it had been estimated that 1,200 tonnes of mercury had been flushed into its tributaries. The gold miners were simply pumping sediment from the rivers and river banks down a sloping sluice crossed at intervals by wooden ridges which slowed down the flow and allowed particles of gold to collect. At the end of the day the miners tipped mercury down the sluice and this formed an amalgam with the trapped gold. The gold was then released by heating the amalgam in a shallow iron dish to evaporate off the mercury.

The European Union was alerted to what was happening in the early 1990s and decided to fund a project to prevent the loss of mercury, a project organized by Imperial College London. With their help, the gold miners were shown how two simple devices developed in Germany could reduce mercury loss by 95%; the first was to slow down the flow of mercury by using a gentler sloping sluice with a trap at the bottom to prevent any loss of it into the river; the second device was a sealed crucible in which the gold amalgam could be heated, and attached to which was a condenser from which the mercury could be recovered. Both devices were easy to make and would pay for themselves in terms of reclaimed mercury within a few days.

The many uses of mercury

Down the centuries, humans have found more than a thousand different applications for mercury, and while most of these have now been abandoned, many persisted well into the second half of the last century. Thermometers, thermostats, barometers, UV lamps, fluorescent lights, contact switches, batteries, medicaments, disinfectants, antifungal agents, agricultural chemicals for protecting seeds, and detonators all made use of metallic mercury or its compounds. One such compound was mercury fulminate* which explodes when it is struck and made an excellent detonator. This was first made by a British chemist Edward Howard in 1799, and he presented a paper on it, and demonstrated it, at the Royal Institution in London in 1800. It soon found use as a primer for percussion caps and explosives. Another compound with commercial applications was mercury(II) chloride which John Howard Kyan (1774–1850) patented as a method of preserving wood in 1832. Wood treated this way was used in ship building and harbours as well on famous buildings in London. (The Kyan method was eventually superseded by a cheaper method which used creosote.)

Much less mercury is now used in consumer batteries and fluorescent lighting, but it has not been entirely eliminated. In the case of batteries, mercury is now restricted to button cells for hearing aids and other small electronic devices. The batteries consist of an outer zinc casing, which acts as the positive electrode, filled with a paste of zinc hydroxide and mercury oxide surrounding a small steel negative electrode. The benefit of these batteries is that they provide a steady voltage of 1.35 for all their working life. In the case of fluorescent lights, a metre-long tube now contains only 10 mg of mercury as opposed to more than 35 mg in previous versions.

Mercury is still used in some electrical gear, such as switches and rectifiers, which need to be reliable, and in the chemical industry as a liquid electrode in the manufacture of chlorine and sodium hydroxide. These are produced by the electrolysis of sodium chloride solution, although this method is being phased out in favour of ones that do not require mercury.

* Chemical formula $Hg(CNO)_2$.

Some mercury is still used to treat seed corn to make it resistant to fungus disease, while in developed societies the nearest most people come to the metal is in dental amalgam fillings. Older uses such as in thermometers, felt production, plating, tanning, dyeing and in medicines have all been superseded.

Although many workers were directly exposed to mercury – more than 30,000 in the USA in the middle of the last century – relatively few had to be treated for illnesses related to it. In the UK, where fewer workers were exposed, the number of officially notified cases of mercury poisoning was on average only five a year. Some trades were notorious for the effect the mercury had on those engaged in them and it was among hat makers that it was particularly noticeable, witness the condition known as 'hatters' shakes' in the UK and 'Danbury shakes' in the USA. The town of Danbury, Connecticut, was the centre of the American hat industry, while in the UK it centred on Stockport in Cheshire, where there is now a Hat Works Museum.

Of the 544 hatters who were medically examined in the USA in 1941, 59 showed signs of chronic mercury poisoning and the cause was the air they were breathing. On the face of it, mercury would seem to be an incongruous part of such a trade. Hats were a fashion item for both men and women for hundreds of years until they fell out of favour in the 1960s. Many were made of felt and felt was made from the fur of rabbits, hares, muskrats and beavers all of which produce fur that is smooth, resilient, and straight. That was the problem.

In order to get the fur to matt together, to produce the felt from which hats were made, it had to be treated chemically with an acidic solution of mercury nitrate. From that point onwards, all who were involved in the hat-making trade were liable to be affected, from those who first processed the fur to those who shaped the hats. The felt went through numerous processes (blowing, forming, hardening, sizing, shaping, ironing, planking, proofing, stoving, and pressing) and at each stage the dust from the felt would contaminate the workplace. The air in some workshops was found to contain 5 mg of mercury per cubic metre. At one time, about 40% of the fur cutters in the hat trade were affected by chronic mercury poisoning, and they showed the classical symptoms of mercury poisoning: always irritable, paranoid about being watched, talking incessantly, and prone to irrational behaviour, so much so that the old-fashioned phrase 'mad as a hatter' was thought to have derived from their behaviour.

The centuries-old trade of gilders also involved exposure to mercury, because gilding was done with mercury amalgams of gold, silver, and tin. These were used to gild buttons, mirrors, and even domes. In the early 1800s the cathedral of St Isaac at St Petersburg, Russia, was built and its domes were gilded, which required 100 kilograms of gold to be applied to its copper sheets, and this was done using gold–mercury amalgam. The result was that 60 workmen lost their lives through mercury poisoning.

Gilding of buttons with gold, and mirrors with tin, also exposed gilders to the vapour of mercury. Buttons were gilded by rolling them around in a container with gold–mercury amalgam (one part gold to ten parts mercury), and then when they were coated, they were put in a wire cage and heated to drive off the mercury leaving the gold on the buttons. At least 20% of the mercury was lost to the local environment. Such buttons were very much part of military and naval uniforms and gilding them this way continued into the 20th century. The advantage of the process was that a gram of gold would gild 500 buttons. Those engaged in the trade suffered from 'gilder's palsy', and those living nearby were also exposed to mercury. In Birmingham in the early 1800s the metal collected in gutters in the streets surrounding such factories. Despite technical improvements for trapping out the mercury from the flues of gilding establishments the process only ceased after 1840, when electroplating took its place. However, Britain's Royal Navy and merchant marine continued with the older form of gilding for a further hundred years because it gave a more durable product.

Until the middle of the 19th century, mirrors were made using silver amalgam and those engaged in this trade suffered accordingly, until the German chemist Justus von Liebig (1803–1873) showed that silver mirrors could be made by chemical means that did not require liquid mercury.* Prior to this the gilding of glass was done with silver amalgam or tin amalgam. In the latter case this involved the gilder laying a thin layer of tin leaf on a slab of marble, pouring mercury over it to form the amalgam, laying the sheet of glass on the amalgam, making sure no bubbles of air were trapped by it, then weighting down the glass and leaving it for 3 weeks. Not surprisingly the men who did the work were severely afflicted. In Germany, the gilding of mirrors

---

* He discovered that when acetaldehyde was poured into a solution of silver nitrate and ammonia, a mirror of silver deposited on the glass flask in which the reaction was carried out. This chemical reaction eventually became the way of making mirrors.

was carried out in Fürth and Nuremberg and it was said that not a single man in those towns had a tooth in his head because of mercury poisoning.

The manufacture of thermometers once accounted for several tonnes of mercury a year. Although mercury thermometers were first made in the 1650s, they did not become part of medical practice until clinical thermometers, with their restricted temperature scale, were invented in 1866 by a Dr Thomas Allbut (1836–1925) of Leeds General Infirmary, Yorkshire. These continued to be used in hospitals until the early 1990s and a typical large hospital of 1,000 beds would order about 2,000 such thermometers a year because they were always being broken. In 1985 the UK's Department of Health tested the air in special baby care units and discovered unacceptably high levels of mercury vapour from this source. Sweden was the first to ban mercury thermometers in 1992, and today almost all devices used by doctors and nurses to measure temperature are electronic.

Broken thermometers have played unexpected roles in promoting new industries. In 1830 Louis Daguerre was experimenting with a photographic process that used a thin layer of light-sensitive silver iodide deposited on a copper surface. The images produced were not particularly good. Then one day he stored his exposed photographic plates in a cupboard, which unknown to him also held a broken thermometer. When he took the plates out the following day he was surprised to find the images wonderfully developed. It took him some time to discover what had caused this marked improvement, but eventually he realized it was due to mercury vapour. Daguerre modified his photographic process and began to produce portraits and pictures that became known as daguerreotypes, and which are now highly sought after.

Another broken thermometer was responsible for the manufacture of artificial indigo in Germany in the 1890s. This had been synthesized in the laboratory by Adolf Bayer, but making it commercially was proving too difficult because the first stage of the process – the conversion of naphthalene to phthalic anhydride – could only be achieved using costly or dangerous oxidizing agents, such as hot fuming sulfuric acid. When the thermometer measuring the temperature of this process broke, and spilled its mercury into the reaction mixture, the process suddenly speeded up as if by magic. The mercury acted as a superb catalyst. Cheap indigo dyes and blue jeans were on their way. (The mercury catalyst was eventually phased out in favour of one made of the oxides of vanadium and titanium.)

Mercury thermometers may no longer be much used in hospitals in developed countries but they are still used in developing countries where they play an important role in medical diagnosis and patient monitoring. Their manufacture, however, can create local problems as occurred at the Hindustan Unilever factory in Kodiakanal, South India, which was a major manufacturer and exporter of mercury thermometers. Then, in March 2001, Greenpeace India organized a protest march to highlight the lax state of affairs in the disposal of factory waste, which was being sold to a local scrap dealer. He had collected more than 10 tonnes of broken thermometers, and when Unilever scientists tested the air in his scrap-yard, they detected mercury and measured this at a maximum level of mercury of 0.02 mg per cubic metre, although this is less than the 0.05 mg that EU regulations permit. Workers in the thermometer factory were not affected, as they had been regularly monitored, but production was suspended while an audit of the company's mercury was carried out.

Currently the largest user of mercury is the chlor-alkali industry, which produces the sodium hydroxide and chlorine gas that the chemical industry needs. The new Clean Air Act regulations introduced in the USA in December 2003 is aimed at cutting emission of mercury from the chlor-alkali plants to 220 kilograms per year, and as older plants come to the end of their useful lives then this figure will fall even further. In the past such plants have 'lost' as much as 200 *tonnes* of mercury annually, and at one time there were as many as 35 such facilities operating in the USA. Today there are only nine such plants remaining and these have around 3,000 tonnes of mercury in use, and are still needing to order around 60 tonnes of extra mercury every year.

As the chemical industry in developed countries phases out its older chlor-alkali plants it is faced with the problem of what to do with the unwanted mercury. In Europe this is bought back by the original producers, the Almaden Mining Company, which is owned by the Spanish Government. In the USA there are vast amounts of mercury being released for re-sale. Some argue that rather than sell this to developing countries such as China and India it should be bought by the US Government and added to the 4,000 tonnes stockpile of mercury held by the Defense Department. This was built up in the 1950s and 1960s because it was needed at one time to produce enriched uranium for nuclear weapons.

A rather cavalier attitude to mercury disposal was once the order of the day and as a result some chemical companies found themselves in trouble.

Thor Chemicals based in the UK reprocessed mercury at its plant in Margate but checks there showed its workers were exposed to more than the law allowed. What Thor did was to close the factory down and start up a similar plant in Natal, South Africa, and within a few years there was a spate of mercury poisoning incidents there. Meanwhile in the USA the now-defunct Linden Chemicals & Plastics of Brunswick, Georgia, was operating a chlor-alkali plant in a less than environmentally friendly way. It simply pumped its waste water into a nearby creek that ran into a tidal marsh, and over the years Linden had disposed of nearly 150 tonnes of mercury in this way. In fact the company had contravened almost every act designed to safeguard the environment and it ceased operations in 1994, thereafter filing for bankruptcy. The cost of cleaning up the mess it had left behind was estimated to cost more than $50 million. The chief executive of the factory, 72-year-old Christian Hansen Jr., was sentenced to 9 years in jail, along with six other company officials.

Organo-mercury

In the previous chapter we looked at the effects of mercury as the metal, as its vapour, and as its inorganic compounds. However, its most insidious form is *organic* mercury, better known as organo-mercury, which is a technical term meaning that one or two carbon atoms are bonded directly to the metal, as in methyl mercury chloride or dimethyl mercury. As we shall see, methyl mercury is worryingly mobile, both in its ability to circulate through the environment, move up food chains, and even to travel unchecked around the human body. (A methyl group consists of a single carbon atom and three hydrogens: CH_3.)

The first recorded deaths from dimethyl mercury occurred in London in 1865. In January of that year, two laboratory assistants at the Royal Institution, in the laboratory of Professor Edward Frankland, were reacting sodium-mercury amalgam with methyl iodide with the objective of making dimethyl mercury. They were completely unaware of the dangers it posed. One of the men, a 30-year-old German, Dr Ulrich, was admitted to St Bartholomew's Hospital on 3 February and died ten days later, the other man, 23-year-old Mr T. Sloper, was admitted on 25 March but survived a year before expiring. These deaths were used as an excuse to attack the eminent Frankland in the press but it eventually transpired that the assistants had been working under the supervision of a Professor William Odling. Moreover, Ulrich had told a friend that his exposure had been an accident, caused by his breaking a sealed tube of dimethyl mercury and spilling the contents, with the result that he and Sloper had inhaled a lot of the vapour while clearing up the mess.

Not that this accident prevented others using dimethyl mercury as a possible medical treatment and in 1887 a series of tests began in which injections of it were tried as a cure for syphilis. The dosage was 1 ml of a 1% solution, but no patient received more than two injections because the experiment was stopped when tests on dogs showed how dangerous the substance was.

More recently, dimethyl mercury accidentally caused the death of a woman chemist, Karen Wetterham, who was a professor at Dartmouth College, New Hampshire. She died on 8 June 1997 aged 49. She had used dimethyl mercury when analysing compounds by nuclear magnetic resonance (NMR) – see Glossary. Karen was studying the way in which mercury ions interact with the proteins that repair DNA and she had made up a solution of two materials and was about to study it using NMR. She had taken the usual precautions of wearing safety glasses, latex gloves, and was handling her materials in a fume hood. However, on that fateful day of 14 August 1996, as she removed some dimethyl mercury from its vial with a pipette, drops of the liquid fell on to one of her gloves. She did not immediately remove the glove, thinking that it would protect her, but dimethyl mercury penetrates rapidly through latex and it then went through her skin.*

It was only in January 1997 that the first symptoms of mercury poisoning began to appear: tingling sensations in her fingers and toes, slurred speech, unsteady gait, and visual disturbances. Mercury poisoning was diagnosed on 28 January and blood analysis showed a level of 4 mg per litre, which is more than 50 times the toxic threshold. Two weeks later she slipped into a coma from which she never recovered. Analysis of her hair showed that her body had absorbed the methyl mercury in August. Why it takes so long to exert its toxic effect is still not known, but the body can only excrete about 1% a day of its methyl mercury burden and this is not fast enough to stop its insidious action.

---

* Ideally when handling dimethyl mercury a researcher should wear heavy-duty protective gloves over an inner latex pair.

Methyl mercury is particularly dangerous because it attaches itself to the sulfur atom of a cysteine amino acid and the molecule then appears to be indistinguishable from another amino acid, methionine, at least as far as gaining access to cells of the body which mistake it for the methionine they needs. This deceit by mercury enables it to move freely throughout the body and even to cross the blood/brain barrier and that of the placenta. Once inside the brain, the methyl group may be removed from the mercury atom, thereby converting the mercury to inorganic mercury, but there is no mechanism by which this can then be transported out of the brain. When methyl mercury gets into the brain of a baby or a child, it stops cell division and blocks microtubule formation leading to permanent brain damage.

Although methyl mercury is dangerous, there are other organic groups that can be attached to mercury, such as the *ethyl* group which has two carbon atoms, and while they may be much less dangerous they can still kill. Such compounds began to be produced by the chemical industry in the 1920s for use as antiseptics, seed disinfectants, fungicides, and weed-killers. By the 1960s more than 150 proprietary products were on sale, based on ethyl or phenyl* mercury compounds. They were manufactured as anti-microbial agents, chiefly to control the spread of fungal disease in plants, and were thought to be much safer than methyl mercury compounds which were rarely used. One of the most effective seed dressings was Ceresan, which was a 2% solution of ethyl mercury chloride. It prevented all kinds of fungal diseases that could blight a crop and destroy a farmer's livelihood, and all over the world the treated seeds were eagerly bought because crop yields were greatly improved by their use. Wheat, barley, oats, and corn that are grown from treated seeds absorb very little of this mercury.

Sadly this form of crop protection led to several mass poisonings in developing countries when villagers made bread from the treated grain. In Iraq in 1956 and 1960 it found its way on to the open market and was used to make flour causing many people to be taken ill. A more serious outbreak occurred there in January and February of 1971 when more than 6,500 people needed hospital treatment and 460 people died. Similar, but smaller, outbreaks of

* The phenyl group is the name given to a benzene ring when it is bonded to something else.

poisoning occurred in Pakistan and Guatemala. As long ago as 1942, two secretaries employed at a warehouse in Calgary, Canada, died because of a consignment of 20,000 pounds (9,000 kg) of diethyl mercury which was stored near the desks where they worked. They died from mercury poisoning even though they never actually handled the material. The fumes from the drums were responsible, and later tests showed that the concentration of mercury in the air was almost 3 mg per cubic metre. Warnings had already been issued about the dangers of such organo-mercury compounds, but they had been ignored.

Those warnings came in 1940 as a result of what had happened to a bright 16-year-old English boy Arthur H (the medical records do not reveal his surname) who had done well at the technical school he attended, and who got his first job with a company that made mercury seed dressings. Within 5 weeks of starting work he began to notice his fingers and toes were numb, and as the weeks went by he became clumsy in his movements and unsteady on his feet. His personality changed so that he became confrontational and abusive. His handwriting degenerated into a scrawl. Mercury poisoning was diagnosed and he was hospitalized, but doctors watched helplessly as he deteriorated, until he spent all his time lying listlessly in bed, unable to speak and barely able to eat. Eventually he began to improve, but it was six months before he could walk again and nine months before he could communicate with people. Two years later he had recovered sufficiently to walk upstairs unaided, but his speech was still hesitant and his writing a mere scrawl. Fifteen years later he was still having trouble, being unsteady in his movements and with a tremor in hands, and this was still the case 25 years later. He never held a full-time job again.

Organo-mercury compounds work their malevolent way through the body only slowly and this was clearly demonstrated in April 1954, when a nurseryman used ethyl mercury phosphate to treat tomato plants that had become infested with stem rot fungus. His first symptoms only appeared the following December and consisted of headache and vomiting, progressing to a numbness in his limbs by the following May as his condition progressively deteriorated. He died in July 1955.

Organo-mercury and autism

Autism is a distressing condition and five children in every 10,000 are afflicted with it, and of these the ratio of boys to girls is 4 to 1.* What causes a seemingly normal baby to develop autism is still unknown, but many in the USA believe that mercury is the cause. In the late 1990s it was accused of causing attention deficit syndrome, stammering, and especially autism. Some thought that thimerosal was to blame. This is a mercury-containing antibacterial agent which added to vaccines to preserve them. The US Centers for Disease Controls first reported a link between thimerosal and mental development in children after carrying out an epidemiological study, although they later admitted that their analysis was flawed. Nevertheless, as a result of a Congressional report many US vaccines phased out its use from 2000 onwards and they no longer contain this preservative. Japan and most countries in Western Europe have done likewise.

Thimerosal is an organo-mercury compound and there are several such compounds which vary according to the organic group attached to the mercury atom. Some, such as methyl mercury, have caused widespread illness in the past, as we shall see. Thimerosal is an *ethyl* mercury compound and by virtue of this it is believed to be safe (see Glossary for more about thimerosal). It was introduced in the 1930s by the pharmaceutical company Eli Lilly, and used originally as a preservative for veterinary vaccines, and then it was used to safeguard human vaccines. There are claims that it was never properly tested for safety in humans.

In the UK thimerosal continued to be used because there is no firm evidence that it was anything other than safe, although parents could request that their offspring be injected with mercury-free vaccines. Indeed an analysis of the medical history of 100,000 children born between 1988 and 1997 in the Thames region showed no correlation between autism and exposure to thimerosal. An even larger survey of the 467,450 babies born in Denmark between 1990 and 1996 came to the same conclusion. The amount of thimerosal in a single vaccine is less than 5 micrograms, and is well below the level

---

* In the 1990s it appeared that the incidence of autism was increasing, although in 2004 it was discovered that the increase was mainly due to other childhood problems now being more correctly diagnosed as autism. When Hershel Jick of Boston University School of Medicine analysed various behavioural disorders among American children he discovered that these had declined as the diagnoses of autism had increased.

that some believe to be dangerous, although on a weight-for-bodyweight basis it is near the maximum recommended for pregnant women and those who are breast feeding. The amount certainly exceeds the guidelines set by the Environmental Protection Agency.*

Surprisingly, a study of mercury levels in the hair of babies who were later diagnosed as autistic, showed there to be much *less* mercury present than in the hair of children who do not develop autism. This work was done by a group in Louisiana led by Amy Holmes who expected to find just the opposite when she started the study, believing that mercury levels would be higher in autistic children. She had contacted the parents of some autistic children and asked if they had saved samples of hair from their baby the first time it was cut. Several of them had, and they sent it to her. When she analysed these for mercury she could find very little. She widened her survey to include the first hair of 94 autistic children, and compared what she found with the hair from 45 non-autistic children. The average level of mercury in this latter group was 3.6 ppm, but that of the autistic children was only 0.47 ppm, and indeed she found the more severe the autism the less mercury there was, with the most severely disabled having only 0.2 ppm on average.

So what do Amy Holmes's findings mean? Those who believe that mercury is the cause of autism explain the low amounts in hair by speculating that children who develop autism have a genetic defect that makes them unable to remove mercury from their bodies which then tends to collect in the brain where it does its damage. They might be right. Until we truly understand the causes of autism, which may in fact be due to several factors, such as parental neglect, faulty genes, illness, or physical damage to the brain, then we cannot rule out mercury. Yet all babies come into this world with some mercury in them, because mercury is a natural part of everyone's diet and a pregnant woman cannot avoid taking some in and passing it to her foetus. To avoid vaccinating a baby, because of the minute amount of extra mercury it will receive, it is more likely to put a baby's health at risk than its not having that vaccine.

* The UK Government's Department of Health announced in August 2004 that thimerosal was no longer to be used in vaccines.

Mercury in fish: prelude to a disaster

Vitamin B_{12} has a coenzyme form known as methylcobalamin, and this molecule is capable of adding a methyl group to mercury, and it can do this to
mercury ions (Hg^{2+}) and to the metallic form of the metal as well. In so doing
it forms the methyl mercury ion (CH_3Hg^+), which is water soluble and can
be taken up by plankton, which in turn become food for other organisms
such as small fish, which in turn become food for larger fish, and so on up
the marine food chain. At each stage along the chain the concentration of
mercury increases, by a process known as bio-magnification. Woe betide
those at the top.

Some algae can concentrate mercury to a hundred times the concentration in the surrounding water. Creatures that live off the algae thus build-up
within their flesh significant amounts of the element so that fish may have up
to 0.12 ppm of mercury. In Minamata Bay, Japan, some fish had as much as
200 times this amount at 24 ppm, while crabs from the bay were discovered
to have more than 35 ppm of mercury in their bodies. This did not prevent
them from being eaten, however.

In 2001 the US Centers for Disease Control and Prevention announced
that a survey they had carried out showed that 10% of US women had
mercury levels in their bodies high enough to impact on a developing foetus.
The FDA's Food Advisory Committee, along with the EPA, issued new
warnings about mercury in fish in December 2003. These urged women of
childbearing age not to eat shark or swordfish, and not to eat more than 12 oz
of any purchased fish per week, and only to eat 6 oz of fish caught by anglers.
The most popular fish is tuna and they said that fresh tuna should only be
eaten once a week, although canned tuna could be eaten twice a week.

Whale meat, and particularly that of the pilot whale, was traditionally a
popular source of protein in the Faroe Islands, but in the early 1990s it was
blamed for the relatively high level of mercury in children that were born
there. The World Health Organization tested 1,000 newborn babies and
found that 20% of them had more than the recommended safe limit of
mercury in their bodies as defined by the International Programme on
Chemical Safety. As a result, the health authorities issued a warning that
pregnant women should avoid eating whale meat.

The levels of methyl mercury in fish have been monitored for many years
and in 2002 it was reported that shark had most with 1.5 ppm, followed by

swordfish with 1.4 ppm. Fresh tuna had much less at 0.4 ppm and canned tuna had half this level of mercury at 0.2 ppm. The reason that canned tuna has only half the methyl mercury of fresh tuna is that during the canning process it loses much of its oil and with it goes a large part of the mercury. These levels of mercury should be compared to those of other popular ocean fish such as cod with 0.07 ppm, plaice with 0.06 ppm, and haddock with 0.04 ppm. Warnings have also been given about not eating king mackerel, tilefish, or marlin all of which have more than 1 ppm mercury, although these are not widely available.

A team led by Francois Morel, of Princeton University, published a report in the journal *Environmental Science and Technology* in 2003, in which they claimed that the level of mercury in yellowfin tuna caught off Hawaii had not changed over the past 26 years. The results were based on fish caught in 1998 compared to similar analysis done on fish caught in 1971. The conclusion was that airborne methyl mercury, which has increased threefold during this time, is not part of the marine food chain. Morel thinks that fish get their methyl mercury from sulfate-reducing bacteria living in hydrothermal vents in the deep ocean.

Minamata Bay disaster

The most puzzling mass poisoning by organo-mercury came from *naturally* produced methyl mercury compounds, albeit from mercury discharged from industry, and this occurred around Minamata Bay in the 1950s. Minamata Bay is on the eastern shore of the Shiranui Sea and part of Kyushu, Japan's southern island. In 1907 Jun Noguchi founded a chemical company there that later became the Chisso Corporation* and he built a factory in Minamata, the effluent of which was discharged into Minamata Bay.

In the 1920s these discharges were affecting fishing in the bay and the company paid compensation to local fishermen. In 1932 the Chisso Corporation began to produce acetaldehyde, a raw material of the chemical industry, using a process that required mercury, and while most of this was retained within the plant, some of it escaped into the waters of the bay where it began to accumulate in the sediment. The amount of mercury discharged into Minamata Bay was estimated to be around 80 tonnes in the years

* Chisso is Japanese for nitrogen.

1932–1968. The concentration in the sediment was as high as 2,000 ppm and the bacteria that lived in these sediments began to produce significant amounts of methyl mercury which passed into and up through the food chain.

Signs of the impending disaster came when dead fish were found floating in the bay in early 1952, but what attracted comment was the behaviour of seabirds which fed off them. They were observed to fly around erratically and even to drop from the sky. Next the cats that lived around the bay area were observed to be behaving oddly, staggering around as if drunk, salivating freely, having convulsions, and dying. Some cats that were affected even ran into the sea and drowned. Meanwhile children playing on the shore were able to pull seemingly dazed octopus and cuttlefish out of the sea with their bare hands. Dogs and pigs were also reported to be afflicted with the strange madness, and then it began to affect the human population. They did not know it, but the seafood they were regularly consuming had high levels of methyl mercury; grey mullet had 11 ppm, chinu fish had 24 ppm, and crabs 35 ppm.

The families of local fishermen ate a lot of fish every day, around 300 grams per person in winter and about 400 grams per day in summer. A person eating 300 grams of grey mullet would be ingesting 3.3 milligrams of methyl mercury. An accumulated dose of 50 milligrams of this toxin results in the victim becoming incapable of work and 200 mg results in death. Of course, most of the fish caught in the bay would have mercury levels in their flesh less than the maximum levels recorded and consequently their diet did not affect them for quite some time, but the methyl mercury they were imbibing was doing them no good.

At the start, Minamata disease was called *kibyo*, meaning 'mystery illness', but it was something to be feared because the death rate was 40% of those most seriously affected. The symptoms of *kibyo* began with a numbness of the fingers and toes, lack of coordination in hand movement, particularly noticeable as the victim tried to eat with chopsticks, a wobbling gait when walking, and slurred speech when talking. As the disease progressed, so these symptoms became progressively more severe resulting in paralysis, deformity, difficulty in eating, convulsions, and death, although this could be delayed for many months. Autopsy revealed structural damage in certain areas of the brain.

In April 1956 a 6-year-old girl who was displaying disturbing symptoms was admitted to the Chisso Corporation's Minamata factory hospital; she had been staggering around and talking incoherently. Within a few weeks her younger sister went down with the disease, as did others who lived in the same neighbourhood. On 1 May, Dr Hajimé Hosokawa, the head of the Chisso factory hospital, reported these cases to the Minamata Public Health Department saying that an unidentified disease of the central nervous system had broken out. That summer saw the first large scale outbreak of methyl mercury poisoning when 52 people were badly affected and 21 of them died. As more and more people went down with the disease, an investigation of recent deaths in the area revealed many other cases that had been misdiagnosed as due to other causes, such as encephalitis, syphilis, alcoholism, or hereditary deficiency.

In August 1956, a Minamata Disease Research Group was set up under the auspices of the Kumamoto University Medical School to investigate the outbreak and by October they were able to report that it was caused by heavy metal poisoning due to eating contaminated fish and shellfish from Minamata Bay. Feeding cats on fish caught in the bay made these animals ill and it took about 7 weeks on such a diet to cause *kibyo*. Which metal was responsible they could not say, but they identified the source as the Chisso works whose effluent contained mercury, thallium, arsenic, selenium, copper, lead, and manganese. At first they suspected manganese, then selenium, then thallium, but tests on cats with these poisons did not produce the symptoms observed in those with feline *kibyo*.

It was not until September 1958 that Professor T. Takeuchi deduced that the cause was methyl mercury, when he found a paper in an English medical journal that had reported similar symptoms among workers in a factory producing dimethyl mercury in 1940. The Kumamoto team then began experiments with this compound on cats, and indeed they succumbed to the new disease. The researchers also began sampling the waters of the bay and discovered alarmingly high levels of methyl mercury, with an extraordinarily 2000 ppm near the Chisso factory drainage channel. Some fish taken from the bay had as much as 20 ppm. The organs of cats that died of the disease were also analysed and showed levels of methyl mercury as high as 145 ppm in their liver and 70 ppm in their fur; cats in a control group had values of less than 4 and 2 ppm respectively. The organs of humans who died of the disease showed high levels of mercury also, especially in their liver and kidneys, although brain tissue could have as much as 20 ppm. The hair of patients in the hospital also revealed high levels of mercury, as high as 700 ppm in one case. Hair was a particularly good monitor of exposure and levels as high as 100 ppm were even found in people in the area who were not exhibiting any symptoms of the disease.

Treatment for those afflicted with Minamata disease consisted first of giving chelating agents such as BAL or EDTA (see Glossary) that would bind strongly to the metal and aid its excretion. If and when the patient showed signs of responding and recovery, then large doses of B vitamins were given in the hope of saving undamaged nerve tissue. However, these methods of treatment were of limited success and even those who recovered were severely handicapped, and often badly deformed, for the rest of their lives. When cells of the central nervous system are damaged they never recover. Victims had to resign themselves to a life of trembling, clumsiness, tiredness, sleepless nights, and partial blindness. A few did improve with time, but others slowly deteriorated.

Another outbreak in 1958 seriously affected 121 people of whom 46 died, but by the following year it appeared to be over. Minamata was not the only place where the disease had struck. In 1965 at Niigata a similar outbreak of methyl mercury poisoning affected 500 people and the source of this mercury was the Showa Denko factory 40 miles upriver from the bay area. This was only brought to light when a medical student who had attended a lecture on the Minamata outbreak informed his teacher that there was a patient with exactly the same symptoms in the University of Niigata hospital. Nor was it only in Japan that mercury pollution was affecting marine life. In April 1970 the FDA reported that fish caught in Lake St Clair had mercury levels of up to 1.4 ppm compared to fish from an uncontaminated lake, which would have around 0.01–0.1 ppm. The mercury in Lake St Clair was coming from a chemical plant at Sarnia, Canada. Thankfully, the consumption of fish from that lake was not a significant part of the local population's diet.

Meanwhile at Minamata the management of the Chisso Company refused to accept responsibility for what had happened, and this so incensed the local population that a crowd of around 3,000 fishermen and peasant workers stormed the factory and a pitched battle with the local police ensued. Several of the ringleaders were arrested and sent for trial and punished. As a result of political pressure, the company agreed to pay token damages of £100 per person (£30 for children) provided the recipient signed a document to the effect that they would not pursue an action for legal damages through the courts.

The Chisso works were not completely indifferent to what had happened, and they had installed a purifier to clear up the plant effluent and the sale of fish from the bay was banned, and slowly things returned to normal, so much so that in 1962 the ban on fishing in the bay was lifted. The Chisso plant was still manufacturing acetaldehyde by the same process and almost all of the mercury was being removed at the waste water treatment stage. The danger, however, came not from any extra mercury going into the bay but from the mercury that was already in the sediment. This had continued to release methyl mercury into the waters of the bay and this was still getting into the fish and on to the tables of those who lived in the villages around the bay.

The Chisso plant was still consuming several tonnes of mercury a year, and one estimate was that as much as 600 tonnes of the metal had eventually ended up in the bay. Sadly the company adopted an attitude of non-cooperation towards those researching the disease, even to the extent of not permitting them to take samples of the waste water they were discharging into the bay. They continued to discharge the waste from their acetaldehyde process until 1968, and their reason then for stopping was not on environmental grounds, but because the process was no longer economic. It later transpired that the company had carried out its own research in the late 1950s into the cause of Minamata disease and had exposed cats to the sludge from the acetaldehyde process, causing them to go down with the condition.

However, for those living in the Minamata area the outbreak of mercury poisoning was far from over. Research in the area by environmentalists revealed enormous numbers of people who had been affected by methyl mercury. In some villages about a quarter of the population exhibited symptoms – and that was 10 years after the area had been thought to be safe again. Of the 100,000 people who lived around the Shiranui Sea, about 3,000 had been sufficiently affected by methyl mercury to apply for compensation from the Chisso chemical company.

In 1996, 3,000 Japanese victims of Minamata disease dropped their lawsuit against the Chisso Chemical Company in return for a $80 million settlement, giving them cash payments of around $25,000 each. Charitable organizations working on behalf of the victims were given $45 million. Originally around 14,000 people had applied for compensation but only 3,000 of them were officially recognized as being affected by mercury. In Osaka one group of 58 people, average age 75, refused to drop their lawsuit and fought on, and in April 2001 the Osaka high court ordered the Japanese Government to pay a further $2.5 million to them.

Perhaps the most disturbing aspect of the Minamata outbreak of methyl mercury poisoning was the effect it had on women. Badly poisoned women do not conceive. Less badly poisoned ones may conceive, but have a miscarriage or the baby is stillborn. Those whose babies lived often discovered they were deformed or suffering from brain damage. Congenital methyl mercury victims were born all along the coast of the Shiranui Sea, not just around the Minamata Bay area. These children were both physically and mentally handicapped and about 40 of them lived for several years. The Japanese custom of preserving the umbilical cord allowed analysis of the mercury content at the time of birth to be measured, and on average this was higher that in normal children. But of course the damage to the embryo could have been caused months earlier than this. Whenever the mother ate contaminated fish, the baby would get its dose of methyl mercury since this can cross the placenta and enter the baby. Even after birth, more methyl mercury could have found its way into the baby's body from its mother's milk.

Watching the detectives

Mercury metal can be diluted with solid materials, by carefully grinding them together in a pestle and mortar, and in this way pharmacists used to prepare grey powder and blue pills. The former consisted of finely divided mercury mixed with chalk, and it could be taken by stirring it into milk, while in the latter it was mixed with sugar and could be pressed into tablets. Both were seen as sure remedies for constipation, although they were introduced originally as part of the treatment for syphilis.

Breathing the dust of grey powder presented an occupational risk to a completely unexpected group of workers: detectives. In the first decades of the last century, mercury poisoning became an undiagnosed illness among those whose job it was to search for fingerprints at the scene of a crime. Grey powder made an ideal dusting powder to highlight fingerprints on surfaces and it was brushed liberally on to all areas where fingerprints might be found. The skill was to leave only the faintest film of powder in the surface, which meant brushing most of it in the air and there it could be breathed in.

Mercury that entered the body in this way tended to stick in the lungs and be absorbed.

Detectives and forensic scientists whose job it was to find and photograph fingerprints began to display the symptoms of chronic mercury poisoning, such as excessive salivation, stomach pains, insomnia, tremors, irritability, and depression, although for a long time these were not linked to the real cause, which was mercury. Indeed it was only in the 1940s that it was realized that the detectives were really suffering from chronic mercury poisoning.

The poet and the poison

MERCURY is not a particularly promising homicidal poison, but it is possible to dispose of someone by feeding them mercury(II) chloride provided you disguise its metallic taste. In the 1800s solutions of corrosive sublimate, as it was then called, were used as an antiseptic and as an insecticide against bedbugs, and its very availability resulted in thousands of poisonings being reported to the health authorities, although these were mainly accidents or as a result of its being taken deliberately in order to procure an abortion. Mercury was not a poison to feature in many murder cases because it was so easily detectable by the intended victims, especially if they started to vomit, which they almost always did. Then the metallic taste became particularly noticeable, and the presence of mercury could easily be confirmed by simple analytical tests.

The poisoners who chose mercury had to use a large dose and this would kill within a day or two. Despite these inherent drawbacks, a few poisoners made use of it. Two of the murderers whose cases we are about to analyse opted for the large single dose approach, but the third murderer achieved her ends by targeting her victim with multiple doses. That murder became famous because of whom she killed and the political repercussions it caused. It was also notorious for the manner in which the final fatal dose of poison was administered.

The Yorkshire Witch

Mary Bateman was known as the Yorkshire Witch and she had a plan that she thought would lead her victims into taking a fatal single dose of mercury. Instead it led her to the gallows. At the time of her crime, Bateman lived in

Leeds, Yorkshire, where she earned her living telling fortunes and swindling gullible clients out of their cash and possessions. She claimed to receive her supernatural information from a spirit medium, a Miss Blythe, into whose mouth she put the advice that always seemed to result in her clients handing over their money and saleable goods, with the promise that if they did as Miss Blythe said, then good luck would soon come their way, and they would be more than compensated.

In 1806, the 38-year-old Bateman was consulted by a clothing merchant William Perrigo and his wife Rebecca. He was worried by chest palpitations but was reassured by a message from Miss Blythe that these would soon disappear – and they did. Over the next few months the trusting couple consulted Bateman many times and gave to her most of their money, and quite a few of their belongings as well. On one visit, Perrigo was given in return a heavy, sealed bag by Bateman and was told that this contained the promised wealth, but that it was not to be opened. It had to be hidden and used only in an emergency. Perrigo did as he was told and his wife sewed it into the mattress where they had previously hidden the savings which they had already handed over to Bateman.

It became obvious to the Yorkshire Witch in spring 1807 that there was little more to be milked from the Perrigos and that she risked exposure if the impoverished couple now decided that the time had come to open the sealed bag. So on one of their visits she gave the couple six numbered powders saying that they were to add them to their food over the next six days, beginning Monday 11 May 1807. These magic powders would ward off a terrible misfortune that Miss Blythe had foreseen was coming their way. In fact the sixth powder was to be the cause of that misfortune, because it contained corrosive sublimate. As a result, Mrs Perrigo fell ill and eventually died, but Mr Perrigo, who had taken only a little of the powder, survived.

There is evidence that poor Rebecca Perrigo may have been taking poisoned powders for some time, judging by the state of her corpse, which was recorded as being covered in black spots of gangrene and stinking so appallingly that those who had to deal with it could only do so while smoking tobacco. Despite what had happened, Mr Perrigo continued to consult the Yorkshire Witch, who was full of sympathy for the loss of his wife, suggesting that she must have taken something that nullified the protective effect of the powder. Perrigo accepted this explanation and even gave Bateman his wife's clothes. Nevertheless he was beginning to lose faith in her powers and when she became slightly threatening, telling him he would become blind unless he continued to pay her, he decided enough was enough. He went home, removed the sealed bag from his mattress and opened it, only to find pieces of lead and newspaper. He informed the authorities of what had occurred and as a result Bateman was arrested and her home searched. A bottle containing a solution of mercury(II) chloride was discovered as well as a jar of honey dosed with arsenic.

Bateman was put on trial in March 1809 and was hanged on the 20th of that month before a large crowd, many of whom regarded her more as a martyred mystic than a manipulative murderer. After her execution her body was cut down and the public were allowed to view it on payment of a small charge, the money raised going to the aid of local charities. Finally the corpse was skinned and pieces of her skin were sold as charms.

Poison through the US Mail

Mercury cyanide* would seem to be a particularly lethal combination of poisonous components. Cyanide is a deadly poison which acts rapidly and it might be thought that death would always be due to this rather than the mercury. However, the chemical bond between mercury and cyanide is strong, and what kills the victim may ultimately depend on the acidity of his or her stomach. In a famous American poisoning case this chemical killed two people, one by virtue of the mercury it contained, the other by virtue of the cyanide. The murderer who used it was Roland Burnham Molineux, the 32-year-old son of a famous American Civil War general, Edward Leslie Molineux, and a member of New York's high society. Despite his social advantages, Roland Molineux was a murderer. His method of delivering the mercury cyanide to his intended victims was via the US Mail, a somewhat hit-and-miss method; indeed his first attempt was a hit, but his second was a definite miss, although it did succeeded in killing an entirely innocent victim.

In late October 1898, Henry Barnett of New York received a bottle of Kutnow's powder through the post and took some of it. Kutnow's powder was obtained by the evaporation of Carlsbad mineral water and was believed

---

* Formula $Hg(CN)_2$.

to be good for all kinds of stomach upsets. He was soon vomiting and had severe diarrhoea. Later he displayed the symptoms we can recognize as due to mercury poisoning and his mouth and kidneys were particularly affected. However, the doctor who attended him diagnosed diphtheria, and Barnett died on 10 November 1898, 12 days after taking the free sample of Kutnow's powder. Jealousy was Molineux's reason for wanting Barnett out of the way. The two men were rivals for the affections of a Miss Blanche Cheseborough and she married Molineux only 11 days after Barnett had died.

The Kutnow's powder that Barnett had received was later sent for analysis and shown to be 48% mercury cyanide. His body was exhumed several months later and certain organs analysed, and mercury found in the kidneys (56 ppm), liver (20 ppm), intestines (20 ppm), and brain (trace). That Barnett did not immediately die of cyanide poisoning can be attributed either to the fact that the solution he drank was not acidic, and neither were his stomach contents. This latter state of affairs is very rare, but not unknown, and indeed it had saved the infamous Russian monk, Rasputin, who survived an attempt to poison him with potassium cyanide in 1916. The deadly form of cyanide is hydrogen cyanide (HCN) which requires acid for its formation.

Thinking he had got away with murder, Molineux struck again, this time sending a sample of Emerson's Bromo-seltzer to Harry Cornish, athletic director of the prestigious Knickerbocker Athletic Club at the corner of Madison Avenue and 45th Street. It arrived on Christmas Eve, together with a pretty silver stand to hold it in, and it was this which caused Cornish to give it to his widowed aunt Catherine Adams, because it was similar in style to other silver ornaments she had in her home, located near Central Park. Cornish lived at his aunt's house and he was there when she took some of the Bromo-seltzer for a headache on the morning of 28 December. Immediately she screamed and started writing in agony; 30 minutes later she was dead. The contents of the bottle were later analysed and found to contain 42% of mercury cyanide.

Among its various ingredients Bromo-seltzer contained tartaric acid. When it was dissolved in water this chemical reacted with the mercury cyanide to generate hydrogen cyanide, and it was from this that Catherine Adams instantly died. But who had put the mercury cyanide in the bottle of Bromo-seltzer? It had to be someone with access to the chemical. And on what basis had the murderer chosen his victims? Both poisoned packages had been sent to members of the Knickerbocker Club, of which Barnett had been a prominent member, and officials of that club informed the police that they suspected Molineux of being responsible. He had been its champion gymnast, but he had resigned after rows with Cornish and Barnett in March 1898. Suspicions were confirmed when detectives learned that Molineux worked as a chemist in the laboratory of a paint manufacturer, and so had easy access to chemicals including mercury cyanide.

Police investigations eventually led to his arrest, and he was put on trial on 4 December 1899 and in February 1900 he was found guilty of murder, entirely on the basis of circumstantial evidence. In 1901 his appeal came up before the New York Court of Appeals and they overturned the verdict and ordered a new trial. This took place in October 1902 and the outcome was that Molineux was acquitted. On his release from jail he did not return to his old job but became a writer of short stories and novels, and he even penned a play about prison life. Some of his works were favourably reviewed, but he was eventually certified insane and died in 1917 in a mental hospital. The underlying cause of his insanity was syphilis contracted from one of the many prostitutes he frequented – his wife Blanche had divorced him several years earlier.

So why did Molineux choose mercury cyanide? It seems odd that he selected this poison when he could simply have used potassium cyanide, which was a commonly available chemical reagent. Maybe he wanted to be doubly certain that death would ensue, and indeed had he simply used potassium cyanide it is more than likely that Barnett would have survived.

Ye poysoning of Sir Thomas Overbury

The most famous murder committed with mercury was that of the poet Thomas Overbury. He survived four attempts to poison him and was only killed when he was given an enema containing corrosive sublimate. He was dead within hours. Rapid death from mercury can occur and there are cases on record of people dying of heart failure after being given a mercury-based medicine and clearly some people are particularly sensitive to this metal, but Overbury was not one of them. He had already survived at least one dose of this poison, and indeed may have eaten several meals to which it had been added.

Sir Thomas Overbury was thrown into the Tower of London by order of

Overbury the writer
—◄+►—

Overbury's most famous work was a book called *Characters* which was published after his death and became a best-seller of its day. It consisted of 77 short sketches of different personality types, some of which are still recognizable, such as 'The Arrant Horse-Courser' who was the equivalent of the second-hand car-dealer today, and 'The Mere Pettyfogger' who delights in malicious gossip. Overbury's most admired character was 'The Milkmaid' with her simplicity and natural goodness. Overbury even wrote some of the book while he was imprisoned in the Tower in 1613, hence the entries entitled 'The Sergeant', 'The Jailer', and 'The Prisoner'. The essay that caused most controversy was 'The Wife' in which he summarized the qualities of the perfect partner, but it also had an overtly political message in that all the qualities that a perfect wife was supposed to possess were clearly lacking in the woman who was eventually to become the wife of Robert Carr – and Overbury's murderer.

King James I on 21 April 1613 despite being an intimate friend of Robert Carr who was the King's 'favourite'. All three men are believed to have had homosexual inclinations and there may well have been sexual relations between the King and Carr, and Carr and Overbury. However, both the King and Carr were capable of normal sexual relations, and both of them fathered children, although there is no evidence that Overbury ever had sex with a woman.

Thomas Overbury was born in 1581 at Bourton-on-the-hill in Gloucestershire. He was educated at the local grammar school, and then went to university in the autumn of 1595 to Queen's College Oxford. He graduated in 1598 with the degree of Bachelor of Arts and moved to Middle Temple in London, which in those days was not solely a centre for lawyers as it is today, but was the home of poets and courtiers. Indeed, Overbury wrote prose and poetry, with some success, but he was intent on a career in Government and he secured a position in the Lord Treasurer's Office. In the summer of 1601 Overbury was sent to Edinburgh with important letters to King James, who was then King of Scotland, and it was there that he first met Robert Kerr, a 14-year-old pageboy. Kerr and Overbury returned to London together and thenceforth kept in regular contact. (Kerr changed the spelling of his name to Carr when he moved to London.)

On 24 March 1603 Queen Elizabeth died and King James VI of Scotland succeeded to the English throne. This very much affected the fortunes of the Howard family. They had suffered in the past because of their Catholic leanings but under the new King their fortunes improved greatly, particularly those of the 63-year-old Henry Howard, who was Earl of Northampton, and his 42-year-old nephew Thomas Howard, Earl of Suffolk, who was married to Katherine; they had a beautiful daughter, Frances. In 1606, when Frances was only 13, she was married to the 15-year-old Earl of Essex, a political marriage that was meant to unite two great families. The marriage was not consummated because the bride was too young, and in any case Essex had to complete his education. While they were apart, Frances spent most of her time at Court, and by the time her husband returned to London in January 1610 she had already had an affair with Prince Henry, heir to the throne, and had now fallen in love with Robert Carr the King's favourite.

The relationship between Carr and the King had begun 3 years earlier on 24 March 1607, at a tournament held to celebrate the fourth year of the King's accession to the throne. It was Carr's role to present a shield to the King, but as he rode forward to do this he fell off his horse and broke his leg. The King was immediately attracted to the tall, handsome 20-year-old, and when the games were over he went round to the room where Carr had been taken and sent for his personal physician Mayerne to attend to the broken leg. Over the next few weeks the King made many visits to the young man and even spent time teaching him Latin – or so it was said.

Whatever went on between master and pupil, Carr became the King's favourite. He was now in a position to help his friend Overbury, and he succeeded in arranging an income of £600 per year for him, plus a position as Server at the King's table. This gave him access to the King's after-dinner talk, and Overbury jotted down many of the King's off-the-cuff remarks about religion, state policy, manly virtues, and the obligations of a monarch. These were eventually published many years after Overbury's death as *Crumbs Fallen from King James's Table*. On Christmas Eve 1607, Carr was knighted and made Gentleman of the Bedchamber. (Overbury was knighted six months later on 19 June 1608.) As that year progressed, Carr and Overbury became more and more influential, the former becoming an advisor to the King, the latter often supplying that advice.

Through 1610 and 1611 all went well for Overbury although he managed to upset many people at Court, including the Queen. However, he was protected by Carr, whose influence with the King continued to grow to the extent that he became Viscount of Rochester on 25 March 1611 and was made a Knight of the Garter two months later. In June of that year he was entrusted with the King's Signet, which in effect made him James's private secretary. Important state letters now passed through his and Overbury's hands. Some of these documents dealt with highly confidential affairs and these he showed to Overbury, including those relating to the Queen's enormous debts. Overbury boasted to others of what he knew, and when the Queen heard of this she complained to the Lord Treasurer. The upshot was that in September Overbury was banished from Court, and sent to France on official business, where he stayed for five months, returning in May 1612 when the affair had blown over and the Queen's debts had been paid.

By then Overbury was needed at Court to help Carr draft important letters and documents. Overbury's finest hours were about to begin, but the Court to which he returned had changed in subtle ways that were eventually to ruin him. For the time being though, Overbury revelled in his new found power and influence. He became noted for his arrogance and overbearing ways, not least towards Carr himself as he realized how much the King's favourite depended on him.

Overbury and Carr also made an enemy of the heir to the throne, Prince Henry, but in October 1612 he fell ill with typhoid fever* and died on 6 November, despite being given Sir Walter Ralegh's famous Elixir of Life. This was reputed to cure any fever except that due to poison, and as a result the Queen openly said that she thought Carr and Overbury had killed her son. The King refused to believe her, but the Queen's suspicions were not without some substance because both Prince Henry and Carr were vying for the favours of Frances Howard; indeed it was generally believed that she and the Prince were lovers, but that she had recently transferred her affections to Carr. Overbury had helped his friend woo Frances by composing love letters for him to send to her. He may have imagined that it was little more than a harmless affair, and that nothing would come of it because Frances was already married, and divorce in those days was impossible. Little did he realize that she was thinking the impossible and knew that a marriage might be annulled if it had not been consummated within 3 years of living together as man and wife. So far Frances had kept her young husband at bay with

---

* His symptoms were diagnosed as typhoid fever by a Victorian physician in 1885.

various excuses and when Essex finally insisted on sleeping with her she always managed to ward off his sexual advances.

She had also turned for help to a Mrs Ann Turner who was the wife of a successful London doctor, and who enjoyed a position in society, having patented a yellow starch for ruffs, a must-have fashion accessory of its day. She was officially the Court dressmaker and costume designer, but she was known to be able to arrange other services through her many useful contacts. She introduced Frances to the famous astrologer Simon Forman,* who provided her with various spells, one for bewitching her husband Essex, consisting of a miniature male doll into whose penis she stuck thorns, and one to charm Carr into loving her, consisting of two dolls that would lock together in a sexual embrace. Later on, when the Overbury scandal erupted, Mrs Turner went back to Mrs Forman, who by then was a widow, and retrieved from her many of the compromising letters that Frances had written. Nevertheless, Mrs Forman kept back two highly incriminating letters, one written by Frances and the other by Mrs Turner herself.

Essex finally despaired of ever having sex with his wife and in the summer of 1612 he departed to his country estate, leaving Frances in London. With her husband out of the way, the secret love affair between Frances and Carr became intense and continued until December 1612, when Essex returned determined to have a final try at consummating his marriage. That Christmas the couple again shared the same bed, but again nothing happened, and for Frances it was crucial that it didn't because she knew that the following month, January 1613, it would be possible to annul their marriage on the grounds of non-consummation.

Overbury now realized what this might mean for him. Were Frances to obtain a divorce and marry Carr, then the favourite would be drawn into the Howard circle and no longer be in need of Overbury's advice and help. Overbury began a subtle campaign to vilify Frances and her mother, going so far as to refer to Frances as a base and filthy woman and calling her mother a bawd. Overbury's words had the opposite effect to the one he intended: Carr and he began to quarrel. They had a spectacular row one night when Carr returned to his rooms to find Overbury waiting for him.

---

* Forman was reputed to have forecast his own death on 1 September 1611, saying to those present at dinner that he would die seven days hence, and this he did, while rowing on the Thames he had a heart attack and died.

'What do you at this time of night?' he demanded, knowing full well where Carr had been. 'Will you never leave the company of that base woman!' he shouted and threatened: 'I will leave you to stand on your own legs!' To this Carr replied: 'My legs are strong enough to bear myself!'

They parted that night, but what Overbury had said was true. Without him, Carr would be unable to cope with the mass of official business that now came his way. He had even let Overbury read correspondence destined for the King's eyes alone. Overbury was in a position to threaten Carr and Carr knew it. A few days after this row the two friends made it up and to outwards appearances they were reconciled, but by now Carr was thinking of a way to remove the troublesome Overbury from his life.

Meanwhile Frances had her own schemes for dealing with Overbury. She approached a certain Sir David Wood, whom she knew also had a personal grudge against Overbury. Frances offered him £1,000 if he would waylay Overbury one evening and kill him. Wood agreed to the plan but only on condition that Carr would guarantee his protection when Overbury had been killed, in what he would claim had been a duel. Nothing came of Plan A.

Plan B was more subtle and would also remove Overbury but it required Carr to get the King to offer Overbury an overseas posting, as an ambassador to a remote country. The King agreed, and on Wednesday 21 April the plan was put into operation. Overbury knew nothing of this and indeed that very morning he had said to a colleague that his 'fortunes and ends were never better'. But by six o'clock that evening he was under arrest and on his way to the Tower.

The first Overbury knew about the matter was when he was informed that George Abbot, the Archbishop of Canterbury, wanted to see him. Abbott offered Overbury a special mission to Moscow, but Overbury refused the post, and immediately went to see Carr to get him released from the offer, which technically Overbury could not refuse because it came from the King. Carr said that he could refuse the Moscow appointment and that another offer would be made. That afternoon a deputation headed by the Lord Chancellor, went to Overbury to inform him that he need not go to Moscow but could choose between the embassy in Amsterdam, Holland or that in Paris. He refused both. When his decision was reported to the King, he brought it to the attention of the Privy Council, which was then in session, and they ordered Overbury to appear before them straight away, which he did at 6pm. When he still refused to accept the appointment he was committed to the Tower for contempt.

With Overbury locked up, Frances could now put Plan C into action: to poison him and to do so in a way that would make it appear that he had died of natural causes. After her arrest there was found in her possession a list of poisons which she no doubt intended to use. This read:

> Realgar
> Aqua fortis
> White arsenic
> Corrosive sublimate
> Powder of diamonds
> Lapis causticus
> Great spider
> Cantharides

We can identify these compounds as arsenic sulfide (realgar), nitric acid (aqua fortis), arsenic trioxide (white arsenic), mercury(II) chloride (corrosive sublimate), and potassium hydroxide (lapis causticus), while cantharides is still the name given to the powder of the dried beetle *Lytta vesicatoria*, popularly known as Spanish fly. What 'great spider' was remains a mystery, but may well have been simply a powder made of dried spiders.

Living in the street at the back of the Exchange in the City of London was an apothecary James Franklin, a Yorkshireman. He was a tall, well-built man with ginger hair, a lock of which he allowed to grow long and hang down his back and which he called his elf-lock. His appearance was not improved by his pox-marked face. It was this man of shady reputation that Frances Howard and her accomplice Mrs Turner went to in order to secure supplies of the compounds on their list. (Franklin had already been suspected of poisoning his wife.) Franklin supplied them with aqua fortis, which they gave to a cat that 'languished and pitifully cried for two days and then died'. Clearly that was no use, nor was diamond dust, which they gave to another cat, which showed no ill effects.

Franklin then sold them some realgar. This killed the cat quickly and so they decided to use it on Overbury – but there was a slight problem. Overbury was being kept in close confinement by the Lieutenant of the Tower, who permitted him neither the services of a servant, nor to see visitors, nor even to send or receive letters. That obstruction to Frances's plan came to an end on 6 May when the Lieutenant was replaced by Sir Gervase Elwes, who had paid £2,000 for the post which he secured as a result of Carr's influence. His first act on assuming his new post was to restore to Overbury the services of a servant. He was also allowed contact with a Richard Weston, a man who had been in Mrs Turner's employ, and he told Overbury that he could smuggle letters in and out of the Tower for him. Consequently he wrote to Carr, whose friendship he thought he still had, believing he could secure his release. Carr wrote back that he was going to bring this about, but in fact he was doing no such thing.

The first poisoning

An attempt to poison Overbury was made the very day that Elwes became the new Lieutenant of the Tower. Frances gave Mrs Turner the phial of poison prepared by Franklin from realgar, and she handed it over to Weston's son to give it to his father. On the evening of 6 May Weston senior met Elwes as he was on his way to Overbury's cell. He was carrying a bowl of soup in one hand and the phial of poison in the other and asked: 'Shall I give it him now?' Elwes clearly did not want to be involved and dissuaded him from adding it to the soup. In fact it was not added to Overbury's food until three days later, on 9 May, when Weston stirred it into some broth.

Weston reported to Mrs Turner that the poison had caused Overbury to be very sick with vomiting and diarrhoea, but that he had recovered, whereupon he was told to give Overbury a larger dose, but that did not kill him either, although again it made him very ill. In fact Overbury believed that by being ill it would allow Carr to soften the King's heart and get him released, and he wrote a letter to Carr to that effect. Carr took no action and in his reply he said that the King was very angry and could not be placated. Nor did Carr do anything when he was approached by Overbury's father and mother, who had come up to London to try to obtain the release of their son, although he promised them he would do all he could.

Meanwhile Overbury continued to be very ill and on 14 June a doctor was allowed to visit him. He diagnosed consumption and prescribed *aurum potabile* [drinkable gold] which was an expensive cure-all. During that June, Overbury was also visited by two other doctors, including Mayerne. The latter was called in as a result of Overbury's father directly petitioning the King. Mayerne prescribed various medications for Overbury and these were made up by an apothecary living near the Tower, who delivered them, accompanied by his apprentice William Reeve, on Friday 23 June. We will hear more of Reeve later.

The second poisoning

Despite Overbury's apparent ill health, no release from the Tower was forthcoming. The Privy Council was due to meet on 6 July and Overbury pinned his hopes for a release on this meeting which the King would attend. He wrote to Carr that he planned to make his illness seem even more convincing by taking an emetic powder, and asked his friend to send him one. In fact this gave his poisoners a chance to strike again.

Frances and Mrs Turner sent some white arsenic to Weston to give to Overbury, and we know from Weston's confession that on Thursday 1 July he mixed this with the emetic powder that Carr had sent. For the next four days, Overbury was very ill indeed and as far as we can judge his symptoms were consistent with arsenical poisoning. He was continually vomiting and had diarrhoea and Weston reported that he had evacuated his bowels between 50 and 60 times. A doctor was called in to attend to him, finding Overbury very feverish and he ordered him a cooling bath, despite which he continued to suffer from a high temperature and raging thirst. This, and more, Overbury related in a letter to Carr, which he wrote on Monday 5 July:

> This morning (notwithstanding my fasting till yesterday) I find a great heat continues in all my body and the same desire of drink and loathing of meals, and my water is strangely high. [. . .] The distemper of heat, contrary to my constitution, makes me fear some fever at the last, and such an one, meeting with so weak a body will quickly end it. And in truth I never liked myself worse, for I can endure no clothes on, and do nothing but drink.

Overbury's hope that the Privy Council would release him from the Tower the following day was in vain. That was the bad news. The good news, all though he was not aware of it, was that by surviving five days after a large dose of arsenic he was not likely to die, and during the second week of July Overbury slowly recovered his health. Frances had failed again. Indeed she was forced to suspend Plan C temporarily because she had more pressing problems to attend to that month. On Saturday 17 July she was to be subjected to the indignity of a medical examination to see whether she was, as she maintained, still a virgin. That had to be proved if she was to secure an annulment of her marriage. The inspection of a heavily-veiled Frances was undertaken by midwives in the presence of several ladies (who were too embarrassed to actually look) and they confirmed that she was still a virgin. This fact seemed so at variance with her reputation that few believed it to be true, with some asserting that the midwives had been bribed, others saying that the young girl they examined was not Frances but a substitute. Nor was the divorce commission, set up by the King to look into the Essex marriage, much impressed with this apparently incontrovertible proof of non-consummation. The ten members of the committee were equally divided, with the most prominent member, the Archbishop of Canterbury, against the divorce.

Even though the King addressed the committee the following day, and let them see there was no doubt in his mind that they should grant the divorce, the commission voted five for and five against the divorce. The King thereupon adjourned the commission until 16 September, but not before appointing two extra members whom he knew would be in favour of granting it.

Meanwhile, Overbury's parents were growing more and more alarmed and were doing their best to have him released. Again they went to see Carr who reassured them that he was doing all he could, saying that their son was receiving the best medical treatment possible. Carr did get permission for Overbury's brother-in-law to visit him on Wednesday 21 July. By then, however, Overbury had been poisoned again.

The third poisoning

Arsenic having failed, Frances decided to try corrosive sublimate. This was added to some tarts that were baked by Mrs Turner, and taken by Weston to the Tower on 19 July with a covering letter to Elwes telling him not to taste the tarts or jellies or give them to his family. They must only be given to Overbury, because they contained 'letters' which the conspirators later admitted was their code word for poison. This warning was necessary because it was Overbury's habit to pass on food he did not want to the Lieutenant and his wife. Frances's letter read:

> Sir, I pray deliver not these things till supper. I would have you change the tarts in place of these now come, and at 4 of the clock I will send a jelly to him as it was sent to me; the tarts or jelly taste you not of, but the wine you may drink, for in it is no letters, I know. Do this at night I pray you.

The effect of the poisoned tarts and jellies was dramatic, as one might expect on a body just recovering from an attack of arsenic poisoning. We have independent evidence for the date because a letter written from the Tower by Thomas Bull, and dated 20 July, reported 'Overbury is still here in prison, shut up close and very sick.' The following day Sir John Lidcote, Overbury's brother-in-law, hearing of his turn for the worse, obtained a warrant from the Privy Council for him to be allowed to see Overbury so that he might draw up Overbury's will. On Wednesday 21 July, Lidcote found the prisoner to be 'very sick in bed, hand dry, speech hollow'. A condition of the visit was that the Lieutenant of the Tower must be present at all times. Nevertheless, Lidcote contrived to have a few private words with his brother-in-law, who asked him whether his friend Carr was 'juggling' with him. Lidcote replied that as far as he knew, Carr was being faithful to him. Brief though this exchange was, the Lieutenant noted it and reported it, with the result that Lidcote's warrant to visit was withdrawn.

As a result Lidcote became suspicious of what was going on and he discovered that Carr was being duplicitous in his dealing with his former friend. He managed to get a message to Overbury telling him what he had discovered and he advised him to start writing conciliatory letters to the powerful members of the Howard family, pleading for their help. Overbury acted on Lidcote's advice and wrote to Henry Howard saying that he was very sorry for what he had done in the past and promising to be 'as faithful to you as your lordship's own heart' when he was at liberty again. What is most illuminating about this letter is not what it said but Overbury's handwriting, which had degenerated to a shaky shawl. This is a typical symptom of mercury poisoning.

Overbury also wrote to Lord Suffolk, Frances's father, again in a conciliatory tone and again it is not the message which holds our attention but the footnote: 'Good, my lord, excuse my blotting by reason of my weakness at this time.' But it was all to no avail, because on Friday 27 August, all correspondence to and from Overbury was stopped by order of the Lieutenant, and the prisoner was moved to a smaller cell, with Weston alone being allowed contact with him. A final attempt on his life was about to be made.

The final poisoning

By the beginning of September Overbury was becoming desperate and he resolved to have one last attempt to influence Carr. To this end he wrote a long letter, in which he recounted all that he had done for Carr and threatening to expose him on a matter 'of another nature'. He ended the letter by saying that on Tuesday 7 September he had completed a 'large discourse' setting down everything which had happened between the favourite and himself, and all of it damaging. He also said that he had sealed this discourse with eight seals and sent it to a friend with instructions to open it if he died. Overbury concluded: 'So thus, if you will deal thus wickedly with me, I have provided that, whether I live or die, your shame shall never die, but ever remain to the world to make you the most odious man alive.'

Although Carr received the above letter, we have no knowledge of the discourse referred to in it. Did Overbury write such an exposé? Probably not. In any case there was no way it could be passed to a third party without Weston's help. Almost certainly Overbury was bluffing, but it was now clear to the conspirators that a final attempt had to be made to silence him for ever. Overbury had to be given another dose of corrosive sublimate: this time it would certainly be large enough to kill him and it would be given in the form of an enema.* Weston learned of this plot to poison Overbury from Franklin, whom he met at the White Lion pub on Tower Hill. He told Weston that the apothecary's young apprentice, William Reeves, had been paid £20 to administer the poisoned enema. To a young apprentice like Reeve, £20 was a truly enormous sum, equivalent to around £10,000 (or $20,000) today.

The events of Monday 13 September and Tuesday 14 September 1613 are fairly well documented. On the Monday, the Lieutenant of the Tower was briefed that another attempt was to be made to silence Overbury. Elwes betrayed his knowledge of this in a rather naïve manner, and in a way that told strongly against him at his trial 2 years later. On the death of a prisoner in the Tower, the contents of his cell, and only that cell, became the property of the Lieutenant. Elwes knew this and was also aware that when Overbury had been moved from his larger cell he had left many of his belongings behind, and in particular an expensive suit of clothes. On that Monday, Elwes had this taken to his new cell; the next day it became his.

---

* Chronic constipation is a feature of prison life, as Overbury was to discover and report in his essay 'The Prisoner'. This states: 'Whatsoever his complexion was before, it turns to choler or deep melancholy, so that he needs every hour to take physick to loose his body, for that (like his estate) is very foul and corrupt and extremely hard bound. The taking of an execution of his stomack gives him five or six stools and leaves his body very soluble.'

William Reeves called on the Monday and made up the enema for Overbury which he administered. That evening, Overbury became gradually worse and worse. In the night he was groaning and unable to control himself, and by daybreak it was clear that he was sinking fast. At 6.45am he asked for a drink of beer and died while Weston was fetching it for him.

The Lieutenant immediately sent a note to Northampton telling him that Overbury was dead. Northampton replied that the body should be buried straight away if it was 'foul', but if it would stand viewing then Lidcote should be allowed to see it. Elwes called in the coroner, who convened a jury of six warders and six prisoners. They found the corpse to be nothing but skin and bone and it stank intolerably. They reported that there was a black ulcer between the shoulder blades, an open sore on the left arm, a plaster on the sole of one of his feet, and blisters on his belly the size of peas and yellow as amber. The verdict of the inquest was death by natural causes and Overbury was swiftly buried at 3.30pm that afternoon, less than 9 hours after he had died. Lidcote could not be contacted on that Tuesday and when he arrived at the Tower the following day he was so shocked to find his brother-in-law had already been buried that he refused to pay the burial costs.

The reckoning

As far as Frances was concerned, things were looking up. Not only was the hated Overbury dead, but the Divorce Commission, which reconvened on the Saturday following his death, was about to release her from her equally hated marriage to Essex. The King forestalled a public debate by ordering them to come to a simple yea or nay decision. The members then voted and the result was seven in favour, five against. Frances was free at last. Three months later, on 26 December 1613, she and Carr, who now had the title Earl of Somerset, were married at a lavish wedding ceremony with the royal family in attendance. That same month saw the first edition of Overbury's poem *The Wife* published in London. It was immediately sold out and indeed was reprinted five times within a year.

The year 1614 started well for the Howard family, whose grip on the levers of power was now stronger than ever. But their fortunes were about to be undermined by a series of blows that they had not foreseen. In March, a treasonable correspondence between Henry Howard and the Pope was discovered and that marked the end of public life for him, but in any case his own life was about to end as a result of post-operative infection following an operation on his thigh. He died in March. In the reshuffle of public offices that followed Howard's death, Suffolk became Lord Treasurer and Carr took on the post of Lord Chamberlain. The favourite had now reached the pinnacle of his power. Meanwhile copies of *The Wife* were selling as fast as they could be printed and rumours began to spread about Overbury's strange death.

At this time neither Frances nor Carr was under suspicion of being responsible for his murder, but a more serious threat was on the horizon in the form of a handsome young man, George Villiers who had caught the King's fancy. The remainder of that year saw Villiers slowly oust Carr from his position as the favourite. Then, 18 months later, in June 1615, an event occurred which dramatically reopened the Overbury affair. William Reeve, who had been sent by de Lobell to live in Flushing in The Netherlands, was taken ill and thought he was dying. He decided to confess his crime. The information he revealed was conveyed by the British Consul in Brussels to Sir Ralph Winwood, Secretary of State, in London.

Carr and Frances, who was now three months pregnant,* were oblivious to all this and were staying with friends in the country. Mrs Turner was with them. On their return to London they soon became aware of Reeve's confession and realized the danger they were in. Carr sent his servant round to see Overbury's former servant, and he brought back the letters that Carr had written to Overbury in the Tower, paying him £30 for them. Elwes got drunk at a dinner party in August at which Winwood was present and he revealed that he had known all along that Overbury had been murdered. Winwood informed the King, and the King ordered Elwes to set down in writing his part in the affair, which he did on 10 September. The King set up a commission headed by Lord Chief Justice Coke to look into the death of Overbury, and those suspected of being implicated were arrested.

On 19 October, Weston's trial opened at the Guildhall in London, and at first he refused to enter a plea which meant a jury could not return a verdict. Physical and moral pressure was brought to bear, including being interviewed by the Bishop of London, and eventually on 23 October he agreed to face trial. He was found guilty and hanged two days later at Tyburn, but not before Lidcote and a group of his friends interrupted the execution by riding

* The baby, a girl, was born on 1 December.

up to the gallows and asking Weston to make a public confession of his guilt. (For this they were fined.)

On 7 November, Mrs Turner was tried and this trial brought to light the black arts of Forman, including his model of a couple copulating, and his black book in which he kept details of the love affairs of prominent people. (It was reported that judge Coke's wife was listed in it, and that was why he would not admit it as evidence.) Mrs Turner was hanged on 14 November and the hangman wore a paper copy of the yellow ruffs and cuffs that she had made popular as a fashion item. This piece of mockery was done at judge Coke's suggestion. On 16 November Elwes was tried, found guilty, and hanged on Tower Hill on 20 November. Next it was the turn of Dr Franklin, who supplied the poisons, and he was tried on 27 November and hanged on 9 December.

The trials of Frances and Carr took place on 24 and 25 May at Westminster Hall. She pleaded guilty – in fact she had admitted to her part in the plot in January – and was sentenced to be hanged. Carr pleaded not guilty and his trial lasted all day on the 25 May. Because the King was worried that Carr might use the trial to reveal details of his private life, he had two men with heavy cloaks standing at Carr's side to muffle him if the need arose. However, the prosecution was in the capable hands of Sir Francis Bacon, and he saw to it that the need for silencing Carr never arose. Even so, the trial lasted 13 hours and many in the packed hall fainted, but Carr stood up to his ordeal remarkably well. Carr's defence was no match for Bacon's polished prosecution; he was found guilty and sentenced to death.

Neither Frances nor Carr was to pay the ultimate price for their crimes; instead they were imprisoned in the Tower. Frances was pardoned on 13 July 1616, but Carr was not pardoned until 1625. The two of them lived in the Tower until 1621, when they were released on condition that they lived together. This they did, although they rarely spoke to one another. Frances died in 1632 and Carr in 1645.

And what happened to the youth who administered the fatal dose? William Reeve's fear of dying had been unfounded. He recovered his health and is said to have returned to England, but he was never prosecuted.

So who really engineered the plot to kill Overbury? There were many who wanted to see him dead and various authors have attributed his death to the maliciousness of some of these people. Charles Mackay in his *Extraordinary Popular Delusions* thinks Carr was to blame. Terence McLaughlin in *The*

Coward's Weapon makes a case against the Queen, who he thinks used her influence with her physician Mayerne to ensure that Overbury was given the fatal enema. McLaughlin thinks it was significant that Mayerne was not called as a witness at the various trials, the reason being that he knew too much. William McElwee in *The Wisest Fool in Christendom* even suspected the King of being implicated. The best, and most thoroughly researched account of all, is Anne Somerset's *Unnatural Murder: Poison at the Court of James I*, and she concludes that Frances may have been guilty but that she was certainly acting under Carr's directions. Indeed the only credible suspect was Frances herself, and she confessed as much. Carr was undoubtedly an accessory before the fact, and indeed his behaviour is consistent with his wanting the deed done and of urging his lover to do it. She was only too willing to oblige.

Arsenic

Arsenic is everywhere

For more technical information about the element arsenic, consult the Glossary.

ARSENIC has a long historical and disreputable pedigree; its very name seems to condemn it as something unspeakable. It appears to have been first isolated by Albertus Magnus (1193–1280) although it was not identified as an element until several centuries later. According to the *Oxford English Dictionary* the first recorded use of the English word *arsenic* was in 1310, and certainly it must have been widely known by the end of that century because it was mentioned by Chaucer in his *Canterbury Tales*, written in 1386. The canon yeoman's tale has the words:

> No need to reckon up the lot,
> Rubeficated water, bull's gall,
> Arsenic, sal ammoniac, and brimstone:
> And if I wanted to waste your time I could recite any number of herbs.

Rubeficated means red, sal ammoniac is ammonium chloride, and brimstone is sulfur. Later in the tale he mentions orpiment (arsenic sulfide, As_2S_3) as one of the four spirits of alchemy. And arsenic was known to be deadly, as the following poem supposedly written by the mythical Basil Valentine (see page 23) in 1604 shows:

> I am an evil, poisonous snake,
> But when from poison I am freed,
> Through art and sleight of hand
> I can cure both man and beast.
> From dire disease I can direct them,
> But prepare me correctly, and take great care

That you keep watchful guard over me,
Or else I am poison, and so will be
To pierce the heart of many a person.

The ancient Romans knew of arsenic materials, as did the contemporary civilizations of China and India. The Chinese used them to kill flies and rodents, and the Indians used them to preserve paper from attack by insects. The Roman writer Dioscorides (40–90 AD) wrote *De Materia Medica* [Medical Matters] in which he listed scores of remedies, mainly of the herbal kind, but also of the mineral variety and among these he mentioned orpiment and realgar, both of which are natural arsenic sulfides. The former he called *arsenikon*, which he said could be used to repress 'excrescencies', in other words warts and other skin eruptions, although he did warn that using it might cause the hair to fall out. The latter he called *sandarache* and recommended it for 'spitters of rotten matter' and that it should be mixed with rosin and heated and the smoke that evolved should be breathed in, thereby curing such coughs and also asthma.

Although arsenic rarely threatens our health today, in the past it has affected the lives of many, but that was at a time when it was generally perceived as beneficial, to the extent of being taken regularly as a tonic. Yet while doctors often prescribed it for many ailments, they began to question its widespread use. In 1880 the Medical Society of London published a list of all the products then on sale which were coloured with arsenic pigments, and there were indeed many of them. For example if you were having an evening playing cards, then not only were the cards themselves likely to contain arsenic, but the green baize of the card table certainly did, and the wallpaper of the room would be printed with its pigments, as would the blinds and curtains at the window. The linoleum on the floor might well be coloured with it as would the toys with which the children played, and even the artificial flowers in the vase on the sideboard would have leaves of arsenic green. Arsenic indeed was everywhere.

Arsenic in the human body

Chickens that are fed an arsenic-free diet do not grow properly and it appears this element may have a role in animal development, at least in chickens and maybe in humans. If so, we need very little of it and our requirement each day may be as little as 0.01 mg, although we take in and can tolerate much more than this. The average person, weighing 70 kg, contains around 7 mg of arsenic, which represents a level of 0.1 ppm. Our blood contains less than this, but our skeleton contains more, and the level in hair is generally about 1 ppm.

Even if arsenic is essential in tiny amounts, and unavoidable in our daily diet, we still know that a little too much is harmful – although we may not register symptoms – and that a lot too much is deadly. Thankfully our body recognizes excess arsenic as a toxin and excretes it quickly. But what sort of levels are we talking about? How much arsenic can the body tolerate without any symptoms appearing? How much arsenic does it take to make us ill? How much arsenic does it take to kill us? The answers to these three questions depend very much on the individual. Clearly it takes much less arsenic to affect a child than an adult, and to affect someone who is already in a weakened state than one who is healthy. A dose of 250 mg would certainly be fatal for most adults and doses of half this amount have also been known to kill. Yet for some people a dose of 500 mg can leave them unaffected, provided they have trained their body to tolerate it.

Arsenic is quickly removed by the liver, and at a rate that is more than enough to remove the daily surplus we consume. A sudden larger dose poses a problem and the body's first reaction is to empty the gut as quickly as possible by vomiting and diarrhoea, both of which can turn extremely violent. The person becomes severely dehydrated with their skin feeling cold and clammy, and soon they pass into a coma with heart failure ending their life within a day or two. Smaller doses of arsenic may produce some of these symptoms but the victim should survive and recover, being eventually none the worse for their experience. Even large doses can also be survived, provided an antidote is given – see Glossary.

Arsenic can enter the body via the skin, lungs, or stomach, although the latter is the only way that would-be poisoners could hope to kill their victim. From the stomach arsenic passes into the blood stream and so moves around the body, quickly ending up in the liver, kidneys, spleen, and lungs. Eventually it will permeate all body tissue and find its way into bone, hair, and nails. In the hair it becomes chemically bonded to the sulfur atoms in the keratin molecules of which hair is made, and indeed ordinary hair has a high level of arsenic because of this, and explains why the analysis of hair gives a good indication of the degree to which a person has been exposed.

The first symptom that results from a normal person taking a fatal dose is vomiting. This may start within 15 minutes or be delayed for many hours, depending on the amount of food in the stomach. Unfortunately, vomiting starts too late to remove the arsenic that has been absorbed into the system, although it may expel the remainder of a massive dose before it has time to be digested. The initial vomiting brings no relief and soon starts again. The victim may complain of thirst and a sore mouth and throat, and experience difficulty in swallowing. Drinking, however, does not assuage the thirst and may only serve to accentuate the vomiting. The stomach becomes very painful and sensitive to pressure.

The body next attempts to eject the poison by emptying the bowels, and diarrhoea begins after about 12 hours, eventually becoming watery, and will continue until a condition known as tenesmus is reached. This is a feeling of wanting to evacuate the bowel without anything emerging. The body also rids itself of arsenic through the kidneys and urine but this process of expulsion can be intermittent which is why, when arsenic poisoning is suspected, a negative result from a single urine sample is not enough to rule it out; two negative samples are necessary to be absolutely certain. The arsenic that has entered the blood stream presents the greatest danger to the body and most of this has to be excreted via the urine, and while most of a small dose can be shed this way within two days of ingestion, larger doses tend to accumulate around the body. A notable effect is to produce muscle spasms, especially in the calves.

In acute poisoning, the person deteriorates rapidly with typical shock symptoms: the pulse is weak and rapid, the skin cold, damp, and pale. Death usually occurs within 12–36 hours but some poor souls have lingered for four days. Rapid death can leave a lot of arsenic in the liver, up to 120 ppm, but survival beyond a day causes this to drop to much lower values of 10–50 ppm and in Chapter 8 we will see how a victim who lived for three days after being given a fatal dose was found to have relatively little arsenic left in his liver. It will take about 14 days for all the arsenic to be completely removed from the body although most of it will have been expelled within a week.

Less than lethal doses of arsenic produce no single symptom which can be said to be typical of arsenic poisoning. Its features are those of any irritant in the stomach such as occurs with ordinary food poisoning: vomiting, diarrhoea, stomach pains, thirst, and a furred tongue. For this reason it was most unlikely for a doctor to diagnose arsenic poisoning and why to some murderers it seemed the perfect poison – and it was, until chemistry came along and spoilt the fun.

Chronic poisoning by prolonged exposure to industrial arsenic fumes or dust, or by being given small doses medicinally, eventually leads to changes in the skin. In the short term this may be an improved complexion, but later a mottled brown pigmentation results. Thickening of the skin of the palms and soles occurs and this is a symptom unique to chronic arsenic poisoning. Weariness, irritability, a loss of appetite and weight, red and watery eyes, and other symptoms also come from prolonged arsenic intake, although to begin with just the opposite effects are observed, as if arsenic were acting as a tonic. Small doses of arsenic act as a stimulant by speeding up the chemical process that supplies cells with energy; and it can even speed up horses. Unscrupulous racehorse trainers were able to improve an animal's chances of winning by giving it arsenic, but if arsenic is detected in the animal's urine, and exceeds 50 ppb, then it is a sure sign of its having been doped in this way.

Arsenic compounds have this tonic effect by stimulating the body's metabolism. For example the arsenite ion (AsO_3^{2-}) boosts oxidative phosphorylation by forming an arsenic molecule that reacts more rapidly allowing oxidation to proceed at an increased rate. It was observed that those who were given arsenic tended to increase in weight, a fact that was used in the 20th century to increase meat production in pigs and poultry by giving them Roxarsone (phenylarsonic acid*) in their feed. This additive was removed from their diet a week before slaughter so that any residual arsenic would have time to be excreted.

Arsenic in plants and food

According to the US Environmental Protection Agency (EPA) the daily intake of arsenic into the human body that can be tolerated with no observable effects is 14 micrograms per kg bodyweight per day, which for the average 70 kg adult amounts to around 1,000 micrograms. In fact the normal human diet contains much less than this, between 12 and 50 micrograms of arsenic per day, and the amount excreted is within this range. However, in Japan the average intake and excretion is in excess of 140 micrograms per

* Formula $C_6H_5AsO_3H_2$.

day due to the eating of fish and shellfish which are known to be rich in arsenic. Even that is well within the EPA limit.

Seawater contains only 0.024 ppm of arsenic but, despite this, some marine organisms can concentrate quite high levels of arsenic in their bodies. In aquatic environments there are alga and cyanobacteria that can methylate arsenic and this begins a journey up the food chain as these are eaten by other creatures like shrimp, which are then eaten by fish, which may finally be eaten by humans, but as the food chain is ascended the amount of arsenic that is retained decreases. Nevertheless some edible species contain surprising amounts; oysters have around 4 ppm arsenic, mussels, 120 ppm, prawns 175 ppm. Even some varieties of fish that feed upon these creatures can have high levels of arsenic such as plaice with 4 ppm. Most fish, however, absorb only minute amounts. Seaweed is also rich in arsenic and on the remote Scottish island of North Ronaldsey there is a breed of sheep which feeds exclusively on seaweed and they appear to thrive on it.

Some plants thrive on arsenic. The Chinese Ladder fern *Pteris Vittata*, also known as the brake fern, grows rapidly and absorbs arsenic. It can often be seen growing out of brickwork, such is its hardiness. Its fondness for arsenic is an asset that may well come into play one day in cleaning up sites that have been heavily contaminated with this toxin, because of its ability to absorb arsenic, the extent of this comprising 2% of its weight. This remarkable ability was discovered by a group led by Lena Ma, a soil chemist at the University of Florida, who found that it could extract arsenic from soil even where the level was low, for example 6 ppm, which is normal for many soils. When it was grown on soil with 100 ppm not only did it absorb more arsenic, but it grew 40% larger than normal.

A typical area that might benefit from brake fern harvesting is Cornwall in England which has been described as one of the world's major arsenic geochemical provinces, and for good reason as we will discover in the next chapter. In that hapless county there are streams with sediments containing 900 ppm arsenic and similar levels are to be found in garden soils and on farmland. (The UK Government's recommended limit is 40 ppm.) Arsenic also turns up in household dust in homes and in some Cornish villages this can contribute 35 microgram of arsenic to the daily intake, but even in villages where the arsenic content of the soil exceeds 1%, the vegetables grown in this absorb very little and are well below the safety limits for foods. Another region to benefit would be the Repora district of New Zealand, south of the thermal hot springs of Waiotapu, where there is sufficient arsenic in the soil and grass to cause poisoning of cattle which graze there. The Champagne Pool at Waiotapu is also rich in arsenic.

Mushrooms can absorb a lot of arsenic and one species in particular *Sarcosphaera coronaria* was discovered to contain 2000 ppm which is 0.2% (dry weight), but ordinary mushrooms have an arsenic content a thousand times less than this. Some arsenic used to enter the body via the lungs, or at least the lungs of smokers. Lead arsenate was widely sprayed on the tobacco crop to protect it against leaf-eating insects and the average American cigarette contained around 40 mg of arsenic, a lot of which was volatilized when the cigarette was smoked. Copper arsenite was also used as an insecticide to control the Colorado beetle in Mississippi and indeed by 1900 this was being so widely applied that state legislation was introduced to restrict its use. In 1912 calcium arsenate was introduced as an agricultural insecticide and this was found to be particularly effective against the boll weevil that infested cotton crops. Vineyards in France used arsenical pesticides sometimes to the extent that they poisoned the grapes and the wine. This was the explanation of the outbreak of arsenic poisoning among sailors on French merchant ships, 300 of whom were affected in 1932.

What is curious about arsenic, and what makes it so unlike mercury, is that its organic forms present less of a threat than the inorganic form. In the United Kingdom the Food Standards Agency carried out its own survey of arsenic in food and it differentiated organic and inorganic arsenic. A total diet study was completed in 1999 and this assessed the actual intake of arsenic by multiplying the arsenic content of individual foods with the amounts of that food in the British diet, taken from the Government's National Food Survey. The conclusions of the Food Standards Agency report was that dietary exposure to arsenic in food was not increasing, as some had suggested, and that in any case people need not worry about the amount or type of arsenic they were getting in their food. Fish was the main source, providing 3 mg per kg on average, and only 0.03 mg was inorganic arsenic, the kind thought to pose the most risk of cancer. The main arsenic compound present in fish was arsenobetaine, an organic molecule derived from glycine, the simplest of all amino acids.*

Poultry had an average of 0.070 ppm, then came cereals with 0.013, meat

---

* The arsenobetaine molecule has the formula $(CH_3)_3As^+CH_2CO_2^-$.

with 0.005, root vegetables 0.005 (but not potatoes, which had 0.002), and bread 0.004. Many foods had less than 0.001 ppm and almost all of it was organic arsenic. Things like eggs, greens, fruit, milk, cheese, etc. had amounts around the 0.001 level. The upshot of all this was that the daily intake of arsenic by the average person was 50 micrograms of which inorganic arsenic might be only 1 microgram. In terms of dietary regimes there was no difference in arsenic intake between vegetarians and non-vegetarians.

Arsenic in the past has poisoned thousands of people accidentally by being added to their food and drink, mistakenly, unknowingly, or deliberately, or was present without anyone suspecting it was there. Curiously it is the last of these threats, when it has been present in only parts per million and causing no immediately observable effects, that it has caused the most distress, damaging millions of lives even today, as we shall see in the next chapter. At the other extreme, excessive exposure to arsenic is realized within hours.

An example of this latter type of mass poisoning involved arsenic trioxide and it occurred in November 1858 at Bradford, West Yorkshire, when over 200 people were taken seriously ill after eating cheap peppermint lozenges that they had bought from a local market. The proper recipe for such sweets was to take 52 pounds of sugar, 4 pounds of edible gum, and 1½ ounces of peppermint essence, but a certain Mr Neale was a crafty manufacturer and found he could replace 12 pounds of the sugar with 12 pounds of much cheaper powdered calcium sulfate, and this is what he used. Sugar cost 6½ pennies per pound whereas the adulterant cost less then ½ penny per pound.

In the week leading up to the Saturday market Mr Neale sent his errand boy to a pharmacist in nearby Shipley to get the necessary calcium sulfate, but the pharmacist was ill in bed and so his assistant, 18-year-old William Goddard went to the store room to get it. He was told it was kept in a cask in the corner, and indeed it was from such a cask that the young Goddard took 12 pounds of white powder, but it was arsenic trioxide. Both it and the calcium sulfate were stored in similar casks, both without visible labels – in fact they were labelled but only on the ends on which they were standing. From that point on it was only a matter of time before the disaster occurred. Neale manufactured the sweets and the boy delivered them to the market stall of a Mr Hardaker whose special offer of 2 ounces for 1½ pence meant they were almost all sold that fateful Saturday.

On Sunday morning the Bradford police were informed that two young boys, 9-year-old Elijah Wright and 14-year-old Joseph Scott had suddenly died the previous evening under suspicious circumstances, and that the peppermint lozenges were to blame. The chief constable of Bradford took rapid action and sent members of his force around the city ringing bells and alerting people about the poisoned sweets, asking them to hand in all those which had not been eaten, with the result that 36 pounds were reclaimed, some of which were later shown to contain as much as 1,000 mg of arsenic trioxide. But many had been eaten and in the week that followed another 20 people died and more than 200 needed hospital treatment.

A mass poisoning of 6,000 people occurred in Manchester in 1900 when 70 people died. In this incident the agent was a local beer which was eventually discovered to contain 15 ppm of arsenic, so that five pints of the brew would provide a dangerously high dose of around 40 mg, and many men drank this amount of beer every working day. The arsenic originated in the glucose used for brewing the beer. This glucose had levels of several hundred parts per million because it had been manufactured using sulfuric acid containing 1.4% of arsenious acid.* The sulfuric acid had been produced from iron pyrites (FeS_2) which had a high arsenic content. When this ore is roasted, i.e. heated strongly in air, the sulfur comes off as sulfur dioxide which can be used to make sulfuric acid. If arsenic is present it too comes off as arsenic trioxide (As_2O_3), which then forms arsenious acid.

A Royal Commission was set up to investigate the Manchester incident, and it reported its findings in 1902. These led to stringent controls over the amount of arsenic in glycerine, glucose, malt, treacle, and beer, all of which might come into contact with sulfuric acid at some stage during their manufacture. The legal limit set for arsenic in such products was 0.01 grain per pound or gallon, equivalent to 0.14 ppm. What puzzled the Commission at one stage was the finding of arsenic in samples of malt that had never been in contact with sulfuric acid or glucose. An investigation finally discovered that it was being contaminated by dust from the walls and ceiling of the loft in which the barley that produced the malt was dried, and that dust had come from the coke fuel used in the heating system and this contained arsenic.

Even as late as 1952 large-scale accidents with arsenic continued to happen. In that year a French chemist Jacques Cazenive, 59, was convicted of killing

* Formula H_2AsO_3.

73 babies and injuring 270 others through a talcum powder called Baumol that he manufactured. This should have contained zinc oxide, which has known skin benefits, but it contained arsenic trioxide instead. Baumol was eventually traced as the cause of the children's illnesses and deaths because there was an outbreak of sores and damaged skin among those who had bought the powder, and when this was realized and the powder analysed, its high arsenic content was discovered.

The arsenic-eaters

In the 1800s it was rumoured that the peasants of the Styrian Alps, near Graz on the border between Austria and Hungary, consumed arsenic trioxide as a tonic and in more than lethal doses. To many it seemed inconceivable, and doctors generally disbelieved such tales despite assurances from their colleagues in the Graz region that they really were true. The men ate arsenic to help them to breathe better at high altitudes, while the women ate it to make them plumper – a desirable female feature in those days – and to give them fresh complexions. (It did indeed result in a rosy cheeked complexion – regarded as a sign of good health – because it damaged the blood vessels in the surface of the skin.) The men also claimed it gave them more energy, aided digestion, prevented disease, made them more courageous, and increased their sexual potency.

It appears that the Styrian peasants first developed a taste for arsenic in the 1600s when mining began in the region. They got the arsenic trioxide from the chimneys of the small huts in which minerals were smelted and from which fumes of white arsenic were often observed to be emitted. Their name for the arsenic was *hittrichfeitl* referring to the white smoke, and they ate the arsenic trioxide like salt, sprinkling it on bread and bacon.

The arsenic-eaters were brought to public notice by a Dr Von Tschudi in 1851 who wrote about them in a medical magazine. This was repeated by Charles Boner in *Chambers' Edinburgh Journal* and was so sensational that it was reproduced in more than 30 other journals around the world. It received even more publicity when J.F.W. Johnston wrote about it in his book *The Chemistry of Modern Life* published in 1855 and even got academic acceptance when Professor Henry Roscoe spoke about it at the Manchester Literary and Philosophical Society in 1860 and included it in the influential textbook *Treatise on Chemistry* which he co-authored with Carl Schorlemmer in 1877.

The Styrian arsenic-eaters started with doses of around half a grain (30 mg) taken two or three times a week and then slowly increased this to a grain and then two grains, and they took such doses for most of their working life. Some men were known to be able to eat five grains (300 mg), well above the fatal dose for a normal adult, and one Styrian poacher was reputed to be able to eat almost a gram (1,000 mg) of arsenic trioxide. The peasants were skilled in measuring the dosage, cutting the necessary arsenic from a larger chunk, and some continued with this regime for as long as 40 years without coming to any apparent harm. Nor was it only humans who benefited from the *hittrichfeitl*; those who worked with horses also gave it to their steeds claiming it improved their health and appearance, and increased their stamina too.

Many regarded tales of arsenic-eating in the way we now regard urban myths. Clearly such natural scepticism could only be disarmed if arsenic-eating was proved by a scientific demonstration, and one done publicly. Such a demonstration was conducted in front of a large gathering at the 48th meeting of the German Association of Arts and Sciences at Graz in 1875. A Dr Knapp presented two arsenic-eaters to the audience, one of whom then consumed 400 mg of arsenic trioxide, the other 300 mg of orpiment (arsenic sulfide). The following day the two men appeared again still in the best of health and it was announced that samples of their urine had been collected and analysed and proved that they were indeed excreting large amounts of arsenic. There could now be no doubt that it was possible to eat arsenic and become almost immune to it by gradually increasing the dose.

Arsenic-eating had its downside. It interferes with the essential element iodine, which the thyroid gland needs to produce the hormone thyroxine that governs several metabolic processes, not least of which is keeping the body temperature steady. The iodine-deficiency disease goitre was prevalent among the Styrian peasants, as was cretinism among their children.

Following from the discovery that the peasants of Styria really were able to make themselves immune to arsenic poisoning, came the Styrian Defence, used by barristers to defend their murdering clients when they came to trial. It was a two-pronged defence which maintained that the arsenic in the victim's body might well have been there because he – and it generally was a he – was a regular taker of arsenic, while the arsenic found in the accused's possession was likewise explicable because she – and often it was a she – was dosing herself with arsenic to improve her complexion. The effects of arsenic in improving the complexion led to several patent cosmetic treatments appearing in the late 1800s, such as Dr Simms Arsenic Complexion Wafers which were said to produce a 'beautiful transparency', remove wrinkles, brighten the eyes, and raise the spirits. We shall hear more of the Styrian Defence, which was used at the trials of Madeleine Smith in 1857, Florence Maybrick in 1889, and even at the Maierhofer trial in Austria as late as 1937.

Arsenic in medicine

The idea of 'eating' arsenic may have horrified the Victorians but they were not averse to taking arsenic in smaller doses for its supposed curative properties. Indeed arsenical treatments for disease had been around for a very long time. Hippocrates (460–377 BC) said that the red ore realgar was a remedy for ulcers on the body. Orpiment and realgar were used to treat abscesses and scrofula in ancient China and were mentioned in medical texts of 200 BC. In Europe, arsenic trioxide was used as early as the 1110s as an antimalarial ('tertian fever') medicine. Other physicians down the centuries favoured the more common yellow orpiment, which was taken to treat all sorts of conditions, such as arthritis, asthma, malaria, TB, diabetes, and venereal diseases, for which it could have had only marginal benefits.

These natural arsenic minerals were part of the stock-in-trade of pharmacies and physicians for millennia and even today some Chinese medicines still include arsenicals among their ingredients. Chinese herbal balls, also known as tea balls, contain them and these are taken for a wide range of ailments, such as to reduce fever, ease rheumatism, and reduce tension. They are taken dissolved in warm wine or water. Their arsenical ingredients came to light when some tea balls being imported into the USA claimed to contain rhino horn, and it was this which made US customs seize a cargo of them in 1995. In fact they contained no rhino horn. What the forensic chemists discovered were high levels of arsenic. Some were found to contain as much as 35 mg of arsenic trioxide so that a recommended dose of two per day would in effect deliver 70 mg.

In the UK, Pakistan and India, ethnic remedies sold as powders have been found to contain as much as 100 mg of arsenic trioxide (along with mercury sulfide). In Singapore in 1975 there was an outbreak of arsenic poisonings which was eventually traced to traditional remedies, some of which contained as much as 10% arsenic trioxide.

Arsenic in small doses is not noticeably detrimental to health and may even have short term benefits. Indeed it is present in several spring waters, some of which were reputed to have tonic effects that might be due to arsenic. The famous Vichy water contains as much as 2 ppm of arsenic, and other spa waters have been found to contain even more. However, most bottled mineral water today contains very little or none at all.

Arsenic really came into medical use with the introduction of Dr Fowler's Solution in the 18th century. Thomas Fowler* was working at the Stafford Infirmary, in the English Midlands, in the 1780s where he was impressed with the curative properties of a patent medicine that went under the name of Thomas Wilson's Tasteless Ague and Fever Drops. Together with the Infirmary's apothecary, a Mr Hughes, he analysed it and discovered that its active ingredient was arsenic. Dr Fowler then devised his own version of the medicine and publicized this in a pamphlet entitled *Medical Reports of the Effects of Arsenic in the Cure of Agues, Remitting Fevers and Periodic Headaches*, which was published in 1786.

Dr Fowler's medicine was a solution of potassium arsenite with a little lavender water, this being added to prevent its being taken accidentally. The recipe for making it was to dissolve 10 g of arsenic trioxide and 7.6 g of potassium hydrogen carbonate in a litre of distilled water; then add a little alcohol and lavender oil. The maximum single dose recommended in the British Pharmacopoeia was 0.5 ml of Fowler's solution which would deliver 5 mg of arsenic. This solution was prescribed for all manner of ailments such as neuralgia, syphilis, lumbago, epilepsy, and skin disorders. Twelve drops of the medicine were to be taken three times a day for a week, the total arsenic amounting to around 120 milligrams. The medicine could be added to a glass of water or taken with wine. Dr Fowler's solution first appeared in the London Pharmacopoeia in 1809 and throughout that century it was regarded almost as a cure-all. It was prescribed as a restorative tonic, and it even acquired the reputation of being an aphrodisiac. It was often taken to aid convalescence. Charles Darwin's mysterious illness may well have been due to arsenic poisoning, and it may be that he sometimes overdosed himself with Fowler's solution which he was known to take regularly to treat a tremor in his hands.

No less a person than James Begbie, Vice-President of the Royal College

* He died at York in 1801.

of Physicians of Edinburgh, endorsed its benefits, and he was Queen Victoria's doctor when she was resident in Scotland. Such an endorsement ensured its continued popularity although he did warn that continuing with the treatment for more than a week would eventually lead to adverse side effects such as dry throat, tender gums, nausea, and diarrhoea. Those who were regular takers of Dr Fowler's Solution – and some took it for years – were putting themselves at risk of an even worse side effect: cancer.

Analysis of 262 patients who took Fowler's solution for many years – often self-administered – found half had thickening of the skin on hands and feet (hyperkeratosis) and about one person in ten had skin cancers. Eventually its use as a therapeutic agent was disapproved of and in the 1950s it was no longer permitted to be sold. One 39-year-old man who reported to his local clinic in 1955 because of the thickened and cracked skin on the palms of his hands and soles of this feet was found to have been taking Fowler's solution three times a day for 12 years, having first been prescribed it for 'nervous debility' and continuing with it via an automatic repeat prescription which his pharmacist never thought to question.

Although Fowler's Solution is no longer used in modern medicine its main ingredient, arsenic trioxide, is making a come-back as we shall see. Arsenic trioxide itself was once used by dentists, who applied it as a paste to the cavities of decayed teeth as a way of destroying the nerve before filling the tooth. Leakage from these cavities, however, would sometimes destroy the underlying jaw bone or kill the tissue of the gums. However, there was a role for arsenic in more sophisticated drugs, and some of these were used until recent times.

The man whose vision led to remarkable advances was Paul Ehrlich. He was born of well-to-do Jewish parents on 14 March 1854 in the Silesian village of Strehlen which was then in Germany but is now in Poland. He studied at medical schools in Strasbourg, Freiberg and Leipzig and was anxious to secure a post in a university teaching hospital. In 1878 he accepted a position at the Charité Hospital in Berlin but in 1885 he discovered he had tuberculosis and spent the next 2 years convalescing in Egypt. In 1890 he returned to Berlin to work with the illustrious Robert Koch and then was given his own research unit albeit housed in a old converted bakery.

In 1899 he was appointed the first director of the Institute for Experimental Therapy in Frankfurt-on-Main and it was there that he was to make the discovery for which he became world famous. He hoped to find a chemical that would be toxic to microbes in the body without being toxic to their human host. His target microbes were the trypanosomes, which are flagellate protozoas transmitted by the tsetse fly and which cause sleeping sickness. The compound which looked most promising was an arsenic compound called atoxyl. This had first been made in 1859 by a French chemist, Antoine Bechamp, and had been tried on patients with the disease by an English doctor H.W. Thomas in 1899 with some success, although it had side effects that were too severe to be worth the risk of using the drug: it damaged the optic nerve causing blindness in some of those he treated.

Ehrlich began work on modifying atoxyl in 1906 and he was helped by the fact that its chemical formula had been worked out.* Knowing exactly what the molecule was, Ehrlich then set about devising means of making similar compounds, hopefully less dangerous. He was a difficult man to work with and insisted on his instructions being followed to the letter. What he needed were compliant co-workers who would perform the necessary chemical experiments but that was not what his colleagues thought they should be doing, and some of them resigned when he demanded they follow his exact instructions. However, one man was prepared to go along with Ehrlich and that was Alfred Bertheim, and others were recruited to assist him.

Every new compound that Ehrlich devised he gave a number to, and told his laboratory assistants how it could be made. When he was satisfied they had produced the right molecule it was tested. This was now done on rabbits infected with *Treponeam pallidum*, the spiral-shaped bacterium that causes syphilis. The Japanese bacteriologist Sahachiro Hata had found a way of infecting rabbits with syphilis and Ehrlich persuaded Hata to come and work with him. Compound number 418 had some success in controlling the syphilis but this first glimmer of a treatment turned into a blaze of light when Bertheim made compound number 606.

The momentous day was 31 August 1909. Ehrlich watched as Hata injected a male rabbit which had large syphilitic ulcers on its scrotum, with 606 and it survived unharmed. A day later when fluid from the ulcers was tested not a living *Treponeam palladium* could be found and within a month the ulcers had completely healed. Ehrlich now agonized over whether they should inject the new drug into a human patient, remembering the blindness

---

* It is sodium *para*-aminophenylarsonate, $Na[H_2NC_6H_4AsO_3H]$.

that atoxyl had caused, and he was relieved when two young assistants in the institute volunteered. They survived unharmed.

Still Ehrlich hesitated, but he offered 606 to a Dr Julius Iverson of St Petersburg where there was an outbreak of relapsing fever, which is spread by bites from ticks. Iverson treated 55 patients with the new drug and reported back to Ehrlich that 51 of them had been completely cured after a single injection. Tests on patients with syphilis soon followed and on 19 April Ehrlich announced the new drug at the Congress for International Medicine at Wiesbaden in Germany. He gave it the generic name arsphenamine and it was marketed as Salvarsan. It created a sensation and was to make news around the world. Soon people were talking about the new cure even to the extent of calling it Ehrlich-606 or just 606.

There were side effects with Salvarsan as one might expect with any arsenic-based treatment, and Ehrlich continued his search for a better version which he achieved with compound number 914, launched as Neosalvarsan and which had fewer side effects, and even better derivatives were eventually developed. By 1937 the number of arsenical compounds that had been made and tested added up to 8,000 and some were used such as arsthinol (Balarsen), acetarsone, tryparsamide, and carbasone. Tryparsamide became widely used to treat sleeping sickness in parts of Africa although some strains of the disease were resistant to it, but it too had its side effects, including dermatitis and damage to the optic nerve. More than 40,000 people were treated with it in the years in which it was used, and the death rate from the disease fell dramatically from about 35% to around 5%.*

The side effects of Salvarsan sometimes resulted in permanent deafness and gangrene, necessitating amputation of limbs, and as a result Ehrlich found himself being attacked in the press with scurrilous accusations that he had forced prostitutes to act as guinea pigs. One editor who made such accusations ended up in prison but the affair depressed Ehrlich and may have contributed to the stroke he suffered on 20 August 1915 and from which he died. Nevertheless he died a millionaire because he had patented Salvarsan and he even won the Nobel Prize for Medicine in 1908, which he shared with the Russian embryologist and immunologist Elie Metchnikoff (1845–1916). Bertheim, who made the first sample of 606, was not to be so lucky with

---

* Hollywood paid tribute to Ehrlich with a biopic *Dr Ehrlich's Magic Bullet* (1940) in which Edward G. Robinson played Ehrlich.

either wealth or international honours, and indeed his life came to a rather tragi-comic end. He volunteered for the army at the start of World War I in August 1914, tripped over his spurs, fell down a flight of stairs, and broke his neck.

Arsenic and cancer

In the USA in the 1990s, claims were made that arsenic caused cancer of the skin, lungs, bladder, and prostate, and that it was linked to diabetes, heart disease, anaemia, and disorders of the immune, nervous, and reproductive systems. The last of these followed from the assumption that it was a potent endocrine disrupter, in other words it could interfere with the working of hormones. Most of these scares proved to be unfounded, but not all. Arsenic is classed as a cancer-forming chemical and it was commonly believed that those in industry who were exposed to the element ran a significantly greater risk of contracting the disease. Evidence that it was so was inconclusive, but it was certainly the cause of lung cancer in workers exposed to the arsenic trioxide fumes given off when metal sulfide ores containing arsenic were roasted. Those who were prescribed Fowler's solution over a long period of time were also at risk of cancer.

The link between arsenic and cancer was first made in 1888 by Jonathan Hutchinson who noted that patients who were being treated for psoriasis with arsenical mixtures developed cancers of the skin. Prolonged exposure caused chronic dermatitis and he suspected this was what eventually developed into skin cancer. We now know that he was more or less correct and the evidence now points to *inorganic* arsenic as the villain; in other words to the compounds based on arsenite (AsO_3^{2-}), which is arsenic in oxidation state (III), and arsenate (AsO_4^{3-}), which is oxidation state (V). There is also epidemiological evidence of links between exposure to inorganic arsenic and bladder, lung, and liver cancer. Research by John Ashby and co-workers at the Central Toxicology Laboratory in Cheshire, England, in 1991 showed that arsenite was capable of causing genetic damage in mice, and thereby possibly triggering cancer, but that orpiment was inactive.

When animals were deliberately exposed to high levels of arsenite it proved impossible to induce cancers this way, even in rats specially bred to be ultra-sensitive to anything that might cause cancer. Some see this as proof that arsenic is not carcinogenic and this appeared to be confirmed in 2001 when Toby G. Rossman of New York University Medical Center demonstrated that arsenic did not react with DNA to cause mutations. She carried out researches on mice that are particularly used to model skin cancer. They were given water containing 500 ppb of arsenic to drink and were then exposed to intense UV light, and while they suffered from sunburn, no tumours formed.

In 1988 the EPA rated arsenic a class A human carcinogen, mainly on the basis of what had happened in Taiwan where there were high levels of arsenic in drinking water and increased incidences of cancer. The effects of such exposure will be looked at more closely in the next chapter, but it has been possible to put numbers to the risk of getting cancer and the National Research Council of the USA has estimated that at a level of 3 ppb in drinking water the lifetime risk of bladder and lung cancer would be one person in 1000 which is much higher that the one person per 10,000 level that is normally acceptable for cancer risks.

Inorganic arsenic may cause cancer but it can also help in the fight against it. It was known many years ago that arsenic stimulated the formation of blood and it was once a recognized treatment for anaemia. While this medication was eventually discontinued it was not forgotten, which is why Trisenox (arsenic trioxide) has recently been approved for use by the US Food and Drugs Administration for treating acute promyelocytic leukaemia. It works by stimulating the production of normal bloods cells which have become crowded out by cancerous white blood cells. It was reintroduced into the USA by Drs Steven Soignet and Raymond Warrell at the Memorial Sloan Kettering Cancer Center in 1997 after they had learned that Chinese scientists had administered it intravenously to those suffering from this form of cancer. It had been given at a rate of 10 mg per day for 45 days and shown to lead to remission of the disease, a fact they reported in the June 1996 issue of the journal *Blood*, which recounted its success on 16 human patients who had not responded to other anti-cancer treatments.

The sources and uses of arsenic

Most arsenic minerals in the Earth's crust are sulfides: red realgar (As_4S_4), yellow orpiment (As_2S_3), silvery arsenopyrite (FeAsS), and iron grey enargite (Cu_3AsS_4). None of these is mined for its arsenic because the world economy has more than it needs and that comes as a by-product from the refining of ores such as copper and lead. World production of arsenic tri-oxide amounts to 50,000 tonnes per year, and that which is needed for industrial processes comes mainly from China, which is the chief exporting country. The amount of arsenic in copper and lead ores is more than 10 million tonnes, but the problem with arsenic has been overproduction and at times surplus arsenic has even dumped in the sea.

There have been many uses found for arsenic over the years. An early one was the discovery that adding a little (0.4%) to molten lead ensured the formation of perfectly spherical pellets of shot when the melt was poured down a shot tower. Copper arsenite (Paris green) was used as a horticultural spray especially to kill the codling moth on apple trees. Lead arsenate was also used, and in 1941 in the USA when 546 orchard workers were tested, seven of them were found to be suffering from clinical arsenic poisoning. Such uses have now been phased out. At one time sodium arsenite solution was sprayed on potatoes to defoliate them prior to harvesting, but this was phased out in 1971.

For many years in the USA wood has been pressure treated with chromium copper arsenate (CCA) to prevent it rotting and being eaten by termites; indeed 70% of treated timber products were protected with CCA and the industry was a $4 billion per year one using $150 million of CCA at 350 facilities around the nation. As of December 2003 the use of this preservative started being phased out in favour of a copper boron compound, although other countries still permit its use.

As recently as 1997 the USA imported 1,200 tonnes of arsenic metal and 30,000 tonnes of arsenic trioxide. Of the latter, 16,000 tonnes were turned into wood preservative, while 5,500 tonnes went to making agrochemicals such as herbicides and animal feed additives, where it is given to pigs and poultry to control various diseases and especially to promote growth. Glass manufacture accounts for 800 tonnes of arsenic trioxide, while 700 tonnes of metallic arsenic went into metal alloys. There may be less industrial poisoning by arsenic today but there are still cases and in the USA in 1998 there were around 1,400 incidents reported to the Toxic Exposure Surveillance System, of which about a third were due to arsenic-containing pesticides.

As some arsenic-using industries found alternatives, newer industries found newer uses, albeit not on anything like the same scale. The electronics industry is one of the minor users of arsenic, which it requires for the manufacture of diodes, lasers, and transistors. It is added to silicon and germanium semiconductors where it acts to supply electrons to the crystal lattice. A new and growing use of arsenic is the semiconductor gallium arsenide which has the ability to convert electric current to laser light. Gallium arsenide is an important product for microelectronics and the arsenic to make this comes as ultra pure arsine gas (AsH_3). This gas is also used as a dopant to add a few arsenic atoms to other materials to provide electrons. Arsenic is now used to make the more sophisticated indium arsenide (InAs).

Arsenic is everywhere in the environment and the amount that is added each year continues to increase. There are natural sources, for example volcanoes emit about 3,000 tonnes per year and micro-organisms release volatile methylarsines to the extent of 20,000 tonnes per year, and this will all eventually be washed out of the atmosphere to land or sea. The burning of fossil fuels releases 80,000 tonnes per year. Although such numbers seem threatening they add only marginally to the planetary burden and in any case environmental arsenic is rendered safe once it meets up with a sulfur atom and forms an insoluble, immobile sulfide.

Arsenic in warfare

The forces of the Byzantine Empire, the successor to the Eastern Roman Empire, had at their disposal a wonder-weapon: Greek fire. According to one account it appeared in the reign of Constantine IV Pogonatus (who ruled from 641–668 AD) and its invention was credited to a refugee from Syria who fled to Constantinople (now Istanbul) after his native land was conquered by the Arabs. Others say that it was really a development of an existing weapon that the Byzantines had used in the 500s but, however it was discovered, it certainly had a profound effect.

Greek fire was invaluable in fighting off the Arab fleets that attacked Constantinople in 673 and 717, and was even used against a Russian fleet in the 900s. In these attacks Greek fire was ejected under pressure from tubes mounted on the prows of their ships, rather in the manner of a modern flame-thrower and it was reputed to catch fire spontaneously and to be impossible to extinguish. Such was the power of this new weapon, and the fear it engendered, that it is thought to have been a significant factor in enabling the Byzantine Empire to flourish for almost a thousand years.

The secret of Greek fire was carefully guarded and with the fall of

Constantinople to the Turks in 1453 all knowledge of how it was made was lost. We can hazard a guess as to what it contained and whether the necessary chemicals were available. The chief flammable component must have been a volatile hydrocarbon and indeed naphtha, which is similar to kerosene, was collected from natural pools of oil that came to the surface in parts of the Middle East. While this would burn well it required something extra to make it burn out of control and what better for this purpose than a combination of arsenic sulfide and potassium nitrate (KNO_3), whose chemical reaction would indeed release a great deal of energy. The former was available as the mineral realgar and the latter as nitre, which was easily collected from the walls of latrines or wherever dung was stored; it is formed by bacteria working on the nitrogenous material in human and animal waste. In Victorian times realgar was used to produce so-called Indian fire which burned with a brilliant white light and which was made by mixing it with potassium nitrate in the ratio of 1 part realgar to 12 parts KNO_3. Such a mixture could easily have been made by the Byzantines.

Arsenic as an agent in warfare languished for many centuries until it was revived in World War I. In that war various chemical agents were used in an effort to break through the lines of trenches that stretched for hundreds of miles along the Western Front. The Germans tried chlorine gas on 22 April 1915 and this had a devastating effect on the unprotected British soldiers as it rolled over no-man's-land and into their trenches. Five thousand men died and more than 15,000 were permanently lung-damaged. In September of that year the British retaliated with mustard gas, a sulfur compound, but the attack was totally ineffective. The disadvantage of these types of chemical agents was that they made the target area unsafe to occupy and it hindered rather than helped the attacking forces. The search was on for 'better' weapons. Several arsenic-based chemicals were found such as Lewisite, Sneeze Gas, and Adamsite, their chemical names being 2-chlorovinyldichlorarsine, phenyldichlorarsine, and diphenylaminechlorarsine. Of these the only one used on a large scale in World War I was Sneeze Gas which was capable of penetrating gas masks and producing unbearable irritation of the respiratory tract.

Lewisite was much more potent and this was developed for use as a chemical weapon, but the war was over before it could be used. Lewisite is an oily liquid with the odour of geraniums and it boils at the relatively high temperature of 190°C, which means that it is not very volatile and so cannot be used as a gas as such, but it could be spread as a vapour – the 'dew of death' – and while it could kill it was more likely to incapacitate because breathing the vapour would cause the lungs to fill up with fluid. The reason for using Lewisite was to disable troops by penetrating their clothing, including protective rubber suiting, causing a violent reaction on the skin forming large painful blisters. On unprotected individuals the chemical would attack eyes, lungs, and skin and eventually lead to liver damage and maybe death.

Lewisite was first made by Julius Nieuwland in 1904 in the chemistry department at Notre Dame Catholic University, Indiana.* He was researching the reactions of acetylene gas and as part of this research he looked at what would happen when he put it with arsenic trichloride ($AsCl_3$). Nothing happened, but he knew that did not necessarily mean that they would not react under the right conditions and so he tried adding aluminium chloride, a commonly used catalyst. The reaction took off and the liquid later to be known as Lewisite was produced. Because he had taken no special protective measures, some of the vapour was breathed in by Nieuwland who was taken ill and spent the next few days in hospital. He never performed that reaction again – but it was not forgotten.

Once the Germans had used poison gas in World War I, the United States, aware that it would eventually be drawn into the conflict, began its own programme by sponsoring work at universities, one of which was the Notre Dame Catholic University. There Nieuwland's PhD supervisor told the head of one of the chemical weapons units, Winford Lewis, about the fateful reaction and its toxic effects. Lewis began to study it, finding a way to make it under controlled conditions and, by November 1918, just as the war came to an end, the USA was shipping Lewisite to Europe. They had made 150 tonnes of it at a plant in Willoughby, Ohio, where military personnel were producing 10 tonnes a day by the time the plant closed. The lethal cargo was dumped in the deep Atlantic Ocean.

That did not mean the end of Lewisite, because other countries began to manufacture it and indeed it was used by the Japanese against the Chinese in Manchuria in 1940. The Americans also began to manufacture it again, and by the end of World War II in 1945 they had amassed 20,000 tonnes of the liquid, which was again dumped in the ocean once that conflict was over. US servicemen in that war were equipped with tubes of an antidote ointment

---

* Father Nieuwland died of a heart attack when visiting his old laboratory in 1936.

known as British Anti-Lewisite or BAL (see Glossary) as a precaution against such an attack. Thankfully they were never needed.

The dictator Sadam Hussein used Lewisite in the Iraq–Iran war of the 1980s, and he may have supplied it to the Sudan government to use against the Sudan People's Liberation Army, which was active in the south of that country. On 23 July 1999, planes dropped 16 exploding canisters loaded with it on the towns of Lainya and Kaya. The Norwegian People's Aid reported that civilians who were exposed to it showed all the symptoms known to be produced by Lewisite. A group of UN World Food Programme workers who visited Lainya a few days later were also affected. Casualty figures among the civilian population are not known although two people are believed to have died, and the same fate befell goats, sheep, dogs, and birds.

Although it would have the capacity to cause untold human misery, it seems unlikely that a chemical warfare agent based on arsenic would today ever be used in a major global conflict, although as an easily manufactured material it might well be employed on a small scale to terrorize people, as happened in the Sudan.

Insidious arsenic

THE journal of the Royal Society of 1671 carried a review of a paper by a Dr Caroli de la Font entitled 'The nature and causes of the plague' in which he put forward the theory that it was due to 'arsenic exhalations' that were polluting the air. Of course he was wrong, but the idea that such emissions could pollute the atmosphere was not wrong and 150 years later, in 1821, they may well have contributed to the death of one of the great figures of history: Napoleon.

Arsenic had, and still has, its uses as we saw in the previous chapter but it is an insidious element and is much more mobile than earlier generations appreciated. When it diffuses into the air we breathe, and gets into the water we drink it causes problems and in this chapter we will look at two ways that it led to – and is still leading to – mass poisonings. The historical story concerns its leakage from wallpaper, the modern story concerns its leakage from underground rocks. The first of these leakages contaminated the air of millions of homes in the Victorian age, the second contaminates the drinking water of millions of people in Bangladesh and neighbouring states of India today. The first of these tragedies was eventually controlled, the second one remains to be dealt with.

Colouring with arsenic

In days gone by the palette of a painter might well have held three arsenic compounds because they could provide brilliant shades of yellow, red, and especially green. The first two of these came from the natural pigments yellow orpiment and red realgar, both of which are arsenic sulfides; orpiment has the formula As_2S_3, and realgar has the formula As_4S_4. Orpiment got its name from the Latin words *auri* (gold) and *pigmentum* (paint) and was popular in the ancient world, especially in the Middle East. Its association with gold is probably what made it attractive to alchemists. Orpiment only became widely used in Europe when synthetic orpiment was manufactured and then it was known as royal yellow or king's yellow and was the preferred source of yellow until it was displaced by chrome yellow (lead chromate) and cadmium yellow (cadmium sulfide).

Within the veins of naturally occurring orpiment there could be found the red mineral realgar whose name comes from the Arabic words *rahj al-gar* meaning dust of the cave. As a pigment it was given names such as red orpiment, red arsenic, and arsenic orange, the name depending on the depth of its colour. Realgar was used by artists at the time of the Pharaohs and was still being used by Dutch artists in the 1600s. The synthetic version was known as ruby sulfur and while it was much favoured, it was prone to convert to the more stable orpiment, especially when exposed to sunlight. Then its red would fade first to orange and subsequently to yellow.

Orpiment too had its problems. Its bright yellow was particularly favoured by Dutch artists of the 17th century. Not all artists were enamoured by it because it caused other pigments to turn black if it was painted over, and this was especially so if they were using white lead which slowly reacted with the sulfur in the orpiment to form black lead sulfide. And there was yet another disadvantage, although it was never apparent to the painters who used it. The passing of the centuries caused the yellows to fade by the slow oxidation of the arsenic sulfide to white arsenic oxide, and as the process of oxidation advanced so it tended to lift the pigment from the canvas.

There are some naturally occurring green minerals such as malachite, which is copper carbonate, but artists and painters who required this colour often used verdigris, the copper pigment that forms slowly on copper when it is exposed to the air. Others mixed blue and yellow pigments to get green they wanted. All this changed with the appearance of a beautiful green chemical called Scheele's green, named after Karl Scheele (1742–1786). This was copper arsenite* which he first made in 1775 and he realized he could make money from the manufacture of it as a new green pigment. In 1778 it went into production, although in a letter he expressed his concerns about

* Formula $CuHAsO_3$.

its toxic nature and felt that purchasers should be warned that it contained arsenic. In the end he accepted that there was no need to worry, presumably thinking that its vivid colour was a guarantee that it could not be misused. The new pigment was soon in demand from artists across Europe – Turner was using it in 1805 – and it continued to be popular for the next 50 years; Manet was still using it in the 1860s.

Its chief rival was emerald green, which had first been made in 1822 – Turner was using this by 1832. Emerald green was a combination of copper acetate with copper arsenite and gave different shades of green. Manufacture of this began in 1814 at the Wilhelm Dye and White Lead Company of Schweinfurt. It was more popular than Scheele's green and was soon being used to print paper, cloth, and even confectionery. It was called by a variety of names such as Schweinfurt green, Paris green, Vienna green, and emerald green. Emerald green was made by a secret recipe until this was published in 1822 by the German chemist Liebig and its poisonous nature was revealed. Manufacturers then changed the recipe, adding other ingredients to lighten the colour, and changing its name accordingly in an effort to disguise its true nature. Emerald green was manufactured and used to colour paint, wallpaper, soap, lampshades, children's toys, candles, soft furnishings – and even cake decorations.

The leaves of artificial flowers in particular were coloured with various arsenic greens and they were very popular in Victorian households. The industry making them employed hundreds of young girls, who suffered accordingly from chronic arsenic poisoning. What is even more surprising is that these greens were sometimes used in ways that seem fraught with danger, and at a banquet held by the Irish Regiment in London in the 1850s the table decorations were sugar leaves coloured with them. Many of the diners took these home for their children to eat as sweets, and several deaths ensued. At another dinner in 1860 a chef was eager to produce a spectacular green blancmange and sent to a local supplier for green dye. He was given Scheele's green and three of the diners later died.

When such toxic pigments were used in oil paintings they posed little threat to those whose job it was to grind them, unless they breathed in the dust, even less threat to the painters who used them, unless they licked their brushes, and no threat at all to those who bought the finished works of art because these were generally given a coat of varnish to protect them. However, the walls on which they hung were a threat because these were covered with other arsenic pigments and these constituted a real threat, because they could release arsenic vapours at a debilitating rate.

Arsenic in wallpaper

Scheele's green was ideal for printing wallpapers, especially those with floral motifs. Wallpaper production rose steadily throughout the 1800s and in the UK it reached 1 million rolls a year in 1830 and 30 million rolls by 1870. When tests were then carried out it was found that four out of five wallpapers contained arsenic. Arsenic in wallpaper had a habit of diffusing into the air of a room and thereby affect its occupants. This had been suspected as long ago as 1815, but the mechanism by which it occurred was not correctly deduced until the 1890s, and what exactly was being released was only solved in 1932.

In 1815 there was a report in a Berlin newspaper to the effect that arsenic pigments in wallpapers were dangerous and this had been written by Leopold Gmelin (1788–1853) who was to become one of the best-known chemists of his day. He was perhaps too far ahead of his time, but people in Germany had already begun to suspect that arsenic wallpapers could poison the indoor atmosphere. Gmelin had noticed that rooms which had been covered with wallpaper printed with Scheele's green gave off a mouse-like odour when the paper was slightly damp. He suspected the vapour was an arsenic compound, and he said that it was unhealthy to spend too much time in such a room. He even went so far as to advocate stripping off all such paper and banning Scheele's green. No one took any notice. Had they done so, and taken action, then much illness and not a few deaths would have been prevented in the decades ahead.

In articles published in the *Dublin Hospital Gazette* in 1861, a Dr W. Frazer of the Carmichael School of Medicine reported that every sample of wallpaper that he had been asked to test contained arsenic, and not just in trace amounts but at levels he thought might be injurious to those who came in contact with them. He imagined that the threat came from breathing the dust from such papers, especially flock wallpapers, but he realized that this was not the only way in which arsenic could get into the air of a room and he postulated that arsine gas (AsH$_3$) or even cacodyl* might in some way be emitted from wallpapers that were damp, and he added that the wallpaper paste might be partly to blame. Frazer noted that often one layer of wallpaper would be pasted over another and that the lower layers often became ripe for 'putrefactive change' and that it was a gaseous arsenic emanation from such walls that was responsible for the peculiar smell that was often noticed in such rooms. He was right about it being due to arsenic vapour, he was right to speculate that the wallpaper paste had a part to play, and he was nearly right when he speculated that it might be cacodyl.

In 1864 there were reports in the press of children actually dying as a result of the vapours given off by mouldy green wallpaper, and the medical journal *Lancet* warned of the dangers of arsenic pigments. A typical wallpaper would contain around 700 mg of arsenic per square metre so that an average sized living room would hold around 30,000 mg of arsenic, in theory enough to kill more than a hundred people. Most of this arsenic would remain on the walls of the room unless they became damp. The nature of the aerial poison was unknown at the time but this did not stop concerned individuals launching campaigns against the use of arsenic-based pigments even though these flew in the face of most medical opinion which regarded arsenic as a potent medicine and good for treating all kinds of afflictions of the human body. In addition the general public had discovered that when arsenic-based papers were used on bedroom walls there was a noticeable disappearance of bed bugs, a major benefit that led to increased sales. Moreover, arsenic cigarettes were popular and reputed to cure nervous complaints, and arsenic-based cosmetics were supposedly good for the complexion. How could the tiny amount that was emitted by wallpapers be dangerous? It seemed illogical to claim otherwise and so Scheele's green and emerald greens continued to be used.

In the last quarter of the 1800s the most famous wallpaper designer of all was William Morris (1834–1896) who was one of the leading lights of the Arts and Crafts Movement, which brought a new style to interior design. He was also a tireless campaigner for left-wing causes, even producing his own socialist newspaper *Commonweal*. Not that his concern for a better life for all

---

* This is a methyl derivative of arsenic and the word cacodyl comes from the Greek word *kakodes* meaning stinking. It consists of two arsenic atoms bonded together, each with two methyl groups attached: (CH$_3$)$_2$As-As(CH$_3$)$_2$.

caused him to question the source of his wealth and its unhealthy origin in the West of England. That region had been mined since the Bronze Age because its rocks were a rich source of copper, tin, and lead. Another mineral that was to be found there was arsenopyrite, which is iron arsenic sulfide (FeAsS) but this was generally seen as a nuisance because if arsenic contaminated copper and tin it made them very brittle. Arsenic was removed by roasting the ores in furnaces whence it was emitted as arsenic trioxide which tended to collect in the flues. This was removed periodically and thrown on to the heaps of mine waste.

Morris's father was a successful broker in the City of London and he lived in a large mansion near Epping Forest, north east of the capital. He speculated in all kinds of ventures but nothing was as successful as his investment in the Devonshire Great Consolidated Mining Company which struck lucky on 4 November 1844 when it discovered one of the greatest copper lodes ever found in Britain. The Morris family held 304 of the 1024 shares issued by the company and they were to see these rise in value from £1 per share to £800 per share within a year and pay a first dividend of £71 a share. (The mine eventually paid out more than £1 million in dividends, equivalent to more than £100 million today.)

Morris's father died in 1847, shortly after one of his investments went spectacularly wrong, and the family had to move to a more modest home in Walthamstow. They retained most of their shares in the copper mine and Morris's widow gave 13 of these to each of her nine children thereby securing each of them a comfortable income of around £700 a year, equivalent to around £100,000 today.* In 1853 Morris went to Oxford University with the intention of training for the ministry, but it was there that he was drawn into the circle of left-wing intellectuals and artists that became the Arts and Crafts Movement.

William Morris's first wallpaper design was called *The Trellis* and consisted of a climbing rose on a wooden trellis. There was a lot of green in the pattern and that green was copper arsenite. Morris was to become a strong advocate of traditional pigments and dyes, and used them whenever he could. He was also aware that the buying public wanted vibrant colours and these could only be produced with synthetic dyes like Scheele's green.

---

* During his lifetime William Morris earned almost £9,000 from this investment, equivalent to £1.5 million now.

Indeed at the time *The Trellis* appeared on the market the production of Scheele's green in England was in excess of 500 tonnes a year. Morris's wallpapers became very popular and this was doubly profitable for him because he was to become a major shareholder in the largest arsenic-producing mine in the world. Maybe he closed his mind to those who claimed that such wallpapers could poison the air of the rooms in which they were hung. He was wrong to do so because they were right: it could.

In 1871 Morris joined the board of Devon Great Consols when it had been priced out of the market for copper by cheap imports but it now realized that its mountains of waste could be processed to extract arsenic which became its main source of income. Up to then arsenic had been imported from Germany where the mines of Saxony produced it as a by-product. Now Britain became the world's largest producer and user. Year by year arsenic production increased as they began to reprocess the slag for its arsenic, generally to local disapproval because of the pollution such work caused. West Country arsenic was of high quality and much in demand from glass and enamel manufacturers (a little arsenic would neutralize the iron which gave glass a green tint). Soon arsenic was being exported as more and more uses were found for it, for example in pesticides, and the price rose from £1 a ton to £20 by the mid-1870s. Production at the Devon Great Consols soared to 3,500 tonnes per year in the 1890s but the company was ultimately to decline as its attempts to mine for other metals met with no success.*

Morris was on the board of the company for 5 years, but his other interests made heavier demands on his time. It would be nice to think he resigned because he was aware that arsenic was coming under increasing criticism for its deleterious effects on those who worked to produce it, those who lived in the vicinity of where it was produced, on the workers in industries which used it, and even on the general public who innocently bought the products of which it was part. Such was the concern about arsenic that the Factories and Workshop Act of 1895 made it a legal requirement to notify the authorities of cases of arsenic poisoning, although the act was mainly designed to deal

---

* The Devon Great Consols mine had a revival in the next century when arsenic was needed for making war gases such as chlorovinyl dichlorarsine a blister gas more unpleasant than mustard gas. In World War II this was made in a tin works at South Crofty mine. This was the last working tin mine in the UK, finally closing down in March 1998.

with the industrial diseases arising from lead and phosphorus which were of particular concern.

While the new act improved working conditions in some industries it did little to relieve those who worked for the arsenic manufacturers. The men whose job it was to scrape the arsenic laden soot from the flues of the furnaces were particularly badly affected and the Government was eventually forced to set up an inquiry into conditions in the industry. Its report of 1901 acknowledged the terrible effects that exposure to arsenic was causing, yet it did not recommend measures to control the industry, mainly because the industry itself was now in rapid decline and indeed Devon Great Consols itself went out of business a year later. It had produced more than 70,000 tonnes of arsenic trioxide in the previous 30 years.

In the 1870s it became possible to use synthetic green dyes that were not based on arsenic and indeed wallpaper manufacturers, including those made for Morris, began to advertise that their products were free from arsenic, although privately Morris remained sceptical that wallpaper pigments were the cause of all the illness being blamed on them. The weight of scientific opinion was against him, however, and as the Victorian era drew to a close it was clearly established that wallpapers were capable of poisoning people by the gas they gave off.

Arsenic in the air

So what was the toxic vapour given off by wallpaper decorated with a green arsenic pigment? The Victorian chemists knew of one deadly arsenic gas and that was arsine in which three hydrogen atoms were bonded to the element (AsH_3), and that this they could make by treating a solution of arsenic trioxide with zinc and sulfuric acid. The German chemist A.F. Gehlen (1775–1815) discovered how dangerous arsine could be when he made some in 1815. Within an hour of sniffing the gas he became ill, displaying symptoms of vomiting and shivering. He took to his bed but became more and more ill and died nine days later. This gas had first been made in 1775 by Scheele but he had avoided breathing it and so chemists had been unaware of its toxicity until Gehlen's accident.

Industrial accidents have been caused by the release of arsine when metal alloys containing arsenic have been treated with acid. The gas can be given off from zinc dross, and in Yugoslavia in 1978 a gang of eight men who were handling this while it was raining were overcome by it. The rain reacted with the dross to release arsine which affected them and they had to be rushed to hospital where one of them died. Arsine gas has a special affinity for the haemoglobin of red blood cells and this is what makes it so dangerous. It could even be formed inside lead accumulators, the electrodes of which were made of lead plates alloyed with a little arsenic to strengthen them, and these were used inside the storage batteries of British submarines in World War I. The batteries were sealed units but on board one submarine there was an escape of arsine gas and the crew of 30 men were all exposed to the same concentration of this dangerous gas for the same length of time. Yet they did not all react in the same manner and it was this that made the incident interesting from a medical point of view. One man was not affected in any way. Seven men suffered anaemia and jaundice and were hospitalized, while the rest showed symptoms of anoxemia, which is abnormally low levels of oxygen in the blood, although not to the extent that they required hospital treatment.

However, arsine was not the gas given off by Victorian wallpaper. Chemists at the time knew there was no possible way in which it could be generated because to do so it would require a chemical reaction between arsenite ions in the dyes and a strong reducing agent. But chemists did know of other arsenic compounds that were quite volatile, and potentially dangerous on this account, such as cacodyl.

In May 1891 an Italian chemist, Bartolomeo Gosio (1865–1944) started a year-long research project with the intention of proving that arsenic dyes could be a health hazard because they gave off a poisonous gas. Gosio's research was motivated by a problem that had been worrying the Italian authorities for many years: the large number of young children who were sickening and dying of inexplicable causes. They too were aware that doctors were sure that arsenical pigments were to blame and decided something must be done. Gosio's technique was to grow various micro-organisms on a mixture of mashed potato containing a little arsenic trioxide and he simply placed samples in damp cellars where indeed colonies of bacteria and fungi quickly grew on them and produced a garlic odour. Some research was also done by adding Scheele's green and emerald green to the culture medium and then growing the common bread mould on it and again the same smell developed. When he took some of the mould from a piece of rotten paper he noted that the intensity of arsenical gas was such as to be dangerous, and indeed a rat exposed to the vapour quickly died, and a small mouse put in a vessel with the fungus expired within a minute.

Others were also experimenting with micro-organisms and arsenical materials and with contradictory results. In 1893 Charles R. Sanger of the Washington University of St Louis, in the USA, did a study of 20 cases in which the poisoning was attributed to wallpaper and showed that indeed this was the cause, the wallpapers all containing arsenic with the ranges being from 15 to 600 mg per square metre. Sanger even noted that the *less* arsenic in the paper, the *more* severe was the poisoning and suggested that when the level of arsenic was very high it was too toxic for the moulds to grow. While Sanger in the USA was able to confirm Gosio's findings in 1893, the German biologist Otto Emmerling could not and in 1897 he published a paper saying that he found no arsenical gas given off with his cultures. This was not really a contradiction of Gosio's findings because he had used different micro-organisms which were unable to make methyl arsenic and release it into the air.

No one really doubted that Gosio had discovered the reason why there was so much ill health among those living in houses whose walls were decorated with arsenical paints and wallpapers. His findings were such that the sickness associated with breathing the air in such rooms began to be called Gosio's disease. Now it was possible for doctors not only to diagnose the condition but to treat those affected by removing them from the cause and advising that the room be treated rather than the patient. Nevertheless it took many years before Gosio's disease disappeared. Microbes found damp wallpaper a fertile place to grow. In those days fresh wall plaster was sealed with gelatine (known as size) prior to being painted or wallpapered and the latter was generally stuck on the wall with a flour-based paste. This combination of proteins and carbohydrates made an ideal nutrient for the microbes, especially when it was damp. As the microbes grew they needed to remove the arsenic from their environment and this they did by converting it to the compound trimethylarsine.

As late as 1932 two children in a house in the Forest of Dean on the English–Welsh border died of Gosio's disease. It was only then that an English chemist Frederick Challenger (1887–1983), who was professor of organic chemistry at the University of Leeds and who researched the biological methylation of arsenic, correctly identified trimethylarsine as the cause. This compound had first been made in 1854 when it was shown to be a colourless, oily liquid with a boiling point of 52°C. Improbably as it appeared at the time, we now know that there are enzymes capable of replacing the three oxygen atoms attached to the arsenic in Scheele's green with three methyl groups.

Although trimethylarsine had been identified as the deadly vapour in the 1930s, it was not until 1971 that it was finally revealed how micro-organisms could add methyl groups to arsenic. Many microbes can perform this little trick, including two kinds of wood-rotting fungi that can grow on wood treated with arsenical preservatives, although 63 other wood-rotting fungi were unable to do so. Natural gas also contains traces of trimethylarsine, a fact only discovered when technicians working on a gas pipeline in California in 1989 discovered an inlet pipe choked with a deposit which turned out to be pure arsenic. Further research showed that natural gas could contain as much as one microgram of arsenic per litre and that this was mainly as trimethylarsine.

There was a strange case of arsenic poisoning in the US Embassy in Rome in the 1950s when a female ambassador, Clare Boothe Luce, was affected. She was the US Ambassador to Italy and she had to resign because of her illness. It was obvious that she had been poisoned, but by whom? Was it by agents of the USSR? This was at a time when the Italian Communist Party was particularly powerful. So serious might have been the repercussion that the CIA sent a team to Rome to investigate. Eventually the source of arsenic was traced. She slept in a room that had not previously been a bedroom and which had a decorated ceiling that contained a lot of arsenic pigments. Was this yet another example of Gosio's disease? In fact it wasn't. On the floor above was a washing machine that caused the floor to vibrate and this caused the room to become loaded with arsenic-laden dust and it was this which she breathed in and which caused the unpleasant symptoms.

Arsenic in the hair

In his last will and testament Napoleon Bonaparte (1769–1821) wrote: 'I died before my time, murdered by the English oligarchy and its hired assassin.' Indeed he did die before his time – he was only 52 – and indeed he was probably right in thinking that the English would like to see him dead, but did they hire an assassin to carry out the job? Some think they did, and know who it was. Others take a less contentious view of his death, maintaining it was a perforated stomach ulcer and that this had become pre-cancerous, which is what the doctor who did an autopsy found.* Yet others maintain that he was indeed poisoned, not by the British Government but by the arsenic from his wallpaper.

Following Napoleon's disastrous retreat from Moscow in 1812, with the loss of 500,000 of his troops, his enemies were able to push him back to the borders of France, and in March 1814 the allied forces reached Paris. On 6 April Napoleon admitted defeat and abdicated. As part of the peace treaty he was given the governorship of the island of Elba, an annual income of 2,000,000 francs, a guard of 400 volunteers, and permission to retain the title of emperor. Two years later he was an exile on the remote island of Saint Helena, situated in the middle of the South Atlantic Ocean and a thousand miles from the nearest continent. This was to ensure that there would be no repetition of his escape from the island of Elba, which had occurred on 1 March 1815, and of his return to France where again he was able to assume power and raise an army. That escape began what was to be known as the Hundred Days and ended with his defeat at the Battle of Waterloo on 18 June 1815.

For his remaining years Napoleon lived comfortably at Longwood House on Saint Helena. He died on 6 May 1821. When famous political figures die unexpectedly there are bound to arise theories that their deaths were not natural but murder, and the question is who wanted them dead? Conspiracy theorists step into provide answers. The death of Napoleon is no exception. One theory is that he was poisoned and that the poison was arsenic because it appears that the great man was suffused with this element and that this explained what they found when they dug up his body. Napoleon was originally buried in Saint Helena and there his body rested for 20 years before it was finally transferred to Paris. When it was exhumed his corpse was found to be almost perfectly preserved which has generally been taken as a sign of poisoning by either arsenic or antimony.†

Analysis of Napoleon's hair shows that it had an arsenic content 100 times the normal level, which according to some is positive proof that he was

---

* Napoleon had digestive pains all his life which was why he was often seen with his hand inside his jacket rubbing his stomach, a posture that was characteristic of the man, and shown in some of his portraits.

† The explanation is given in chapter 11.

poisoned. Ben Weider is a Canadian and President of the International Napoleonic Society, and he submitted two tiny samples of Napoleon's hair to Roger Martz, chief chemist of the FBI, for analysis in 1995. Matz used graphite furnace atomic absorption spectroscopy to determine the amount of arsenic present in them and reported that the first sample, 1.75 cm long, contained 33 ppm of arsenic, while the second sample, 1.4 cm long, contained 17 ppm and he concluded that both were consistent with arsenic poisoning. This suited Weider, who believes Napoleon was deliberately poisoned and in his books *The Murder of Napoleon* and *The Assassination at St. Helena Revisited*, co-authored with Sten Forshufvud of Sweden, he identifies the murderer as Count Montholon.

When Napoleon fell ill in March 1821 we know he was given tartar emetic (potassium antimony tartrate) by his doctors to check his fever. Then on 3 May he was given a dose of calomel (10 grains), but without his knowledge, this being a strong purgative and the reason for this was that Napoleon admitted to not having evacuated his bowels for many days. The dose was well in excess of the normal amount given to treat constipation and Weider believes that this was the agent that finished him off, and explains how his stomach lining was found to be heavily corroded when it was opened after death. (The arsenic in his hair is consistent with his theory that this was given to weaken his constitution and disguise the fatal poisoning.)

The evidence provided by Napoleon's hair is not so straightforward as it first appears. There are authentic samples of locks taken from his head dated 1805, 1814, 1816, 1817, and 1821, the last being cut from his head soon after he had died. Various research institutes have tested these, and the specimens tested in 2002 by Ivan Ricordel of the Paris Police Laboratory showed high levels of arsenic in *all* of them. The values ranged from 15 to 100 ppm, but all were well above the 3 ppm that is today regarded as the maximum safe limit. (The more normal level is 1 ppm.) The value for arsenic in Napoleon's hair dated 14 July 1816 was 77 ppm but a year later it was down to 5 ppm for hair dated 13 July 1817. Clearly Napoleon had been given, or had taken, a dose of arsenic the previous year. Was this deliberately given by a murderer? Or by his doctor as Fowler's solution? Or even taken as self-medication? There have even been suggestions that Napoleon was so fearful of being poisoned that he was taking arsenic in the belief that this would protect him by building up his resistance to poisons. However, this seems an unlikely theory.

Another sample of hair was taken by Napoleon's faithful valet, Marchand, who went into exile with him. This he cut from the emperor's head the day after he died and which he sealed in an envelope that he took back with him to France and where it was treasured by Marchand's descendants. A few strands of this were analysed in the 1990s and they revealed the presence of arsenic at various points along each strand and this varied from 51 ppm down to 3 ppm then up again to 24 ppm over the period of six months that they had grown from the emperor's head. These variations seem inconsistent with a steady intake of arsenic from some environmental source, such as the water supply, but they might be correlated with release of arsenic from wallpaper which would depend on the weather. When this was wet and damp then conditions were right for mould to grow and release trimethylarsine, but when the weather was warm and dry then they would be less likely to do this. Nevertheless, the results showed that at various times in the last months of his life, Napoleon had been exposed to high levels of arsenic. Indeed it appears that his symptoms during this period were consistent with arsenic poisoning.

In September 1994 at a debate of the Napoleonic Society in Chicago the results from other tests on the emperor's hair were revealed. These had been carried out by the FBI on hair owned by a Jean Fichou of Rennes, France, and they too were reputed to have been cut from Napoleon's head at the time of his death. Their provenance is far from reliable and we can question their authenticity because they showed an average arsenic level of only 2 ppm, which is totally incompatible with the hair that Marchand's descendants had kept.

Robert Snibbe, who is President of the Napoleonic Society of America, dismissed the theory that Napoleon was deliberately poisoned with arsenic in order to weaken him, as Weider believes, as 'complete hogwash' but there is still the puzzle of how he came to have so much arsenic in his body at different times. Could it really have come from his wallpaper? That was the question asked on national radio by Dr David Jones of Newcastle University, England, in 1980 in a talk called 'A Touch of the Vapours'. The answer was that it might well have come from that source because wallpapers printed with Scheele's green were then being manufactured and if these had been hung in Longwood House they might well have emitted arsenic because the climate of Saint Helena is particularly damp.

The only problem with the theory was that there was no way of knowing what the interior decorations of the house were like while Napoleon was a resident – or so Jones thought. A resident of Norfolk knew otherwise. She was Shirley Jones who had listened to the programme and it reminded her of a scrapbook that she had. This had been handed down in her family for generations and all its pages were dated. On a page for 1823 there was recorded a visit that one of her ancestors had made to Saint Helena and he had both taken a leaf from the tree that stood above Napoleon's grave, and torn a piece of the wallpaper from the room in which the great man had died. These he stuck in his scrapbook. Another listener to Jones's radio talk was Karin Cross of Weston-super-Mare who had a picture of the drawing room in Longwood House to which Napoleon was moved during his final illness. This clearly demonstrated that the wallpaper from the scrapbook was that on the walls during Napoleon's final days.

Tearing a piece from the wallpaper in Longwood House would not have been difficult to do, as a team of BBC investigators found when they went there in 1992 to make a programme based on Jones's theory. They noted that the current wallpaper was already coming away from the wall because of the damp climate. Clearly Shirley Jones's ancestor could have removed some and it is unlikely that the rooms in Longwood House had been redecorated so soon after Napoleon's death. The paper in the scrapbook showed it to be of a floral design with a star pattern of green and brown like that in Karin Cross's illustration. But was the green an *arsenic* pigment? A tiny fragment was analysed by the same non-destructive methods used to measure the arsenic in Napoleon's hair, and indeed it showed the element to be present. The fragment of wallpaper contained 120 mg of arsenic per square metre, with most of this being in the star-like pattern which had 1500 mg of arsenic per square metre.

Damp wallpaper might well explain the arsenic in Napoleon's body, but would the trimethylarsine emitted from the walls of his rooms have been enough to kill him? Just possibly, and at least it would have made him ill and we know it was present because some of those who accompanied Napoleon in his final exile complained of the 'bad air' in Longwood House. During his stay there, Napoleon complained of many symptoms such as shivering, swelling of the limbs, sickness and diarrhoea, and stomach pains. Because the symptoms of arsenic poisoning are so easily confused with other causes then it might have been that Napoleon suffered from too much arsenic during his years at Longwood.

Weider goes much further than this and attributing his death to murder, and he makes a strong case against the Comte Charles-Tristan de Montholon, who was head of the Longwood household, especially as his wife Albine was having an affair with the emperor and even had had a child by him. Moreover Montholon was a former aristocrat who, somewhat suspiciously, had only come to favour with Napoleon after his defeat at the Battle of Waterloo. Montholon is not the only suspect. There was Dr Antommarchi, the emperor's personal physician; was he bribed by the French Government to kill him because the restored monarchists were still afraid he might yet return? Or was it the British Governor of Saint Helena, Sir Hudson Lowe? He got on badly with Napoleon and was so worried that he might try to escape that he posted guards around Longwood House to keep a watchful eye on him. (Napoleon responded by having deep paths dug so that he could walk round the garden unobserved.)

Of course Napoleon was not the only permanent resident of Longwood House, and if he were affected by arsenic then so would others who lived and worked there. The emperor had a retinue of 20 people who accompanied him to Saint Helena, and some of these also died under rather suspicious circumstances, such as the major domo Franceschi Cipriani, one of the maids, and a young child who lived in the house. Cipriani fell ill on 24 February 1818 with violent stomach pains and cold chills. He was given hot baths to treat the latter but two days later he suddenly expired and was buried the following day. Others were suspicious of the death, but when his grave was later opened in order that an autopsy might be carried out there was no body to be found. The maid and the young child died a few days after Cipriani, which to some suggests they were deliberately poisoned as well, although it might well have been that they succumbed to Gosio's disease at a time when Longwood House was particularly badly affected.

Whatever was happening to the interior décor of Longwood House, there was no doubt that the wallpapers contained Scheele's green and it is not unlikely that Napoleon himself chose some of them, possibly ones with a green and gold theme which had been his imperial colours. If so, then he did himself no favours and though they reminded him of his glorious past, they slowly added to his present agonies.

Somewhat surprisingly, arsenic has recently been discovered in the hair of the man who ruled the nation that was eventually to defeat Napoleon: King George III of Great Britain. Strands of his hair, which were discovered in a long-forgotten envelope in a little-used display cabinet in the Science Museum London, to which they had been sent by the Wellcome Trust. They were passed to Professor Martin Warren of Queen Mary College London in 2003 and they were analysed for mercury, arsenic, and lead, the metals most likely to affect anyone with a predisposition to porphyria the condition from which the King suffered. The analysis showed there to be 2.5 ppm of mercury (a typical level would be around 1 ppm), arsenic 17 ppm (normally this would be around 0.1 ppm), and lead 6.5 ppm (normally this would be less than 0.5 ppm).* Warren argues that the exposure to arsenic was the most damaging and that this was probably a contaminant of the tartar emetic that was often prescribed, and sometimes forcibly given, to George III at times when he was considered to be mad. The madness of King George III is explored more fully in Chapter 13.

Arsenic in drinking water

When arsenic contaminates drink at levels like that in the beer once brewed in Manchester (see Chapter 5) it will eventually be discovered and dealt with, because it leads to toxic symptoms in those who consume a lot of it. When arsenic contaminates drinking water at similar levels it may go unnoticed for several years because much less water is drunk at any one time. And therein lies its opportunity to inflict widespread damage. The symptoms which appear are generally skin lesions and skin cancer, although there have been claims that other conditions are linked to it such as bladder cancer, kidney cancer, lung cancer, diseases of the nervous system, high blood pressure, and diabetes.

The villagers of San Pedro de Atacama live in an isolated region of Chile and their drinking water contains 500 ppm arsenic, yet they show no signs of arsenic-related diseases. Newer residents to the area are much more likely to suffer from skin disease and cancers. Those whose ancestors have lived a long time in that region appear not to be affected by it, whereas among the newcomers it appears that 10% of all deaths over 30 are due to bladder and lung cancer and that is linked to their arsenic intake.

A curious arsenic poisoning of drinking water occurred at Yellowknife on the North Shore of Great Slake Lake, Canada, in the 1940s. A gold mine in

---

* This information was kindly provided by Professor Warren.

the area had a smelter to process the ore and this removed arsenic in the usual way by releasing it as arsenic trioxide to the atmosphere where it fell to earth well away from the town itself. In winter of course it fell on the snow that covered the area, and that's how it ended up poisoning two elderly fur-trappers who had built themselves a log cabin in the area and in winter they used the snow outside as their main source of drinking water. Slowly, and without realizing it, they were poisoned and eventually died, only to be discovered when the thaw came and their bodies were found.

Regions around the world where the level of arsenic in drinking water is high are the Lagunera region of Mexico, the Codoban region of Argentina, Inner Mongolia, Taiwan, Finland, the West Bengal region of India, and Bangladesh. Even in the USA there are regions where the natural level of arsenic in drinking water was once high. In Taiwan an epidemiological study in the 1960s showed that the prevalence of skin cancer was related to the amount of arsenic. In one region where the level of arsenic was 500 micrograms per litre, which is 500 ppb or 0.5 ppm. (For the rest of this chapter I will use ppb because these are the units preferred by those who deal with arsenic in drinking water.) It was found that 10% of people aged 60 or over had cancers, mainly on their skin. Thankfully these types of cancers are not usually fatal and can be removed. Internal cancers, however, are not so easy to diagnose or treat and in one part of Taiwan where the level of arsenic was 800 ppb there was a high incidence of bladder cancer. Another study of 40,000 people found more than 400 cases of skin cancer in one region where the drinking water contained 600 ppb. Even when measures were taken to purify the water, the incidence of cancer continued high for many years because of the long latent period before which the disease appears.

The largest mass poisoning by arsenic has involved more than 30 million people. The arsenic was in their drinking water and that came from tube wells that were drilled in the 1970s in West Bengal, India, and in Bangladesh. The wells were installed in a drive by the United Nations Children's Fund (UNICEF) to provide safe drinking water for a population that had traditionally taken its water from contaminated streams, rivers, and ponds and which had suffered accordingly from water-borne diseases such as gastro-enteritis, typhoid, and cholera.

Tube wells are 5 cm in diameter and tap the abundant ground water about 200 metres below the surface and a few had first been installed in the 1930s when the region was part of the British Empire. Those villages that had them showed the hoped for decrease in illness, especially among children. Analysis of the water showed it to be free of heavy metal contamination although it was not tested for arsenic because this was not suspected to be a contaminant, an oversight that was to lead eventually to litigation. On the face of things the wells were of great benefit and by 1997 UNICEF could report that it had already surpassed its goal of providing 80% of the population in those regions with 'safe' water. Such had been the success of the programme that many villagers installed wells privately and today three out of four wells are indeed privately owned.

The first cases of chronic arsenic poisoning were identified by Dr K.C. Saha at the Department of Dermatology, School of Tropical Medicine in Calcutta, India, in 1983. Patients were suffering from characteristic skin lesions and pigmentation changes on their upper chest, arms, and legs, and a horny thickening of the skin on the palms of their hands and the soles of their feet, a condition which goes by the medical name of hyperkeratosis. He recognized these symptoms as due to exposure to arsenic, but his patients denied every using any arsenic-based products. So where was the arsenic coming from? Tests soon showed that their drinking water was to blame and as more and more cases of arsenic poisoning came to the attention of the authorities it became clear that the problem was widespread and soon thousands of wells were being tested and high levels of arsenic revealed – up to 4,000 ppb in some places in the Indian state of West Bengal. Things were even worse in Bangladesh where 70 million people have been exposed to low level arsenic poisoning over many years, and this has caused widespread arsenicosis, in which the skin erupts in disfiguring, leprosy-like lesions. After many years of such exposure cancerous growths begin to appear.

The World Heath Organization (WHO) says that arsenic in drinking water should not exceed 10 micrograms per litre (10 ppb). This sets a gold standard that many countries believe is not realistic and in many parts of the world including Bangladesh and the USA, a statutory limit is set at 50 ppb, which is a level at which arsenic is deemed to have no effect. As more and more wells were tested in Bangladesh it became apparent that large numbers of them had arsenic levels above 50 ppb and a significant fraction had levels in excess of 300 ppb. More than 20 million people were drinking water of the former kind and as many as 5 million were drinking water of the latter kind. In one study carried out under the supervision of Kimiko Tanabe of Miyazaki University, Japan, water from 282 wells in the village of Samta in the Jessore

District were analysed, of which 42 had arsenic concentrations in excess of 500 ppb, and one well was even delivering water with 1400 ppb. Of the other wells, 39 had concentrations between 300 and 500 ppb, 114 wells had between 300 and 100 ppb, 57 wells had between 100 and 50 ppb, and only 30 wells had less than this last concentration of which a mere 10 wells produced water with the WHO standard of 10 ppb. What the Japanese workers found was that where a tube well might yield as much as 1,200 ppb, a deep tube well on the same site would see this fall to only 80 ppb.

In 1997 the Bangladeshi Government instituted a Rapid Action Programme and it focused on a group of villages where contamination was severe and this found that almost two thirds of 30,000 tube wells were delivering water in excess of 100 ppb. A team from the Dhaka Community Hospital visited 18 affected areas and examined 2,000 adults and children and concluded that more than half had skin lesions due to arsenic. In the Indian state of West Bengal, there are fewer people using the water from tube wells but of the 1.5 million who do use it there may be 200,000 cases of arsenicosis, with many more on the way because symptoms do not appear for 10 years or more and cancers may not appear for 20 years.

So what can the people of this region do? One answer is to dig deeper wells that go below 200 metres or shallower wells that reach down only 20 metres. In this way the contaminated layer of ground water can be avoided. Once the population has access to uncontaminated water then the arsenic in their body is quickly lost; indeed giving them anti-arsenic chelating agents is not necessary because the body will detoxify itself. Those who display the skin conditions associated with arsenic can be treated with lotions and antibacterial ointments and slowly the condition will improve. The Indian government has issued chlorination tablets that will oxidize the arsenic in water, converting it from AsO_3^{3-} to AsO_4^{3-}, which forms an insoluble salt with the iron which is also present in the water.

In February 2004 the Bangladesh government approved four simple technologies that could remove arsenic by treating the water with various agents, advising that the best one would be to use an iron-based material that could be produced within the country. Meanwhile research led by Jonathan Lloyd of the University of Manchester, UK, has discovered that bacteria are responsible for the release of arsenic from rocks in which it would otherwise remain indissolubly bound. As ground water is extracted from tube wells it is replaced by surface waters seeping down and carrying organic matter and bacteria with it.

So why had no one suspected the water from tube wells might be dangerous because of arsenic? The British Geological Survey (BGS) is part of the British Government's National Environmental Research Council and in the summer of 2002 an attempt was made to take them to court by lawyers representing hundreds of Bangladeshi villagers. The claim was that when they had undertaken to survey water quality in their country in 1992 they had failed to test for arsenic, and that if they had tested for this element then thousands of people would have been spared the suffering of arsenicosis.

In effect the BGS had given the well waters of Bangladesh a clean bill of health, and they did the same for the drinking water in Hanoi, Vietnam, in 1996, but again they had not tested for arsenic which turned out to be very high at 3,000 ppb when tested by a group of researchers from Switzerland in 2001. Moreover it was pointed out that when the BGS had undertaken a survey of water in the UK it had tested for arsenic although it did not find any source that breached European guidelines. However, the BGS claimed that they did not think it necessary to test for something they did not expect to find, an opinion endorsed by UNICEF whose Bangladeshi representative said they only tested for known contaminants, of which arsenic was not one.

The BGS's later investigations started in 1996 and its findings were published in 2002, and it did indeed test for arsenic. Water from more than 100,000 wells was tested. Nadim Khandaker, a former head of UNICEF's arsenic mitigation project in Bangladesh says that the second report produced by the BGS and the Bangladeshi Government is the baseline that everybody accepts, and it looked both at water from shallow wells and deep wells. Among the former, 27% of wells had water with more than 50 ppb. However, it appears that 80% of the Bangladeshi population has access to safe drinking water. In May 2003 the High Court in London said that the villagers' case could go to trial but the Court of Appeal made a ruling in February 2004 that overturned that decision saying that the relationship between the BGS and the villagers was not sufficiently close as to impose a 'duty of care' on the BGS to protect the claimants from exposure to arsenic. Indeed the Court ruled that to allow such a case would go far beyond any case ever decided in UK law.

In the 1990s environmentalists in the USA ran campaigns to have the limit on the amount of arsenic in drinking water reduced from 50 ppb to

10 ppb. In the dying days of his presidency Clinton agreed to this as part of the Safe Drinking Water Act which required the Environmental Protection Agency to implement the new standard by January 2006. This would affect about 12 million Americans living primarily in the West, plus a few in the Midwest and New England. This decision was overturned in March 2001 under President Bush, much to the disgust of environmentalists but to the obvious delight of the smaller water companies. The 50 ppb limit had been set in 1942 by the US Public Health Service in the belief, wrong as it turned out, that arsenic in drinking water was the cause of heart disease. Not that removing arsenic is all that difficult and it simply requires it to be passed over alumina (aluminium oxide) which absorbs it and forms an insoluble aluminium arsenic salt. It is the expense of installing such filtration systems that is costly.

Remediation steps have to be taken when drinking water is contaminated and yet it is a problem that is costly to solve. When half a million people in New Mexico were exposed to arsenic levels in their drinking water that were too high, action was taken, but at a cost of $100 million. Taking similar action in Bangladesh would be ruinously expensive. Even replacing a shallow well with one that draws water from the lower, safer water table, costs $1,000. (Shallow tube wells cost only $100 to drill which is why they have proved so popular.)

What puzzled some investigators, such as Khandaker, was the way in which arsenic affected only some people who were exposed to it. He has pointed out that in Kachua, where almost all wells are contaminated with arsenic, only one person in 100,000 shows visible symptoms of arsenicosis, while in Sonargaon, where only around 60% of the wells are contaminated the condition occurs in 70 per 100,000 of the population. He contends that there must be some other factor involved in causing the disease and says he can point to extended families where only one adult was affected and yet all had been drinking water from the same well for many years.

In 2002 a paper was published by a group of US, Canadian, and French researchers led by Bibudhendra Sarkar, and this listed many other metals that were contaminating the well waters of Bangladesh. This paper made the point that while undoubtedly arsenic was the most worrying contaminant – they even found wells in which the arsenic level was in excess of 6,000 ppb – other metals and especially antimony, should not be overlooked. The other worrying metals that were exceeding WHO guidelines were manganese, lead, nickel, and chromium. Arsenic, however, remains the contaminant that must be addressed. Clearly the authorities should strive to meet the WHO guideline of 10 ppb but at what cost? Could arsenic in drinking water be safe at a level of 50 ppb, which is the guideline used in many countries until recently? This level produces no debilitating symptoms in those who drink that water because they take in what might be regarded as a threshold minimum that the human body can easily tolerate. As for the effects of long-term exposure then the risks of cancer appear to be higher, but in a country where the chances of living to an age when such diseases might appear are relatively low, then addressing the more immediate injuries of hyperkeratosis would seem to be the number one target.

How do you measure the amount of arsenic in drinking water when it is at the parts per billion level? The answer is with a great deal of difficulty and by using sophisticated technical apparatus. What is really needed is a method of testing that will immediately indicate whether water is safe to drink and such a method has been developed by Jan Roelof van der Meer of the Swiss Federal Institute for Environmental Science and Technology. He has modified strains of *E. coli* bacteria so that they emit a green fluorescence when they come into contact with water that has more than 4 ppb of arsenic.

Malevolent arsenic

W<small>E</small> can never know who committed the first murder with arsenic or even who discovered the deadly nature of arsenical compounds. Although the natural arsenic minerals orpiment and realgar are poisonous, they are not particularly effective as murder weapons because they are insoluble and highly coloured, so that feeding them undetected to the intended victim would not be easy. The most reliable form in which to administer arsenic, knowing that it would succeed in killing someone, would have to be as the oxide. This is not a naturally occurring substance but it was easily obtained. When copper ores that had arsenic as an impurity were smelted, the arsenic was oxidized and emitted as white smoke, some of which sublimed on to the walls of flues and chimneys of the smelter and from where it could be gathered. When people talk of 'arsenic' they are almost invariably referring to its oxide, whose chemical formula is As_2O_3, with arsenic atoms bonded to oxygen atoms. Over the centuries this has had many names such as white arsenic, arsenious oxide, arsenious acid (because it dissolves in water to form an acidic solution), arsenic trioxide, and its proper chemical name, arsenic(III) oxide. I shall call it by the name which is still in common use even among chemists: arsenic trioxide.

Some murderers used solid arsenic trioxide, stirring it into foods like stews, porridge, or rice pudding to disguise it, but the more usual method was to dissolve it in something that the victim would drink. Not only is arsenic trioxide soluble, but the solution in which it is dissolved does not betray its presence because it is colourless and almost tasteless; if anything it imparts a slightly sweetish taste to the water. Yet even with such advantages favouring the would-be poisoner, it was still possible to fail to kill, either by not understanding its simple chemistry or misjudging the dose required. Sometimes the ignorance and incompetence of murderers worked in their favour because repeated small doses of the poison gave the impression that the victim was suffering from some deep-seated illness, so that when a final fatal dose was administered the end was not unexpected.

Stirring arsenic trioxide crystals into a glass of cold water will leave most of them sitting on the bottom, and while the amount that goes into solution will make someone very ill it might not be enough to kill them. It takes time for arsenic to dissolve but this can be speeded up by heating the mixture, and then the amount which goes into solution may be such that a mere mouthful (25 ml) will deliver 450 mg, which is double the fatal dose; indeed a lethal drink of arsenic trioxide could be dissolved in only 10 ml of water. The smallest dose of arsenic trioxide to which death had been attributed was 130 mg, but such a dose would not usually be fatal. On the other hand, some people have survived much larger doses.

In earlier centuries it was almost impossible to prove that someone had died of arsenic poisoning. Indeed arsenic trioxide was the perfect murder weapon in that its symptoms of vomiting and diarrhoea were very like those of many common ailments, so that even doctors were unlikely to diagnose arsenic poisoning as the cause of an illness or death, and surgeons who carried out autopsies were unlikely to find evidence that arsenic trioxide had been administered. All this changed once chemists began to study arsenic and its compounds and devise tests to detect its presence.

The first conviction based on forensic investigation did not occur until the mid-1700s and the first conviction based on unchallengeable evidence of this kind did not occur until James Marsh came up with his test for arsenic in the 1830s. Even then, it was possible for the guilty to walk free because juries were loathe to convict on forensic evidence alone, and if the defence lawyer played on their lack of scientific knowledge regarding chemical analysis it would create confusion in their minds, and such confusion would lead them to disregard it. Sometimes when a conviction was secured there were those who argued that the verdict was wrong. One such murder trial was the famous Maybrick case in which the young American wife Florence Maybrick was accused of poisoning her rich British husband James. Many have argued that she was innocent, but in the next chapter we will re-examine that murder and with our knowledge of the way arsenic behaves, we can work out exactly how she did it. In this chapter I will deal chiefly with those cases in which the interest focused on arsenic, where its detection, or lack thereof, was a key feature.

Arsenic murders down the ages

The ancient Assyrians of the 8th and 9th centuries BC were familiar with yellow orpiment, and the Greeks and Romans knew that it formed a white compound when roasted, which would be mainly arsenic trioxide. It was also known from an early date that heating orpiment with natron (natural sodium carbonate) produced a product that was deadly and that when it was dissolved in water it gave a clear solution. This reaction would form the soluble salt sodium arsenite, which would indeed have been very poisonous. Thus from the very earliest days there were those who knew the deadly nature of arsenic trioxide and its salts and how to make them. Such knowledge was both dangerous and politically useful, and there were some unexpected deaths that seem likely to have been caused by it. Laws of ancient Rome, dating from around 100 BC were specifically designed to cover cases of death by poisoning.

One of the more notorious poisoners of ancient Rome was Agrippina. She disposed of those who stood in her way and she almost certainly used arsenic trioxide because it was so effective and it enabled her to escape detection. Agrippina undoubtedly murdered her husband in order to be free to marry her uncle, the emperor Claudius, and thereby gain political power and promote her son Nero into becoming Claudius's successor. To bring that about Agrippina first eliminated her opponents among the palace advisors, and then poisoned Claudius's wife Valeria. Once she and Claudius were married she persuaded the emperor to allow his daughter Octavia to marry her son Nero. All that remained was to poison the emperor's son Britannicus, who would undoubtedly have succeeded him, and persuade the emperor to name his step-son Nero as his successor. When he did that he sealed his own fate. She poisoned Claudius in 54 AD and Nero became emperor at the tender age of 16. Sadly poor Agrippina soon fell out of favour with her son and he had her murdered in 59 AD, although not with poison. Or so the story goes.

The use of poison in the furtherance of political ends is supposed to have reached a fine art in Italy in the 1500s and 1600s. The most notorious practitioners were Cesare Borgia (1476–1507) and his sister Lucrezia (1480–1519) whose names are still synonymous with such depravity. (Lucrezia may even have borne a child as a result of an incestuous relationship with her father the Pope). The pair employed a white powder they referred to as

La Cantarella and which was almost certainly arsenic trioxide. It was said they got the recipe for making it from the Spanish Moors, and indeed their father was a Spanish cardinal called Rodrigo Borgia who became Pope Alexander VI in 1492. He died in 1503 after attending a banquet with his son Cesare and it was even rumoured that his death was caused by his eating poisoned food and wine that was destined for someone else. This seems unlikely because Cesare was also taken ill, although he recovered. It is impossible now to know whether *La Cantarella* was arsenic trioxide although it seems likely that it was. Lucrezia died in 1519 at the age of 39, apparently in a state of grace, having given up her scandalous life for one of religious devotion. Her brother died in a skirmish in 1507 aged 31.

The use of arsenic trioxide was not restricted to the upper echelons of the Roman Catholic Church; it also circulated among the gentry and the most notorious supplier of it was a woman known as Toffana of Sicily. She sold a solution of arsenic trioxide under the name of 'Manna of St. Nicholas' calling it after a holy water of the time; it was ostensibly bought as a cosmetic and indeed some women may even have used it as such to improve their complexion; others used it as an effective poison and referred to it as *Aqua Toffana*. It has been estimated that at least 500 people were sent to early graves through its administration.

Toffana began her career in Palermo about 1650 and moved to Naples in 1659, where she built up a clandestine organization for the distribution of the *Aqua* and she operated successfully through a network of agents for more than 50 years. The Governor of Naples finally put a stop to her trade in death when he discovered that *Aquetta di Napoli*, the name under which her product was being sold in his city, was so deadly that a mere six drops in a glass of wine would kill.* When Toffana heard that a warrant had been issued for her arrest she fled to a convent, from which she had to be forcibly removed and thrown into prison. Under torture she confessed all and was strangled to death in 1709.

Nor was it only in Italy that arsenic was misused. In France it acquired the name *poudre de succession* [inheritance powder] and there were clandestine organizations distributing it to those who wished to dispose of relatives. It is

---

* If this were true than it seems likely that *Aqua Toffana* was probably a solution of sodium or potassium arsenite because these can pack much more arsenic into the same volume of water than arsenic trioxide itself. Her recipe for making it has not survived.

now difficult to be certain to what extent arsenic was used but there were times when its deployment reached scandalous proportions.

The Marquise de Brinvilliers (1630–1676)

The beautiful Marie Madeleine D'Aubray was the daughter of the Mayor of Paris and Councillor of State. She became the Marquise de Brinvilliers when she married the aristocrat Antoine Gobelin de Brinvilliers, Baron of Nourar, in 1651. Antoine naturally had a series of mistresses and he appeared not to mind when his wife became the lover of a gambling friend of his, Gaudin de Sainte-Croix. However, her father was not so understanding and had the young man committed to the Bastille for 6 weeks in 1663. While in prison he met a notorious poisoner and when Sainte-Croix was released he too began to dabble in poisons, purchasing the necessary ingredients from the famous Paris chemist Christopher Glaser, who was Apothecary Royal and tutor to the sons of Louis XIV. Sainte-Croix experimented with ways of dissolving arsenic trioxide so that it could be administered easily and without causing suspicion. His mistress tested his concoctions on the patients in a local hospital to whom she brought charitable gifts of food and wine. Sadly these much appreciated additions to their meagre diets made them not better but worse, and some of them even died.

What the Marquise de Brinvilliers had researched was the amount of arsenic trioxide that would kill and the amount that would simply make someone ill. The first person to suffer at her hands was her father whom she first made ill in May 1666 and who she finally despatched in October of that year. Her aim was to inherit his wealth. She rather resented the fact that her actions had also benefited her two brothers, so she introduced a fellow conspirator into their home as a servant. He was known as La Chaussée and it was he who poisoned them the following year. Meanwhile the Marquise de Brinvilliers led a life of debauchery bestowing her favours on various men in addition to Sainte-Croix to whom she had written incriminating letters offering to pay him for his deadly concoctions, although he appears to have accepted payment in kind rather than in cash. That state of affairs she sought to put on a permanent basis by murdering her husband with a view to marrying Sainte-Croix, but it was not to be.

Sainte-Croix suddenly died in July 1672 in his laboratory, and his secret cache of incriminating letters was discovered. The Marquise de Brinvilliers made several attempts to regain them and when these were unsuccessful she fled first to England, then to Holland, and finally took refuge in a convent near Liege. In February 1673 La Chaussée was arrested, tortured, confessed, and was broken at the wheel. A youthful police agent from Paris was sent to Liege where he eventually conned the Marquise de Brinvilliers into leaving her refuge in order to become his lover and he arrested her on 25 March 1676. She was tried, tortured, confessed, and was beheaded.

Guy Simon, an apothecary, was called in to try and identify the poison that had been discovered in Brinvillier's house. He dissolved it in water and added a few drops of the liquid to oil of tartar and to sea water but nothing precipitated. He heated some of it but no acidic fumes were emitted and after intense heating it had all evaporated. Its toxic nature was proved when it was fed to a pigeon, a dog, and a cat. They all quickly died, and when he inspected their bodies he noted that there was clotted blood in their hearts. Simon also tested the white powder that deposited from a solution that was left to evaporate and he did this by giving it to another cat which promptly vomited for half an hour and then died. All his observations were consistent with the poison being a solution of arsenic trioxide. The one man who could have resolved the issue, Christopher Glaser the Apothecary Royal, had prudently departed from France and was never heard of again.

Arsenic also featured in the so-called 'Affair of the Poisons' which rocked high society at the Court of Louis XIV of France in 1679–1680 and which fascinated all of Europe with its tales of love potions, sorcery, witchcraft, and arsenic powders. At the centre of this web of intrigue was Catherine Deshayes Monvoisin, known as La Voisin, who was nominally a mid-wife and married to a silk merchant and jeweller by whom she had ten children. When her husband went bankrupt it fell to La Voisin to support the family and this she did by carrying out illegal abortions, performing satanic rites, and supplying love potions or deadly poisons as required. As a result she became wealthy and even entertained on a grand scale at her villa in the fashionable Villeneuve suburb of Paris in the 1670s. La Voisin was eventually arrested in March 1679 along with various accomplices, and she was burned alive at the stake on 22 February 1680. In her heyday La Voisin was even consulted by Madame de Montespan, a long-time mistress of Louis XIV, who turned to her for help when she wanted to regain the King's waning affections. Help in this case consisted of a satanic mass performed over the courtesan's naked body at the altar, and love potions to put in the King's

Old habits die hard

The tradition of using 'inheritance powders' to enrich oneself by the death of a near relative may never have been entirely lost in Italy, and indeed it surfaced again in the 1930s among the immigrants of an Italian neighbourhood of South Philadelphia. The press dubbed it The Great Arsenic Murder Ring, and indeed many people died suspiciously and much life insurance was collected on their deaths. Altogether 30 people were put on trial for murder, and 24 were found guilty. The distributors of the inheritance powders were two brothers Herman and Paul Petrillo along with Morris Bolber who was also known as Louie the Rabbi. They persuaded women to take out life insurance policies on their husbands and then to give them a white powder which would, in the words of the gang, 'send them to California'. It was arsenic trioxide. Some women clearly connived in the plots, handing over part of the money they collected to the Petrillos and Bolber, while others later claimed they were innocent, among whom was a Stella Alfonsi. She was still able to tell her story to George Cooper for his book *Poison Widows: a true story of witchcraft, arsenic and murder*, published in 1999. The three ringleaders of the affair were sentenced to death and two were electrocuted, the third being given a life sentence. Twenty-one of the willing poisoners were sentenced to life imprisonment, or given long jail sentences.

drinks.* What makes it somewhat unrewarding to try and unravel La Voisin's crimes and those of others in the 1600s is that we can never be certain of the poisons that were used, because chemistry was still in its infancy. That situation was to change in the 1700s as that branch of science slowly developed. Two women of that century who were eventually executed for using arsenic were found guilty with the help of the emerging science.

Mary Blandy (1720–1752)
Mary Blandy was the only child of Francis Blandy, attorney-at-law and town clerk of Henley-on-Thames in Oxfordshire. Although it was known that she

* Readers wishing to know more should consult Anne Somerset's fascinating book *The Affair of the Poisons*.

would bring a dowry of £10,000 to a future husband, she was still unwed at 26. Then she met lieutenant William Henry Cranstoun who charmed her and with whom she fell hopelessly in love. When the Blandy's were told that he already had a wife and child in Scotland he replied that the woman was lying and that she had no claim on him, and that this would be confirmed by a court case he was bringing against her. Indeed he even persuaded his wife to write a letter under her maiden name of Murray to the effect that they were *not* married. She did this thinking to help her husband who had told her that his being married was a bar to his promotion in the army, and that for the time being he wished to appear to be single. He showed this letter to Mary and her parents who took it at face value and even allowed Cranstoun to live with them. Unfortunately for Cranstoun, the Scottish courts in March 1748 upheld the woman's claim to be married to him although he managed to keep this from the Blandy's for two more years, during which time Mrs Blandy died.

Eventually Francis Blandy decided he wanted nothing more to do with Cranstoun, and he returned to Scotland, from where he sent Mary packets of arsenic trioxide, which he labelled 'powders to clean Scotch pebbles' and telling her to put some in her father's food because he said it would change her father's opposition to their relationship. She may well have first done this in November 1750, putting some in his tea, but it only made her father ill, although in June the following year, on receipt of more powder from Scotland, she tried again, making her father very ill but he again recovered. In August she added a larger dose of the powder to his gruel (oatmeal and milk) and again he was taken ill, this time fatally so; a servant who also ate some of the food was also made very ill. Francis now realized that his daughter had poisoned him and that he was dying; he expired on the 14th of that month but not before he tried to save his daughter by saying that he forgave her.

When the townspeople of Henley learned what Mary had done a mob appeared menacingly outside the Blandy residence and she had to flee to the safety of the Angel pub where the landlady gave her shelter. Mary was arrested on 17 August and imprisoned in Oxford Castle. She was put on trial on Tuesday 3 March 1752, and this lasted almost 12 hours. At the end of that day the jury did not feel the need to retire to reach a verdict but delivered it there and then: guilty.

Evidence that she had used poison was proved by having some of the white powder identified by Dr Anthony Addington. He showed that four tests performed on a sample of the powder gave exactly the same results as tests carried out on white arsenic. The forensic evidence helped to convince the jury that Mary had murdered her father and she was duly convicted and hanged, but not before pleading for extra time in order to write her version of events: *Miss Mary Blandy's Own Account of the Affair Between Her and Mr Cranstoun.* Clearly she was an odd young woman and maybe she was always a little naive. She seemed unnaturally cool and calm even as she approached the gibbet, asking the hangman not to hang her too high 'for the sake of decency' and indeed a large crowd had turned out to witness her execution. But hang her high he did on 6 April 1752.

As for Cranstoun, he fled to France, first to Boulogne, then to Paris, and finally to Furnes in Flanders where he stayed with a distant relative. He changed his name to Dunbar but he had not long to live and in November 1752 he went down with a fever and died on the 30th of that month, having converted to the Roman Catholic faith, a sign perhaps that he felt the need to confess his part in the murder of Mr Blandy and the execution of his daughter.

Anna Zwanziger (1760–1811)

Anna Schonleben was born near Bayreuth in Bavaria in 1760, and was orphaned when only 5 years old. She was then passed from one family member to another until she was sponsored by a wealthy guardian who paid for her to be educated. At 15 she was married off to a 27-year-old lawyer Herr Zwanziger, but it was not a happy marriage, because her husband was an alcoholic and could not get work, and Anna's giving birth to two children only made matters worse. Her answer to the family's need for income was somewhat unconventional but it was one way she could cash in on her undoubted good looks and ability to hold an educated conversation: she became a high class call girl, confining her client base to judges and other men in powerful positions.

Soon she was providing for her family and drunken husband with whom she was clearly deeply attached, in that having decided to divorce him she did so, but promptly remarried him the day after her divorce came through. Finally in 1796 he died aged 48, and Anna then tried to earn a more respectable living by opening her own shop but that business was unsuccessful, so she resorted to prostitution again but that lasted only a short time because she became pregnant. Although she gave up her baby for adoption, it soon died. By now Anna was losing her looks and her preferred profession was no longer an option so she offered her services as a housekeeper and was employed by a succession of families. It was a position for which she showed a certain flair although she often resented being told what to do by the mistress of the house, a situation she sometimes resolved by poisoning the wife and then making a play for the husband. While they may have allowed her to console them in bed, none of them found her particularly suitable as a future wife.

Then Anna had a stroke of luck. Justice Wolfgang Glaser asked her to be his housekeeper. He had recently separated from his wife, and on 5 March 1808 Anna took up residence with him. Her good fortune was short lived because Frau Glaser reappeared on the scene on 22 July 1808. Her return was short lived because the following day she was suddenly taken ill with vomiting, diarrhoea, and stomach pains. She had recurrent bouts of this sickness and died a month later. For some reason Anna left Glaser's service and went to keep house for 38-year-old Justice Grohmann who was a chronic invalid. Under Anna's care he began to recover until he was suddenly stricken with vomiting, diarrhoea, and stomach pains in April 1809 and died on 8 May. Next she moved to the home of Justice Gebhard and his wife who was in the final stages of pregnancy and who was delivered of a baby boy on 13 May. A few days later this frail woman lay dying and she accused Anna of poisoning her but no one took any action and after the funeral Anna continued to keep house for the judge.

On 25 August 1809 Gebhard invited two friends to a meal and soon after they were taken ill, as was a messenger who arrived at the house and was given a glass of wine, clearly indicating that this was how Anna had poisoned them. A porter who also called at the house was offered some but noted it had a white sediment at the bottom and only took a little. Even so it was enough to make him very ill. No one died and possibly they accepted what had happened was due to food poisoning, but what occurred the following weekend could not be dismissed so lightly. On that evening, 1 September 1809, the judge entertained five friends for an evening of beer and skittles. All became ill and blamed Anna, and at their urging the judge dismissed her, but not before she mixed a large dose of arsenic trioxide with the salt. As a parting gesture she made coffee for the two housemaids, and gave the 5-month-old baby some milk. Then she bid them goodbye and left. All became very ill and finally the police were called. The salt box was taken away for analysis and found to contain a lot of white arsenic. When Anna was arrested she was found to have two more packets of the poison in her possession. Frau Glaser's body was exhumed and arsenic was found in it. Anna confessed her guilt and was beheaded by sword in 1811.

Analysing for arsenic

In the 1700s chemistry emerged as one of the sciences, and chemical analysis was an essential part of it with the result that it became possible to *prove* that arsenic had been used to murder someone. Yet deaths deliberately caused by arsenic went undiscovered, nor was this surprising. Doctors naturally diagnosed the symptoms of arsenic poisoning as being due to natural illnesses, and even when they were puzzled by a sudden death and carried out an autopsy, there was nothing to indicate that arsenic was the cause. No doubt many murders went unrecognized and their perpetrators went unpunished. However, when suspicions were aroused the outcome might well be a dramatic trial that would be gleefully followed by the growing number of newspaper readers.

The biggest step forward in the forensic analysis of arsenic came as a result of research by James Marsh (1794–1846) who worked at the Woolwich Arsenal in London. He was particularly annoyed by the acquittal in 1832 of John Bodle, charged with the murder of his 80-year-old grandfather George Bodle who owned a farm at Plumstead. A maid who worked at the farm testified that John had said he wanted his grandfather to die because he stood to gain a handsome inheritance when it happened – the old man's estate was worth around £20,000 (equivalent to £2 million today). A local pharmacist confirmed in court that he had sold John some arsenic trioxide.

Marsh had been asked to prove the presence of arsenic in some suspect coffee that John had given the old man and in organs removed from his body. Marsh used the standard test of the time, which was to show the formation of a yellow precipitate of arsenic sulfide when hydrogen sulfide gas was bubbled through a solution containing arsenic. However, the samples of precipitated arsenic sulfide which Marsh presented to the court had become discoloured by the time the trial was held and the jury were not convinced by them. As a result Bodle walked free. Ten years later he was found guilty of fraud and sentenced to 7 years in jail after which he was ordered to be deported to the colonies on account of a blackmail offence. He then confessed to murdering his grandfather, in the sure knowledge that he could not be tried twice for the same offence.

Meanwhile, Marsh had been working to devise a test for arsenic that would provide unquestionable proof of its presence and the results of which could be exhibited in court and would impress jurors. In 1836 he published his results in the *Edinburgh Philosophical Journal*. The first requisite of the test was that all the arsenic should be brought into solution and this could be achieved by heating samples of suspect tissue with strong acids that destroyed all the organic matter and at the same time dissolved the arsenic they contained. The next step in the Marsh test, as it became known, was to convert the arsenic in solution into the gas arsine (AsH_3), and this could be achieved by adding pieces of metallic zinc to the acidified solution.

Any arsine gas that was evolved was passed through a heated glass tube which caused it to decompose into hydrogen gas and metallic arsenic which condensed on the cooler part of the tube as a mirror-like film. The amount of arsenic was then assessed by comparing the mirror with standard mirrors of known weight. (That the mirror was arsenic could be shown by heating it in air when it was converted to arsenic trioxide which sublimed as a white deposit further along the tube.) Some of the tubes with arsenic mirrors could be sealed off and shown to the jury. The extreme sensitivity of the test brought its own problems because it would also detect traces of arsenic in the test reagents, especially in the zinc. Even when great care was taken to ensure that there was no such contamination it still revealed the minute traces of arsenic that are always to be found in the human body, and this was a problem when bones were all that remained for testing.

Sensitive as it was, the Marsh test had one disadvantage: it could only be carried out in a laboratory. It was not suitable for on-the-spot analysis. In 1842 Egar Hugo Reinsch (1809–1884) found that a better test for arsenic was to dip a strip of brightly polished copper foil into the suspect liquid. (If the sample were a solid then it was first boiled with hydrochloric acid in a test tube to dissolve the arsenic.) Any arsenic present deposited on the copper foil, and if this was then heated the arsenic would be vaporized again. This test was capable of detecting as little as a tenth of a microgram of arsenic (0.0001 mg). The amount of arsenic on the copper strip could be measured later by converting it to arsenic trioxide and weighing that. In theory it was ten times more sensitive that the Marsh test but it required a certain skill to perform and there were pitfalls as we shall see. The Reinsch test, as it was soon called, could show a positive result within minutes, although if there was only a little arsenic present in the solution then it could take an hour or more for the deposit to form, a fact that had important implication in the Maybrick trial – see next chapter.

In the 1900s even better tests for arsenic were developed but the chemical methods devised by Marsh and Reinsch were more than adequate to expose the murderers of the 1800s, although not always to get them convicted and punished.

The golden age of arsenic murders

Despite the advances in forensic analysis, there were serial poisoners who managed to evade detection for many years and many murders.

Hélène Jegado (1803–1854)

It is still not clear how many people this pious young woman disposed of before she was arrested in July 1851. Hélène Jegado was orphaned at the age of 7 and was then looked after by the curé of Bubry, M. Raillau, who employed two of her aunts as his servants. When one of them went to work for another curé at Seglien in 1826, Hélène went with her and it was there that she was first accused of tampering with the food, although no one died as a result. It was only when Hélène went to work for a priest in Guern, M. Le Drogo, that people began to die suddenly. Within three months of her taking up her new post, there was an outbreak of sickness so severe that it carried off six of those who lived in the house, as it did Hélène's sister who came to visit her. Hélène was a model of care and attention while all this was going on, and seemed genuinely distressed when people died. Not only that, but she was clearly an intelligent and religious young woman.

So began the trail of deaths that followed the faithful servant as she moved from job to job. Wherever she went, people who ate the food that she prepared became ill and died, and for a time she even gained sympathy from the fact that she appeared to be forever dogged with such bad luck. 'My masters die wherever I go' she sobbed at one funeral. And so it went on from 1833 to

1841 as she moved around Brittany from town to town poisoning young and old and stealing their belongings. During these 8 years she disposed of her employers, members of their families, and their servants – probably 23 in total. (The Statute of Limitations of French law meant that these crimes were excluded from the indictment when she was finally brought to trial in 1854.)

For some reason Hélène appears to have stopped her murdering ways in 1841, only to begin again in 1849 when she obtained a position as the sole domestic servant to a married couple called Rabot. Their son Albert was already ill when she took the job in November that year, and he died the following month after eating porridge cooked by Hélène. It was then that she was discovered to have been helping herself to the family's wine store and she was sacked but not before she was able to serve them some arsenic-laced soup which they, and a guest, survived eating although it made them very ill. Hélène moved on and began again to dispense arsenic, leaving a trail of at least five deaths and many illnesses.

The best documented evidence at Hélène's subsequent trial was her murder of a fellow servant when she went to work for M. Theophile Bidard, Professor of Law at the University of Rennes. He took her on after she told him that she needed the job to support her two small children and her poor old mother. The other servant working for the professor was Rose Tessier and it was not long before Rose was taken ill, on Sunday 3 November 1850, after a dinner that the three of them ate together. Hélène offered to nurse Rose and sat up with her all night although she was again convulsed with violent vomiting after drinking a cup of tea prepared by Hélène. A Dr Pinault was called but diagnosed nothing more than a nervous affliction. Over the next few days poor Rose waxed and waned and died on the following Thursday. As usual, Hélène seemed overcome with grief at the funeral.

The Professor then hired a Francois Huriaux, but she was taken ill and soon had to leave, to be replaced in her turn by Rosalie Sarrazin. To begin with she and Hélène got on very well but this phase did not last and, in May 1851, she too began to be seized with vomiting and diarrhoea. By this time there was such bad blood between the two women that in the end the professor decided it was Hélène who must go. Her response to this was to intensify her poisoning of Rosalie while at the same time appearing to be reconciled with her and the professor, faced with having no domestic help at all, agreed to let Hélène stay on. She even went with Rosalie to see Dr Pinault who diagnosed an upset stomach and prescribed Epsom salts.*

The harmony was not to last and on 22 June Rosalie was again taken ill after drinking a glass of Epsom salts prepared by Hélène. Rosalie's mother then took a hand in looking after her daughter and while drinks prepared by her seemed to be fine, those prepared by Hélène made her very ill. Professor Bidard now wrote to Dr Pinault asking for a second opinion. The second doctor confirmed that the treatment she was being given was the correct one. Meanwhile the professor questioned Hélène who satisfied him that she was blameless and for a few days Rosalie appeared to recover but when he went to the country on 27 June he was quickly called back to Rennes because Rosalie was worse than ever.

When the professor saw how sick she was, and how she could not stop vomiting, he was so alarmed that he ran to get Dr Pinault, only to meet him in the street conferring with another doctor about Rosalie's illness. The three men went back to the professor's home and managed to stop the vomiting but it was too late. They had concluded that Rosalie was being poisoned and ordered Rosalie's vomit to be saved for analysis. The vessel containing it was locked away by the professor, despite Hélène's protestations, and when Rosalie died on 1 July it was handed over to the authorities. The professor of chemistry at Renne University, M. Malagutti, carried out the necessary forensic analysis of the vomit and various organs of the dead girl and showed they contained large quantities of arsenic.

Hélène was arrested and put on trial, charged not only with Rosalie's murder but those of many others, although it was the details of her latest murder which provided the most convincing evidence in court. The jury took only an hour and a half to find her guilty on all counts and she was sentenced to death. Hélène pleaded her innocence even when she was on the scaffold and in a last desperate attempt to stay her execution she named another woman who she said was the real murderer. It was a fruitless gesture but, after the execution, the Procureur of Rennes felt obliged to interview the woman whom he found to be old and entirely blameless.

It is difficult to be certain how many people Hélène Jegado murdered but it was probably around 30, making her one of the most prolific of arsenic

---

* Magnesium sulfate.

poisoners. Certainly it put her well ahead of Mary Ann Cotton, who was Britain's most consistent arsenic poisoner.

Mary Ann Cotton (1832–1873)

Again it is not possible to know exactly how many people Mary Ann Cotton murdered, but they almost certainly included her mother, three husbands, a lover, eight of her own children, and seven stepchildren, making a grand total of 20.

Mary Ann was born Mary Ann Robson at Lower Moorsley, County Durham, the daughter of a young coal miner and his wife. Her childhood was uneventful but her father died in a pit accident when she was young and her mother soon remarried. It appears that her stepfather treated Mary Ann kindly. Not only did she turn into a pretty young teenager, but she was a bright girl and even taught at the local Methodist Sunday school. When she was 15 she obtained a position as nursemaid to an ageing colliery manager. At 19 she married a miner called William Mowbray on 18 July 1852 and went with him to Cornwall where she bore him four children, three of whom died rather suddenly of what the doctors called gastric fever.

The Mowbrays moved back to Murton in County Durham in 1856 with their remaining child, a daughter, but she too went down with gastric fever and died in 1860 aged 4. Nevertheless Mary Ann continued to procreate giving birth to four more children, two of whom died within a year. In 1863 the family moved to Hendon, near Sunderland, because William Mowbray no longer wished to be a miner and had taken a job on a steamship, at the same time taking out insurance for himself and his remaining children, one of whom died in September 1864 and whose death brought them a small sum from the Prudential Insurance Company. Things became difficult for the Mowbrays when William injured his foot and was laid off work. While he was convalescing at home he fell ill of gastric fever in January 1865 and soon joined his children in the next world. Mary Ann collected the £35 insurance, equivalent to six months salary of her dear departed.

It was not long before Mary Ann had a new love, one Joseph Nattrass, but he seemed unwilling to marry her and saddle himself with two stepchildren. Though one of these died unexpectedly – of gastric fever – he still seemed reluctant to make Mary Ann his wife so she fostered her sole surviving child with her mother. Even so, Nattrass's ardour had cooled and with no prospect of a quick marriage to solve her financial problems, Mary Ann had to seek employment. She took a job at the local hospital and it was there that she met 32-year-old George Ward and they were married on 28 August 1865, but the marriage was not to last. When Ward became unemployed he quickly went down with a fatal case of vomiting and diarrhoea and he died in October 1866, surprisingly without having made the fecund Mary Ann pregnant.

Mary Ann now took a position as housekeeper to a widower called Robinson whose wife had recently died of tuberculosis leaving him with five children to look after. That burden was lightened a little when one of them died the week after Mary Ann moved in. Soon Mary Ann was pregnant again and then in March 1867 she was called away to look after her sick mother, who sadly died nine days later, leaving Mary Ann to reclaim the daughter that her mother had been looking after, and take her back to live with the Robinsons. Mary Ann married James Robinson in August 1867 and their child, a girl, was born at the end of November, but died three months later – of gastric fever.

The following year another child was born to the couple but Mary Ann and her children were thrown out of the house in the autumn of 1869 when Robinson discovered that she had been milking his bank account and selling his things. That action no doubt saved his own life. Mary Ann now reverted to her old name of Mrs Mowbray, forged a set of testimonial references and got another job as a housekeeper to a doctor in Spennymore, who also noticed things disappearing and dismissed her. Next she moved to Wallbottle where she formed an intimate liaison with a miner, a widower Frederick Cotton, whose wife had recently died. They were married (bigamously as far as Mary Ann was concerned) in September 1870 and Mary Ann gave birth to a boy in January 1871. Mary Ann was already six months pregnant when she married Frederick at Newcastle and when the baby, a boy, was born they went to live in West Aukland taking Frederick's two remaining children by his first wife with them. By a strange coincidence they lived in the same street as her former lover Nattrass. In September 1871, Cotton unexpectedly died and by Christmas Nattrass had moved in with Mary Ann.

However, she had her eye on a man of higher social standing, an excise officer, with whom she had taken a job as a nurse and by whom she was soon pregnant. To clear the way for marriage she disposed of Nattrass and two of the children. The only offspring she now had was 7-year-old Charles Edward Cotton, and something had to be done to remove him from the scene. First she tried fostering him on an uncle, but he was having none of it, then she turned to the local workhouse but they refused to accommodate him, whereupon she warned the superintendent that if they did not take him in the boy would soon be dead. On 12 July she poisoned Charles, and when the superintendent heard of the boy's sudden death he went to the police. They contacted the doctor who had attended the child in his final illness and he now refused to issue a death certificate and carried out a post-mortem. Because he could find no evidence that the child had been poisoned, the coroner's inquest returned a verdict of natural causes.

However, the doctor took samples of tissue from the boy's body, did a Reinsch test on them and discovered arsenic. On 18 July Mary Ann was charged with murder. The bodies of her other recent victims were exhumed and also shown to have high levels of arsenic. Mary Ann was sent for trial but when she appeared in court in December 1872 it was clear that she was in the last months of pregnancy so the trial had to be postponed. She gave birth to a baby girl in January 1873. Her trial started in March of that year and lasted three days. She was charged with the murder of Charles Edward Cotton and, to support the charge, details of other deaths attributable to Mary Ann were given in evidence, with the result that she was found guilty. She was hanged in Durham gaol on 24 March. (Her defence lawyers argued that Charles Edward had died due to the arsenic vapours from the green wallpaper in his bedroom.) Mary Ann's chosen murder weapon was thought to be a proprietary de-worming compound.

While Jegado and Cotton were serial killers, there were many others who used arsenic to further their ends by murdering only one person and their names are famous in the annals of crime, although their *modus operandi* and detection adds little to the arsenic story. But there are some cases that were unusual such as those of Dr Smethurst, Madeleine Smith, and the Reverend Herbert Hayden, which shows that it was possible for the innocent to be convicted and the guilty to be acquitted.

Dr Thomas Smethurst (1803–date of death not known)

Dr Thomas Smethurst was 24 years old when, in 1828, he had married a woman who was 20 years older than himself. In 1858 they were living in a boarding house in London's Bayswater district when the now 54-year-old doctor met 43-year-old Miss Isabella Banks, and it was not long before she too moved into the same boarding house. It soon came to the landlady's notice that the newcomer and the doctor were having an illicit love affair and she asked Isabella to leave, which she did – and Smethurst left with her. They went to live in Richmond, where they passed as man and wife, the two having gone through a bigamous marriage ceremony at Battersea Parish Church.

In March 1859, Isabella fell ill and suffered 3 weeks of vomiting and other symptoms such as bloody diarrhoea, and she finally passed away on 3 May. The two doctors whom Smethurst had called to attend her believed she had been given an irritant poison and went to a local magistrate to report their suspicions; in fact Isabella was about 6 weeks pregnant which might have accounted for some of her symptoms. However, she had made a will leaving everything to Thomas, and while her estate amounted to only £1,700 it was not insignificant at a time when the average annual wage was £100. Not that Smethurst needed the money; he had been a successful entrepreneur in promoting the water cure and he ran a hydropathic clinic at Moor Park in Surrey. He also wrote on the benefits of water in his book *Hydrotherapie*, published in 1843, was editor of the *Water Cure Journal*, and was a great advocate of the benefits of sea-bathing. What brought him into the public eye was not these treatments but the treatment of his mistress: he was accused of poisoning her and was arrested.

Among the many bottles of medicine that were found in Smethurst's possession was one, labelled bottle number 21, which contained a mouth-wash and which tested positive for arsenic. Examination of Isabel's faecal stools also revealed arsenic. Or at least that is what the eminent forensic expert Dr Swaine Taylor said when Smethurst first appeared before the magistrate, and he repeated this allegation at the coroner's inquest, where the jury naturally found that Isabella had died of arsenic poisoning wilfully administered. Smethurst was consequently sent for trial at the Central Criminal Court in London. There Dr Taylor had a confession to make: the arsenic he had detected by means of the Reinsch test came from the impure copper strip he used in the test. That arsenic was liberated in the tests on bottle 21 was explained by its contents which was a solution of potassium chlorate, a chemical that had the ability to release the arsenic that was present as an impurity in the copper. Moreover, Dr Taylor had to admit that when he tested Isabella's organs he could find no arsenic in them. (In fact it may have been antimony that had killed her, and examination of her intestines and one of the kidneys yielded between 15 and 30 mg of antimony.) Neverthe-less there was enough circumstantial evidence produced by the prosecution to convince the jury that Smethurst was guilty and he was sentenced to be hanged.

Almost immediately there was doubt expressed about the verdict and the Home Secretary, Sir George Cornewall, commissioned a leading surgeon, Sir Benjamin Collins Brodie to review the data from the trial and give his opinion. He concluded that there was no convincing evidence that Smethurst was guilty and the Home Secretary overruled the jury and granted the condemned man a pardon. He was released but immediately re-arrested and charged with bigamy, of which he was convicted at a second trial, and sentenced to a year's hard labour. When he emerged from jail, Smethurst then sued Isabella's relatives who claimed her will was void because she was of unsound mind. He won his case. Meanwhile Dr Taylor had had to face the derision of the press, including calls for his resignation, but he weathered the storm and continued to be the prominent forensic scientist for another 18 years.

The symptoms displayed by Isabella before death were not due to arsenic but probably due to irritable bowel disease (IBD), also known as Crohn's disease.* Details of the post-mortem carried out on Isabella by Sir Samuel Wilks were described in a letter to the *Medical Times and Gazette* in 1859 and in it he described the severe ulceration of the colon that is now associated with Crohn's disease and which causes stomach pains and severe diarrhoea. Wilks had interpreted the acute rawness and ulcers in her intestines as due to arsenic poisoning.

Madeleine Smith (1838–1912)

Madeleine Smith was the 19-year-old daughter of a wealthy Glasgow architect. She had fallen head over heels in love with 26-year-old Emile L'Angelier, a shipping clerk from Jersey, whom she had met by chance one day while out walking. Although they had had a torrid sexual relationship she knew he was totally unacceptable as a husband, and a new man had come

---

* Although the symptoms of irritable bowel disease had been reported by doctors for hundreds of years, it was a paper published in 1932 in the *Journal of the American Medical Association*, with Crohn as the lead author that led to its being called Crohn's disease.

into her life so she poisoned Emile with a cup of cocoa liberally laced with arsenic. He died an agonizing death shortly after an evening tryst with Madeleine on 22 March 1857. She was brought to trial later that year but saved from the gallows by the curious Scottish verdict of *not proven*. Her lawyers had used the Styrian Defence saying that L'Angelier could have been a secret arsenic eater, and that Madeleine had bought arsenic to use as a cosmetic.

The Victorian public savoured the trial with its prurient details of their relationship although the press portrayed L'Angelier as the seducer and fortune seeker and Madeleine as the victim whose only release was to get rid of him. After the trial Madeleine moved to the south of England eventually ending up in London where she met and married George Wardel, who was one of William Morris's designers, and later his business manager. (At one time Wardel had owned a silk works in Leek, Staffordshire.) Soon Madeleine was working for the company as an embroiderer of tapestries. She became a member of the group of intellectual socialists that included many leading artists and writers of the day, including George Bernard Shaw who knew her well. Eventually her marriage broke up and she ended her days in the USA, dying in obscurity at the age of 74 in 1912.

The Reverend Herbert H. Hayden (1850–1907)

In September 1878, the minister of the Methodist Church in Rockland, Connecticut, enticed a young servant girl, Mary Stannard, to a lonely spot where he gave her a drink consisting of a ounce (28 g) of arsenic in a cup of water telling her it would cause an abortion – he believed he had made her pregnant. When she started screaming in agony he knocked her unconscious with a blow to the head and then slit her throat. In October the following year his trial for murder lasted 15 weeks and God-fearing Americans were enthralled by it. Particularly noteworthy were the extent of the forensic evidence and the disreputable attempt of the defence lawyers to find alternative explanations for the amount of arsenic found in Mary's body.

Mary Stannard was 21 years old and came from a poor family. She worked as a servant and those who knew her found her cheerful and hard-working, but when she was 19 she had an illegitimate child, Willie, by a man whose name she never revealed. In 1878 she was employed by the 29-year-old Reverend Hayden and his wife, Rosa. They had two children and were expecting a third. Hayden was attracted to Mary, nor did she rebuff his advances, with the consequence that on 20 March 1878 they slipped away from the parish oyster supper and had a furtive sexual encounter. Mary thought that it resulted in her becoming pregnant and she confided in her sister that the man responsible for her condition was the minister. Mary said she would confront him and ask him to arrange for her to have an abortion.

On Monday 2 September she managed to see Hayden alone in his barn and he promised to get her some 'medicine' that would bring this about. The following day he left for Middletown, although he told his wife he was going to Durham to buy some oats. He gave himself a reason for visiting Middletown by first calling on a man who was making him some tools. Then he went to Tyler's Drugstore where he purchased an ounce of arsenic for 10 cents, saying he wanted to get rid of rats that were a problem around his house. The poison was wrapped up and clearly labelled and its sale, but not its purchaser, was entered in the sale book.

Somewhat unluckily, as Hayden left the drugstore he was accosted by a man who knew him, and they spoke of the imminent birth of Hayden's third child. On his way back to Rockland, Hayden called at the Stannard house ostensibly to ask for a drink of water, but really to speak to Mary and tell her that he had obtained the 'medicine' and would meet her that afternoon at the Big Rock which was well away from prying eyes. When he got home he transferred the arsenic to an old spice tin and disposed of the packaging. Soon after his midday meal he left home, telling his wife that he was going to stack wood at a woodlot he owned.

He arrived at the Big Rock at 2.30 where Mary was waiting, and there he dissolved the whole ounce (28,000 mg) in water and persuaded her to drink it all down. By 3pm she knew something was wrong and by 3.15 she was screaming in agony and starting to run home. Hayden grabbed a chunk of wood and with two heavy blows he knocked her unconscious. Then, thinking to make her death appear to be suicide, he took out his jacknife and slit her throat. He arranged her body in a peaceful pose, her hat under her head, so that it appeared as if she had deliberately lain down to die. Of course he dare not leave his knife at the scene of the crime and his plan now was to go home and return later with another knife. He then ran to the woodlot, stacked a few logs to show that he had been there that afternoon, and went home, where he washed his jacknife and placed it out of sight on a high shelf. Unfortunately for the minister, his plan began to unfold when Mary's body was discovered by a member of her family who went to look for her, worried that she had not returned home after a heavy downpour late that afternoon.

On Wednesday 4 September the Reverend Hayden rose early and went to the woodlot where he spent a couple of hours stacking wood and loading his wagon in an attempt to make it appear that he had spent most of the previous afternoon gainfully employed. Then he returned home and finding neighbours visiting his wife he made a point of asking for his jacknife and 'finding' it on a high shelf in the kitchen. Meanwhile at the Stannard house an autopsy had been performed on Mary's body which revealed that she had not been pregnant but was suffering from an ovarian cyst. The doctor who performed it did not search for evidence that she had died of poisoning because that was not suspected. At the inquest Hayden was accused of her murder by Mary's sister and he was arrested the following day.

Herbert Hayden appeared before the Justice Court on Tuesday 10 September, which was held in the basement of the Congregational Church in nearby Madison. After a 2-week trial it decided that he was innocent of the charge of murdering Mary Stannard and he left court to the cheers of a large group of supporters. All this was somewhat premature because so far nothing had been said about Mary having taken a large dose of arsenic, but Professor Samuel Johnson of Yale Medical College was about to remedy this defect in the evidence. When he did, and when it was realized that Hayden had been to Middletown to purchase arsenic the day Mary died, then it became imperative to issue a new warrant for the minister's arrest, which was done on 8 October. He was brought to trial a year later, by which time Mary's body had been twice disinterred for further samples to be taken for arsenic tests to be carried out.

The Hayden trial aroused interest across the USA and became known as 'The Great Case'. The forensic evidence against him was damning and indeed was of a much higher standard that normal. Not only Professor Johnson of Yale but his colleague Professor Edward Dana took the stand on behalf of the prosecution. Dana had even travelled to England to the Devon Great Consols mine to obtain samples of arsenic, this being the sole source of arsenic that was sold throughout the USA. He did this because he had discovered that it was possible to identify a particular batch of arsenic trioxide by examining its crystals under a microscope. At the mine he learned that the material could be so identified despite the crystals being ground up before being bagged and shipped. The reason was that the tiniest crystals were unaffected by the grinding and their shape did indeed vary from batch to batch.

Dana's evidence indicated that the arsenic in Mary's stomach was the same as that from the drugstore in Middletown and different from the ounce of arsenic in the spice tin that was eventually handed over to the court. (It was never discovered who replaced the missing arsenic although it would have been easy for Hayden to buy more arsenic anywhere in the state during the 2 weeks before he was re-arrested.) Dana was correctly able to identify different samples of arsenic crystals when he was presented with them in random order, and without being told where they were from. He and Johnson had of course performed the Marsh test on all samples that they had obtained. Two other professors also provided more forensic evidence of arsenic in the organs taken from Mary's body.

However, it was the sheer amount and novelty of the forensic evidence which allowed the defence counsel to so confuse the jury by endlessly challenging it that they appear to have ignored it when reaching their verdict. The defence even maintained that the large amount of arsenic found in Mary's stomach had been planted there deliberately to incriminate the Reverend Hayden. The quantities discovered in the body were exceedingly large: 23 grains (1,500 mg) were extracted from her liver, which is where most of that which had been absorbed through the stomach wall had ended up, although some had penetrated to her lungs and brain. When it was challenged by the defence that this was not possible in the short time that she had lived, further evidence proved that it was possible. (The learned professors took small doses of arsenic and showed that it could be detected in their urine within 15 minutes.) When it was suggested that the arsenic in her brain could have diffused there from the stomach as her body lay in the grave, tests on other cadavers were carried out to show this did not happen.

What perhaps persuaded the jury to disregard the scientific evidence was the defence's deliberate questioning of every statement made by the professors, who were always reluctant to swear that there was *absolutely* no doubt about their findings. More telling was the fact that Mary was not pregnant – thereby removing the motive the minister had for murdering her. Most telling of all was the tearful and emotional performance that his wife Rosa gave in the witness box which reduced many in the court to tears as she showed complete faith in her husband's innocence even to the extent of lying for him about his movements that day.

Finally, on 14 January 1880, the jury retired to consider its verdict. We know from the voting slips which they discarded that originally they voted 10-to-2 for not guilty, then the following day it was 11-to-1, but even after another nine ballots they never achieved the necessary 12-to-0 for an acquittal. A plucky farmer, David Hotchkiss, was not to be browbeaten by the other members of the jury, and of course he was right. It was a hung jury, and that would mean another trial, but knowing how the jury had voted meant that the state was not prepared to finance another long drawn out prosecution that would be likely to fail. The Reverend Herbert Hayden walked free. The family moved to New Haven where he worked as a carpenter and shop assistant. He died of liver cancer in 1907 when he was 57.

Later arsenic murders

In the years following World War I there were some notable arsenic poisoning cases, such as that of Major Herbert Armstrong who was hanged in 1922 for murdering his wife. He would have escaped detection had he not sent poisoned chocolates to a fellow solicitor who was threatening him financially. When these failed to do the trick, he invited the man to tea and fed him scones spread with butter and arsenic trioxide. They made him very ill but he survived and it was these clumsy attempts that led to Armstrong's arrest and the exhumation of his wife's corpse. She had died a year earlier and her body was found to be full of arsenic. Her liver alone contained 138 mg of the poison. The Armstrong case was a carbon copy of the Greenwood case of 1920. Both men were solicitors, both lived in the same part of the country, and both poisoned their wives with arsenic trioxide. The only difference was that Greenwood had been acquitted.

The last large-scale poisoning by arsenic occurred in Britain in 1943 at a student hostel in St Andrews. Many students were taken ill, and two died, after eating sausages made from meat containing arsenic trioxide. Some sausages were found to contain as much as 650 mg of the poison. The case was never solved.

Modern methods of analysing arsenic are capable not only of detecting very small amounts of the element but of measuring the quantity of arsenic very accurately. Techniques such as colorimetric analysis, atomic absorption, and atomic emission are able to measure amounts of arsenic the older tests would not even detect. Neutron activation has been a particularly sensitive method, capable of detecting arsenic in a single strand of hair. In neutron activation the suspect sample is bombarded with neutrons from an atomic reactor and this produces a radioactive form of arsenic called arsenic-76 which can then be accurately assayed by the amount of radiation it emits. The position of arsenic in a strand of hair can show when doses of arsenic were administered, and the length of time between such doses. Hair grows at a fairly constant rate so that, for example, a high arsenic level at a point 5 mm from the hair root shows that the person was given the poison 14 days previously. However, in one of the first major trials in which it was used, neutron activation analysis proved to be of little value.

Marie Besnard (1896–date of death not known)

The strongest reason for a would-be poisoner to avoid arsenic was its easy detection in the victim's body after death. If a corpse had to be exhumed it was still possible to argue that any arsenic found in the remains might be due to contamination from the surrounding soil, if that had been in contact with the body. This line of argument saved the French mass poisoner Marie Besnard, despite the proof of poisoning that came from neutron activation analysis and which was submitted at her second trial.

Marie Besnard's husband Léon died on 25 October 1947, but not before confiding in a visitor that he suspected his wife had poisoned him. Why might she have done this? The locals of the small town of Loudun in the French Department of Vienne suspected she had done it so that she could continue her love affair with her 20-year-old toyboy, a German prisoner-of-war called Dietz who had been working on the Besnard's farm that year and who had become her lover. Indeed she went on several holidays with him in the years that followed. Such were the rumours flying around the town that a court order was obtained for the exhumation of Léon's corpse in 1949 and samples of his remains were sent to Marseille for forensic analysis. There it was discovered that they contained large amounts of arsenic; 39 ppm on average.

Rumour also had it that Marie's mother had likewise died suddenly in January 1949 and her remains were also exhumed and sent for analysis and they averaged 58 ppm of arsenic in the tissues taken from them. Next the remains of Marie's first husband, Auguste, who had died suddenly in 1929, were exhumed and again the arsenic level was high (60 ppm), and then there was a great-aunt to be dug up, and her father, and father-in-law, and even some neighbours. All showed levels of arsenic that strongly suggested they had been poisoned. From some of these victims, and there were 12 in all, Marie had inherited substantial legacies making her a very rich woman.

Marie had been arrested on 21 July 1949, but it was not until February 1952 that her trial began. Wisely, she invested some of her ill-gotten wealth in a brilliant Paris lawyer, Albert Gautrat, and this paid a remarkable dividend. At the first trial Gautrat skilfully exploited a mix-up in the labelling of a few of the exhumed remains to suggest that none of the forensic evidence was reliable, even though the analysis had been carried out by one of France's most respected forensic scientists, Dr Georges Béroud. The trial had to be abandoned. Marie was not released because she had already been found guilty of illegally collecting annuity payments of a deceased relative by forging her signature, and had been sentenced to 2 years imprisonment.

The second trial started on 15 March 1954 in Bordeaux, and by then new analyses had been conducted by leading Paris forensic experts and again the presence of large amounts of arsenic confirmed. Gautrat now changed his direction of attack, questioning the validity of these new results by producing experts who suggested that the arsenic might well have come from the soil of the graveyard of Loudun, and that soil microbes might well have made this arsenic so mobile that it could indeed contaminate the remains of bodies buried there. (The fact that it hadn't contaminated other bodies in the graveyard was only discovered later.) Again this resulted in the trial being abandoned while fresh evidence was gathered. Marie was meanwhile allowed out on bail.

The third trial began in November 1961. New research had been undertaken and by now it was possible to analyse hair samples by the new method of neutron activation analysis. This too showed excess arsenic to be present. Unfortunately the new technique gave the defence counsel yet another opportunity to throw doubt on the results, and Gautrat was able to exploit the natural uncertainties that invariably accompany a new method, playing off one group of scientific experts against another. The man who had carried out these tests had irradiated the hair in a nuclear reactor for 15 hours whereas Gautrat's experts thought that they should have been irradiated for 26 hours. Evidence from the world famous Nobel Prizewinner Joliot-Curie, confirmed that the exhumed hair contained arsenic at high levels, while other experts attested that bodies of animals which had been buried in the

Arsenic revenge

In 1998 a strange tale of mass arsenic poisoning was told by a 73-year-old Lithuanian Joseph Harmatz. He was one of a few who escaped from the Jewish Ghetto in the capital Vilnius, set up by the Nazis after they overran the country in World War II. Harmatz joined the partisans and spent war living in the forests. After the war he and two other fighters vowed to take vengeance on their hated enemy. This was to take the form of poisoning Nazi SS men who had been guards in concentration camps and it was carried out in 1946. The group called themselves Din, Hebrew for 'revenge' and they obtained a large amount of poison in Israel, concealed this in cans labelled condensed milk. Their lethal cargo was discovered by British police on board the ship they were taking to return to Europe and they were arrested and sent back to jail in the then British colony of Egypt. The poison was thrown overboard.

The Din group next planned to kill the SS guards being held in Stalag 13, a former German prisoner-of-war camp, where they were awaiting trial. The would-be assassins had obtained a quantity of arsenic trioxide from a French chemist which they had smuggled into Germany. This they dissolved in water and passed to one of their number who worked in the bakery where the bread for the camp was baked and he used it to paint the outside of 3,000 loaves that were supplied to the prisoners. More than 2,000 prisoners were taken ill with arsenic poisoning and Harmatz believes that as many as 400 died. By the time the American Authorities realized what was happening he and his team had fled to Czechoslovakia. Other more believable versions of the incident claim that while around 2,000 prisoners were affected, only 200 required hospitalization, and none died.

graveyard and exhumed after many months were found to be free of arsenic. Gautrat even had evidence to counter this line of argument. He knew of research carried out in 1952 on a dog that was poisoned with arsenic and then buried. After 2 years its remains were dug up and no arsenic was found. Clearly what happened to a body once it had been buried could be presented as rather unpredictable and clearly a guilty verdict would not be sound. So it was that on 12 December 1961 the prolific arsenic poisoner Marie Besnard was acquitted for lack of proof.

Marcus Marymont (1921–)

As arsenic-based chemicals were replaced with safer alternatives in the 1950s, cases of wilful arsenic poisoning became a thing of the past, although occasionally they did occur.* One conviction for murder by arsenic was that of 37-year-old US Sergeant Marcus Marymont in 1958 (he was based in Britain) who poisoned his 43-year-old wife Mary Helen. She was admitted to hospital on 9 June 1958 in a collapsed state following a meal that had left her vomiting for 24 hours. Despite all that could be done to save her it was clear that she was dying, and when told of this her husband merely shrugged his shoulders. The doctors attending her thought that she had been poisoned and when they said there needed to be a post-mortem he agreed, but later withdrew his consent, saying it would upset his three young children. Nevertheless a post-mortem had to be performed because the medical authorities were suspicious, and samples of her tissues sent to Scotland Yard, where high levels of arsenic were discovered: her liver contained 60 ppm, her kidneys 9 ppm, and her fingernails 120 ppm.

Meanwhile, investigators had discovered that Marymont had a strong motive for wanting his wife out of the way: he was having an affair with 23-year-old Cynthia Taylor, a married woman who was separated from her husband and whom Marymont had met in a night-club in Maidenhead 2 years previously. They began sleeping together later that year, and he even preferred to spend Christmas 1957 with her rather than with his family, who lived in Ruislip in the outer suburbs of London. Enquiries in Maidenhead found that Marymont had tried, unsuccessfully to buy arsenic from a pharmacy run by Mr Bernard Sampson, who had told him it could only be sold to him if he obtained a permit, whereupon Marymont left the shop and never returned. In fact he obtained arsenic trioxide from the USAF base where he was stationed, which was at Sculthorpe, near Fakenham in Norfolk, and it was there that two civilian cleaners remembered his coming to the

---

* The last person to be convicted of murder using arsenic in the UK was 46-year-old Zoora Shah of Bradford, Yorkshire, who in 1992 had murdered Mohammed Azam with it. He was a married man with whom she had a relationship and who had repeatedly beat and raped her, and forced her to have sex with other men. Zoora refrained from telling the Court about his abuse, thinking it brought shame on her family, and she was found guilty and sentenced to life imprisonment. [I am indebted to Paul Board for bringing this case to my attention.]

chemical laboratory one evening and handling a bottle of arsenic trioxide, commenting that it should be locked up for safe keeping.

Neutron activation analysis of the hair of Mary Marymont was undertaken by the forensic chemist Dr I.G. Holden. He found that she had been poisoned with arsenic on at least three occasions, with doses given at weekly intervals, before she was fed the final dose that killed her. Her husband only visited her at weekends and this tallied with the spacing between bands of arsenic at intervals along strands of her hair. The hair nearest the root had 20 ppm which indicated that she had been given a fatal dose of arsenic shortly before she died.

Marymont was tried at a US General Court Martial at Denham in Buckinghamshire in 1958. His defence was that his wife had committed suicide, having become very depressed in April after she found an un-posted letter addressed to Cynthia Taylor and realized her husband was having an affair. Marymont claimed that she must have eaten arsenic-containing rat poison that she had bought because they were troubled with mice under the kitchen sink. The fact that she had been fed doses of arsenic at weekly intervals, coinciding with Marymont's visits to Ruislip, disproved this line of defence and he was found guilty of murdering his wife, and of misconduct with Cynthia Taylor. He was deported to Fort Leavenworth Prison, Kansas to serve out a life sentence.

Michale Swango (1955–)

There is reliable evidence that Michael Swango may have killed as many as 60 of his patients and several of his colleagues during the 20 years he practised as a doctor and paramedic. The trail of deaths started while he was a student at Southern Illinois University School of Medicine, Springfield, Illinois, from 1978 to 1983. It was there that five patients died under mysterious circumstances and his fellow students even gave him the nickname Double-O-Swango based on that of James Bond's 007, with his licence to kill.

Swango later worked at Ohio State University Medical Center, Columbus, Ohio, where an elderly patient accused him of injecting something into her intravenous drip and her story was backed up by the patient's room mate and a student nurse. The senior doctors at the hospital investigated the incident and concluded that patients were confused and the student nurse unreliable. Although they cleared Swango of wrong-doing they told him to leave the hospital at the end of the first year of his residency there. He then moved to

Quincy, Illinois, where he fell out with a fellow doctor and sprinkled an arsenic-based ant-killer on his doughnut and put some in his coffee. This murder attempt was discovered and Swango was tried, found guilty, and sent to prison for 5 years although he was allowed out on parole after only 2.

Swango moved to the University of South Dakota School of Medicine, Vermillion, South Dakota, and obtained a residency there in internal medicine, explaining his previous conviction as due to false allegations by jealous co-workers – and he was believed. Eventually his past caught up with him and he was dismissed. He moved to take up a residency in psychiatric medicine at the VA Medical Affairs Center in Northport, New York state, but it was not long before he had to move from that post. In 1993 he left the USA and went to work in a rural hospital in Zimbabwe where he was again suspected of poisoning and killing five patients and he was dismissed. Swango went to Zambia where more poisonings led to his dismissal, after which he returned to the USA in 1997, in order to obtain a visa so he could take up another job in Saudi Arabia, and he was arrested. In September 2000 he negotiated to avoid the death penalty by plea bargaining, admitting to killing three patients, and he was sentenced to life imprisonment without parole.

The president who wasn't poisoned with arsenic

President Zachary Taylor (1784–1850) was the 12th president of the United States. He spent his boyhood on the family farm in Kentucky and enlisted in the army when he was 22. He soon developed a reputation for honest and plain speaking and rose through the ranks being commissioned first lieutenant in the infantry in 1808 and making it to major general in 1846. It was then that he made his name in the war with Mexico when his army won the battle of Buena Vista against overwhelming odds and despite disobeying orders not to engage the enemy. Taylor was a national hero. He was chosen by the Whig party as their presidential candidate and he beat the Democratic candidate Lewis Cass.

When Taylor was elected he was a fit and healthy man, and nicknamed 'Old Rough and Ready', but early in July 1850, only 16 months after taking office, he was suddenly taken ill after he sat too long bare-headed in the boiling sun listening to a speech. He went indoors and was given a dish of cherries and a glass of cold milk to refresh him, but soon he began to feel much worse and started vomiting. Doctors were called and despite dosing him with various medicines, none of which seemed to help, he died on 9 July. Had he been poisoned? That at least was a possibility and indeed there were some with motives for wanting him out of the way. Taylor had run into problems over the admission of anti-slavery California to the Union, which was strongly opposed by the Southern states, and he had discovered that three members of his cabinet were involved in a financial scandal.

In 1991 the Florida author Clara Rising began researching a book on Taylor and came up with the theory that he had been murdered with arsenic, and she gathered enough circumstantial evidence to enable her to present a strong case for his body to be exhumed and tested. She was prepared to pay the $1,200 autopsy fee and permission was granted for Kentucky officials to remove hair, nails and bone scrapings from Taylor's remains which rested in a crypt in the Zachary Taylor National Cemetery in Louisville, Kentucky.

These samples were sent to the Oak Ridge National Laboratory and to two Kentucky laboratories, and analysed for arsenic. The results were negative. There were traces of arsenic but far less than those needed to cause arsenic poisoning. Taylor had died of natural causes, either cholera or gastroenteritis, whose symptoms are very similar to those of arsenic poisoning.

Murder revisited: the guilt of Florence Maybrick

O F all the arsenic murders, the Maybrick case is the most intriguing. On 7 August 1889 Florence Maybrick was found guilty of murdering her husband James and sentenced to death, only to be reprieved 2 weeks later and her sentence commuted to life imprisonment. There are those who believe she should have been acquitted because she was innocent. There are those who believe that even if she was guilty she did the world a service in that the man she killed was really Jack-the-Ripper. That somewhat dubious claim was made in the 1990s with the publication of an old diary supposedly written by James Maybrick.

In the furore which followed the trial, Florence was seen as a martyr by two groups: the supporters of the Women's Rights Movement, and those who campaigned for a Court of Appeal. The first of these saw her as a victim of a male dominated legal system, and the second saw her as a prime example of injustice which the British legal system as it then stood was unable to rectify. The Women's International Maybrick Society even enlisted the support of three US Presidents, but to no avail because, unbeknown to them, Queen Victoria had taken an interest in the case and believed Florence to be guilty. Until the Queen died, there was no possibility of her release from prison, although she was set free soon afterwards.

Legal problems raised by the Maybrick trial centred on the summing-up of the Judge, Mr Justice Fitzjames Stephens. In its latter stages this became little more than a tirade of moralizing generalizations that dwelt on Florence's admitted adultery, implying that a woman capable of committing such a sin, was indeed capable of murder. (Nothing was said at the trial about her husband's mistress and the five children that she had borne him.) The summing-up was flawed in other ways; for example the judge introduced material that was not produced during the trial and he read accounts of what witnesses had said from newspaper cuttings of their evidence because his own notes were in such a poor state. He made several slips over times and dates, and his directions to the jury would today have automatically led to an appeal.

Judge Stephens was 60 at the time of the Maybrick trial. He was later to be referred to as Mad Judge Stephens and appears to have had some sort of mental illness in 1885 but had recovered sufficiently to continue his judicial duties. Seemingly normal at the time of the Maybrick trial in August 1889, he was behaving so oddly 2 years later that there were questions in the House of Commons about his sanity. He was persuaded in April 1891 to retire from the bench and was given a baronetcy as compensation. Shortly afterward he was discreetly admitted to an Ipswich lunatic asylum where he died in March 1894. Those who have written about the Maybrick case have invariably speculated that Stephens was not of sound mind at the time of her trial, and indeed his behaviour adds piquancy to this remarkable case.

The star of the show, however, was Florence herself. She had married James Maybrick, a man old enough to be her father, and when she fell for a younger man her way out of her marriage was to poison her husband. But she was an incompetent poisoner and only succeeded by giving him several doses of arsenic over a period of 2 weeks. Because of the rapid elimination of this poison from the body, there was not much arsenic in James Maybrick's body when he died.

The marriage of Florence Chandler and James Maybrick

James Maybrick was born in Liverpool in 1839, the second son of a parish clerk. He had three brothers: Thomas, who was older, but with whom he had little contact; Edwin, who was younger and unmarried, and with whom he was in partnership as a cotton broker; and Michael, who was also younger, and who was a well-known composer under the pen-name of Stephen Adams.* At the age of 20 James Maybrick was apprenticed to a shipping

---

* His best-selling song was *The Holy City*, which sold millions of copies and was popular for more than half a century.

broker in London. There he met a young shop girl and they became lovers and set up house together. She eventually bore him five children, although during most of this period James worked in Liverpool where his company was based. The firm also had an office in Norfolk, Virginia, which is where James spent most of his time from 1877 to 1880. It was in March of that year that he sailed on the White Star liner *Baltic*, from New York to Liverpool. Among the passengers were the Baroness von Roques and her 17-year-old daughter, Florence Chandler.

Florence was born on 3 September 1862 in Mobile, Alabama. Her father died early the next year and her mother married again shortly after, but her second husband soon died, and her third husband was a Prussian, Baron von Roques, and this marriage broke up in 1879. While all this was going on, Florence had been fostered out to relatives and educated by governesses, but in March 1880 she set off with her mother on a tour of Europe. During the course of the journey across the Atlantic, Florence fell in love with James, and James was drawn to Florence on account of her beauty. Through the summer and autumn of 1880 James courted her and they were married in the spring of 1881 at St James's Church in Piccadilly, London. They went on their honeymoon to Bournemouth and then took up residence in Norfolk, Virginia. Their first child, James, was born in March 1882.

For 3 years the Maybricks moved between Virginia and Liverpool, eventually settling down in the Liverpool suburb of Sefton Park where their second child, Gladys, was born in June 1886. In 1887 there was a matrimonial crisis when Florence discovered that James had a mistress. About this time James, a chronic hypochondriac, was dosing himself regularly with various tonics, including Fowler's arsenical solution, which was popularly regarded as an aphrodisiac. By now James Maybrick was approaching 50, and his sexual powers were beginning to wane, while Florence was still only 26 and in the first flush of youth.

In 1888 the Maybricks moved to a large 20-room house called Battlecrease House in Riversdale Road, in the Liverpool suburb of Aigburth. To outward appearances all was well between James and Florence: they led an active social life of dances, parties, whist evenings, and race meetings. Behind the façade their marriage was a marriage in name only and Florence's affections had transferred themselves to Alfred Brierley, a handsome, 6-foot, 34-year-old bachelor whom she had first met when he had been a guest at a ball they had held at Battlecrease House in November 1888. Florence had another secret: she was heavily in debt to a moneylender from whom she borrowed money whenever she went to London.

In 1889, relations between James and Florence deteriorated to the extent that they could no longer hide their feelings and they even quarrelled in public. Early in March they engaged in a heated exchange at a hotel in Southport after a race meeting when they were playing cards with another couple. Soon after that row, Florence arranged for her and Brierley to spend a weekend together in London. And so it was that on Thursday morning, 21 March, Florence left Liverpool for London. She told James that she was going at the request of her aunt who was to undergo an operation and wanted her near at hand. She told the children's nanny, 28-year-old Alice Yapp, to redirect any mail to the Grand Hotel, London, where she said her mother the Baroness was staying, and whom she planned to see during the forthcoming week. She had in fact booked a room for herself and Brierley at Flatman's Hotel near Cavendish Square.

Alfred Brierley first made his appearance at the hotel for dinner at 7.30pm on the following day, Friday, having travelled up on the afternoon train from Liverpool. The two of them spent the rest of Friday, all day Saturday, and Sunday morning together, before Brierley returned to Liverpool and Florence went to stay with her friends in Kensington. During the next few days she visited various people, and on one evening she was taken by her brother-in-law Michael Maybrick to the Café Royal and then to the theatre. She caught the train back to Liverpool on Thursday 28 March, to attend the 50th anniversary Grand National Steeplechase the next day, an occasion marked by a visit from the Prince of Wales.* It was also the day of a very public quarrel between Florence and James, which started at the race course and ended with Florence walking off arm-in-arm with Brierley. It was continued that evening at Battlecrease House and concluded with Florence getting a black eye.

After a sleepless night, Florence went round the following day, Saturday, to her friend Mrs Briggs and told her that she could not go on living with James. She complained that not only had he a mistress, but he opened her mail, and kept her on such a tight budget that she was in debt to the tune of £1,200 (equivalent to around £100,000 now). Mrs Briggs calmed her down and together they went round to the family doctor, Dr Hopper, to show him

---

* The race was won by Frigate at odds of 8-to-1.

the results of James's blows. Then they went to Mrs Briggs's solicitor to seek advice about getting a divorce. However, he disillusioned Florence about her chances of securing one, but suggested that she could have her mail sent to a private box at the Post Office, and it was there that the two women next went.

That Saturday afternoon Dr Hopper called as Battlecrease House and in the role of mediator he patched up the quarrel between husband and wife. Florence also confessed that she was in financial difficulties and James agreed to settle her debts when he next went to London. Mrs Briggs was imposed upon to stay at the house for a few days while things blew over. A week later, on Saturday 6 April, Florence visited Alfred Brierley at his rooms and followed this with a letter, to which she received no reply. By this time Alfred's ardour was cooling fast and, unknown to Florence, he had already booked himself on a long cruise around the Mediterranean.

On Saturday 13 April James left for London, where he did indeed pay off some of Florence's debts. He stayed with his brother Michael and on the following day he had a medical check-up by a Dr Fuller. He was suffering he said from severe headaches, constipation, and numbness of the limbs, but was reassured to be told that there was nothing organically wrong with him and was given a prescription for a laxative, a tonic, and liver pills, and told to come back in a week's time. The laxative was Plummers pills which contained antimony sulfide as the active ingredient and James was instructed to take one of these every night.

On Monday, James returned to Liverpool and the next day he had the prescriptions made up at a local pharmacy. His health began to improve and on the following Saturday, 20 April, he returned to London for his second appointment with Dr Fuller. The Plummers pills had clearly worked and James was told to discontinue these and to use sulfur tablets instead. The doctor also handed James a prescription to take with him and told him he would send him a bottle of medicine which he would make up for him. James returned to Liverpool and had the new prescription made up by the pharmacist on the Wednesday and the extra medicine arrived from London on the Friday. The medicines seemed to have the desired effect in that his health continued to improve. James was a hypochondriac and surrounded himself with dozens of pills and potions both at home and at his office, some of which went back to his bachelor days. (An old box of pills was produced by the defence at the trial and these consisted of iron, quinine, and arsenic.)

On Thursday 25 April James made a new will leaving most of his money in trust to his brothers, although to be used for the benefit of his children. Florence knew nothing of this nor that she was only to inherit an insurance policy which would have yielded £2,500 when he died. What James didn't know was that he had only 16 days left to live.

Florence prepares her poison

Florence had decided to dispose of her husband with arsenic. This she first had to acquire and then devise a way in which to give it to him. Putting it in his food would be difficult because all meals at Battlecrease House were prepared by the cook and so she decided to add it to the medicines that James was so fond of taking. In the years following Florence's trial, people swore affidavits to explain the presence of the arsenic found in Battlecrease House. In 1894, one man attested that he had supplied James with both white and black arsenic as trade samples in the February before his death. This seems highly unlikely and certainly only those campaigning for Florence's release from gaol believed it to be true.

A more believable explanation emerged in 1926 when a Liverpool pharmacist revealed that he had supplied Florence with a packet of arsenic trioxide. The packet was labelled 'ARSENIC POISON' and on one side of the packet was written the words 'for cats', this being the reason she had given for needing it. Florence had not wanted to sign the poisons register and the pharmacist did not insist on this, as she was a regular customer at his shop. The arsenic he sold her was black arsenic, so called because it was mixed with soot to prevent its being misused. Florence returned to the shop at a later date and he sold her another packet when she said she had lost the first one. Again the pharmacist wrote 'for cats' on the second packet and this was found in the house after James had died. The pharmacist said he was afraid to come forward at the trial and only contacted the police after Florence had been found guilty. His evidence would indeed have told heavily against Florence if given at the trial but it no doubt convinced the authorities that she had been rightly convicted.

The second weapon in Florence's armoury of arsenic was flypapers. These were on sale at many shops and cost half-a-penny each and they were known to be highly toxic. Indeed flypaper solution had been used in a famous Liverpool murder about the time the Maybricks took up residence in that city. In 1884, two sisters, Flanagan and Higgins, had disposed of a husband, son, stepdaughter, and lodger by poisoning them with flypaper solution. They were convicted and hanged.

Victorian flypapers were activated by placing them on a plate of water, and they killed any fly that sampled the solution, which was made attractive to them by being sweetened with sugar.* To prevent their misuse, the flypapers also contained a little of the bitter tasting quassia and a brown colouring agent. Soaking a single flypaper in water would extract most of its soluble arsenic salts within a few hours to produce a tea-coloured solution. Considering that a typical flypaper contained as much as 400 mg of arsenic, it can be seen that this was potentially lethal for an adult human, provided that it could be administered without the bitter taste and brown colour arousing suspicion. Strong tea, coffee, or meat extract drinks offered the best cover and these were used. Brandy was also a good medium.

Sometime around Easter, Florence bought some flypapers at a local pharmacy, complaining that flies were troublesome at Battlecrease House. She paid cash for them, despite having an account at the shop, and had them sent round to her home. When asked about this in court, the shopkeeper could not recollect the exact date of purchase except that it was mid-April, but he had remembered it because Florence was the first person to buy flypapers that spring. In fact flies were not a problem at Battlecrease House, where there were unused flypapers in the pantry that had been left over from the previous summer. The flypapers bought by Florence were seen by a housemaid who discovered them soaking in a basin covered by a plate and hidden under a towel in Florence's bedroom. The children's nurse, Alice Yapp, when told of this discovery took a look for herself.

At her trial Florence claimed that her reason for soaking flypapers was to use the solution as a cosmetic face-wash. This might explain the presence of a very weak arsenical solution of about 6 mg that was found in a scent bottle in the linen cupboard after her arrest. Florence had used an arsenical face-wash as a teenager and 10 years earlier, when she was 16 years old, a New York pharmacist had given her a prescription for just such a cosmetic solution

---

* The active agents were the salts sodium arsenite and potassium arsenite, which were made by treating a solution of arsenic trioxide in water with sodium or potassium carbonate. The salts so formed are much more soluble than arsenic trioxide itself and just as deadly.

and she carried this around with her on her various travels, but having outgrown its need she had ceased to use it. The prescription came to light when her mother was moving from Paris to Rouen after the trial, and she even tracked down the Paris pharmacist who had made up a bottle of the face-wash many years previously.

Florence was of course an amateur poisoner. She probably over-rated the effectiveness of arsenic and gave James small doses. This misjudgement nearly enabled her to evade detection. Conversely one might concede that she was skilful enough to do this for a very good reason, which was that it made James's symptoms appear to be a case of gastroenteritis. However, her other actions show Florence to be impetuous and emotional rather than careful and rational, so it is more than likely that she bungled her first attempts to poison James because she did not know the kind of dose that would kill him. The several small doses that she gave her husband were not of themselves fatal, but they did undermine his health.

Florence secreted most of her poison in a hatbox in the room where she now slept. This box contained three bottles: one held black arsenic and water, another a saturated solution of arsenic oxide, and the third had the remains of a saturated solution with only a little liquid remaining. This cache of poisons was discovered after James had died. In another hatbox there was a glass containing some milk and a handkerchief, both heavily contaminated with arsenic oxide. The packet, and what was left of the black charcoal (less than 5 g), was found in a chocolate box in her trunk. It is clear that Florence was running a modest poison dispensary from the dressing room that she was using as her bedroom. Her technique seems to have been to add water to black arsenic to dissolve the arsenic trioxide and use a handkerchief to filter out the soot, thereby obtaining an almost clear solution of arsenic.

The poisoning of James Maybrick

On Saturday morning 27 April, James had planned to go to the races at Wirral. Shortly after breakfast, and after taking a double dose of the special medicine sent by Dr Fuller from London he was violently sick. He complained of numbness in his legs and hands but did not feel ill enough to cancel his day out. He called in at his office about 10.30am and left to go to the races at noon. Florence stayed at home. At the racecourse, James was noticeably unsteady while riding his horse. It was a wet day and he got soaked. Later in the evening when dining with friends he was rather shaky and spilled a glass of wine.

The next day, Sunday, James felt very ill and again the likely cause of this was another dose of the poisoned medicine. That morning, James had a glass of brandy, hoping that this would settle his stomach, but by nine o'clock he was feeling so ill that he drank a mixture of mustard and water as an emetic. Florence, ever solicitous of his health, mixed this for him. The emetic worked and James vomited copiously. A Dr Humphreys, who lived only 10 minutes walk away, was sent for and came at 10.30am. James told him that a cup of strong tea which he had drunk first thing that morning, was responsible for his condition. The doctor diagnosed an upset stomach and advised his patient to stick to a light diet and drink only milk or soda water. When James told Dr Humphreys about Dr Fuller's special medicine he advised him not to take any more of it.

For his lunch that day James had arrowroot prepared by the cook. For his tea he had oxtail soup, and for his supper arrowroot again, which the cook started to prepare but which Florence completed making. James left most of this, which returned to the kitchen darker in colour, which the cook thought was due to vanilla essence being added to it. At 8pm James retired for an early night, leaving Florence and brother Edwin talking downstairs. At 9pm, James started vomiting again and rang for help. He complained of being unable to move his legs. Dr Humphreys was again sent for and prescribed potassium bromide and tincture of henbane. James seemed to be so ill that Edwin agreed to stay the night. If vomiting is taken as a sign of recent arsenic poisoning, then James had been fed some for his supper. The colour of the arrowroot suggests that Florence added some of her flypaper extract to it, and perhaps she had put some of this in the morning cup of tea as well.

On Monday morning 29 April, it must have been obvious to Florence that she had failed in her objective. Edwin found his brother feeling much better, as did Dr Humphreys, who confirmed that James was fit enough to get up and go to work. All signs of his upset stomach had gone and all he was left with was a furred tongue. To prevent a recurrence of the trouble the doctor recommended him staying on a light diet for the rest of the week. The doctor stopped all previous medicines and prescribed instead Seymour's preparation, a solution of herbal extracts that was often prescribed as a digestive tonic and liver stimulant. That day James visited his office but stayed for only about an hour. Meanwhile Florence went shopping. At another nearby pharmacy she purchased two dozen flypapers of the arsenical kind. Again she paid for them in cash. Whatever she did with those 24 flypapers, she did in secret. No one at Battlecrease House was aware of their presence and they were never seen again.

On Tuesday, 30 April, cook prepared unsweetened bread and milk for James for breakfast, but he found it had been sweetened and returned it to the kitchen. He went to work after lunch and arrived about 1pm. He sent the office boy out to buy some Du Barry's Revalenta Arabica invalid food and deliver it to Battlecrease House. That evening, Florence went to a private masked ball with Edwin as her escort, and he accompanied her home again and stayed the night.

Wednesday was May Day. The doctor called early and proclaimed James totally recovered. As soon as the doctor had gone, James set off for the office. Edwin was still asleep but James told the cook to prepare some invalid food that Edwin would bring to the office for him. Cook prepared Du Barry's food and put some sherry in it according to James's liking. She gave it to Florence who asked her to get some paper and string to wrap it in. When Cook returned with these items she discovered that Florence had already wrapped the container. Edwin was given this to take with him when he left for the office later that morning.

With the arrival of his lunch, James sent the office boy out to buy a saucepan to warm it up in, and a basin and spoon to eat it with. However, when he tasted a spoonful of the liquid he found it not to his liking and complained to Edwin that the cook had put cooking sherry in it. He ate very little of it, but afterwards felt unwell, although he was not physically sick. He was sufficiently worried to visit the doctor again after he got home from work. That evening, the Maybricks held a small dinner party for some of their friends.

Florence had decided that Du Barry's invalid food was to be the route to poisoning James. On Thursday 2 May, James went to work as usual, taking a jug of prepared Du Barry's invalid food with him. At lunchtime he again ate only a little of it, because of the taste, and again he felt ill in the afternoon. On this occasion when the woman who cleaned the office came to wash up the pan she noticed black and white particles in the bottom of the pan he had heated the food in. This suggests that Florence had used black arsenic. In the evening, James complained of pains in his legs.

Florence was nothing if not persistent. It seems likely that she went back to using flypaper solution for the next day's meal. That Friday morning, James sent for Dr Humphreys, who called at 10am. He found nothing wrong with his patient, except a furred tongue. James asked him if a Turkish bath would do him good, and the doctor agreed, James departed for work, taking with him his jug of Du Barry's and sherry. He went for a Turkish bath before lunch, and then returned to the office where he heated up the invalid food and this time consumed it all.

This was the fatal meal and he ate it later in the day than usual. It had been left standing long enough for a skin to form on the surface of the liquid and a piece of this skin stuck to the side of the jug. It remained there despite the cleaning lady washing the jug. That day the jug was left behind at the office and it was later taken for testing after James's death. The analyst was able to get a positive arsenic reading when he tested the dried skin and also when he tested some hot water that he had used to rinse the pan in which the food had been cooked, and the dish from which it had been eaten. He confirmed that the arsenic he found did not come from the glaze on the vessels by washing them a second time with boiling water and showing that no arsenic was detectable.

After his meal, James began to feel ill and rushed home. There he vomited and blamed the invalid food and sherry, because this was all that he had eaten. He went straight to bed and became progressively worse. At about 11pm Dr Humphreys was called and found James very pale, in great pain, and with gnawing sensations in his thighs. To give James some relief from his pain, the doctor inserted a morphine suppository.

The fatal week

Saturday 4 May 1889: That morning found James no better. He vomited again and when the doctor called he attributed this to the effect of the morphine. James was very thirsty but the doctor advised him not to drink anything and only to rinse his mouth with a gargle or to moisten it by sucking on a wet cloth. When he had gone James and Florence had another quarrel. He told her that he knew what she had been doing behind his back, and that he had started investigations into her movements while she was in London. Florence, alarmed by this, despatched a telegram to Alfred Brierley, telling him that James was making enquiries, and more specifically that he had been advertising in the London newspapers for information about them.

Edwin called to see his brother and found him unable to keep anything down – he vomited after a brandy and soda, and later that day the same thing happened after taking some medicine. In the evening, the cook made James some broth. Florence told the housemaid, Elizabeth, that from now on she, Florence, would empty the slops from the sick room herself.

Sunday 5 May: James was retching but was unable to bring anything back. The doctor called and noted that his throat was red and very sore and his tongue dirty and furred. For this he suggested a mouthwash of Condy's fluid (potassium permanganate solution). He told James to change to Valentine's beef juice in place of Du Barry's invalid food. James was suffering from incredible thirst and ordered fresh lemonade from the kitchen, but Florence only allowed him to gargle with it. Edwin called to see his brother and decided to stay the night at Battlecrease House.

Meanwhile, that same Sunday morning, Alfred Brierley bought copies of all the London newspapers, and scrutinized them for James's advertisement. He found nothing and wrote a letter to Florence, telling her that he was planning to go abroad for a few weeks. He also hinted that it would be best if they did not meet again until the autumn.

Monday 6 May: James started to recover and was sufficiently improved to sit up in bed and read the morning paper, and to dispatch some business telegrams. Dr Humphreys called and prescribed Fowler's arsenical solution, of which James took three doses. The arsenic content of these would amount to less than a milligram. He suggested calling in a second opinion but Florence was against the idea. She was now in complete control of the sickroom; she even changed James's bed clothes, telling the housemaid not to do it.

Dr Humphreys called again that evening and noted that James's diarrhoea of the past three days had progressed to the condition technically known as tenesmus – ineffectual straining to evacuate the bowels. He applied a blistering agent to James's stomach, this being a recognized medical treatment of the time. He prescribed no new medicines, but outlined a light diet of beef tea, chicken broth, Neave's invalid food, and only milk and water to drink. James appeared to be recovering.

Tuesday 7 May: When Dr Humphreys called to see James that morning he was told that he felt 'a different man' and said that the blistering treatment was responsible for an improvement in his health. He still had a fetid mouth and a tickling sensation in his throat which he described felt as if there was a hair lodged in it. He took some food and was not sick after it. Edwin, however, decided to call in a second opinion and on his insistence Florence telegraphed for a Dr William Carter to come with Dr Humphreys on his evening visit at 5.30pm. Edwin went in to the office that day but returned on the 4.45pm train to be there when they called.

Dr Carter examined James and was told that he had been suffering from vomiting and diarrhoea for several days. He diagnosed chronic dyspepsia due to an irritant in the stomach. He prescribed a sedative (antipyrine), a medicine to increase salivation to ease his throat (jaborandi), and a mouthwash (chlorine water). Dr Carter noted the absence of bad breath despite the foul state of James's mouth. He ordered a continuation of the diet recommended the previous day with the addition of that Victorian cure-all: chlorodyne.

On this Tuesday, nanny Yapp saw Florence surreptitiously pouring medicine from one bottle to another – the medicines were kept on the landing near the main bedroom – but said nothing.

Wednesday 8 May: This was the crucial day in the Maybrick affair. James continued to improve and told Edwin, who wanted to send for brother Michael, that he was feeling much better and not to bother Michael. Dr Humphreys paid his morning visit and noted a general improvement, even though James had had a restless night, during which Florence has sat by his bedside.

It was obvious that Florence was the person most in need of rest and at 9am a private nurse was sent for so that she could get some sleep. Perhaps it was exhaustion that made Florence take the next step. She realized that all her efforts to poison her husband had so far failed, and with her lover Alfred cooling rapidly, time was fast running out. During the course of that Wednesday morning she gave James another dose of poison. He started vomiting violently and she wired Edwin at work, and Michael in London. Edwin also cabled Michael before he returned to Battlecrease House on the 12.40pm train.

Michael received yet a third telegram saying 'come at once, strange things going on here' from the Maybricks' friend Mrs Briggs who had visited

James that morning. She had been taken aside by nanny Yapp who told her that peculiar things had been happening lately, specifically mentioning the soaking flypapers. As a result of these three telegrams, Michael hastened to Liverpool.

The private nurse, called Gore, arrived at Battlecrease House to find her patient very ill. James had just been vomiting but did not vomit during the rest of that day. As soon as the nurse had taken over in the sick room, Florence used the opportunity to dash off a letter to Alfred. Too tired to go to the post herself, Florence gave the letter she had just written to nanny Yapp to post on her afternoon walk with little Gladys. The time was 3.30pm. Alice Yapp set off, but seeing the letter was addressed to Alfred Brierley, the temptation to open it proved too strong. (Her excuse at the trial for doing so was that she dropped it in the mud.) What she read confirmed all her suspicions.

Wednesday

Dearest . . . since my return I have been nursing M. day and night. *He is sick unto death.* [This sentence was underlined in the original.] The doctors held a consultation yesterday, and now all depends upon how long his strength will hold out. Both my brothers-in-law are here and we are terribly anxious. I cannot answer your letter fully today, my darling, but relieve your mind of all fear of discovery now and in the future. M. has been delirious since Sunday, and I know now that he is perfectly ignorant of everything, even of the name of the street, and also that he has not been making any inquiries whatever. The tale he told me was a pure fabrication, and only intended to frighten the truth out of me. In fact he believes my statement although he will not admit it. You need not therefore go abroad on that account, dearest; but, in any case, please don't leave England until I have seen you once again. You must feel that those two letters of mine were written under circumstances which must even excuse their injustice in your eyes. Do you suppose that I could act as I am doing if I really felt and meant what I inferred then? If you wish to write to me about anything do so now, as all the letters pass through my hands at present. Excuse this scrawl, my own darling, but I dare not leave the room for a moment, and I do not know when I shall be able to write to you again. – In haste, yours ever, Florie.

All Alice Yapp's suspicions were confirmed by what she read and she returned to the house and showed the letter to Edwin. He realized what Florence had been doing and tried belatedly to undo some of the damage. He instructed Nurse Gore that from then on she alone was to administer to James. Unfortunately Gore had already given him one dose of the medicine prepared by Florence, but she threw the next lot away at 7pm on the pretext of needing the tumbler for something else. Edwin gave Gore a fresh bottle of Valentine's meat juice for James. But it was too late to save him.

James deteriorated by the hour. Eventually Michael arrived and was shocked to discover his brother James only semi-conscious. He told Florence that he was not satisfied with the treatment his brother was getting and at 10.30pm that evening he went round to see Dr Humphreys.

The nurse meanwhile settled James down for the night but he slept only 3 hours in all. Florence, in the adjacent room, had a full night's sleep, her first for several days.

Thursday 9 May: By morning, James had improved slightly and both Michael and Edwin agreed he was looking better. Despite this progress, James vomited at eight o'clock and his diarrhoea was still prevalent. Both doctors paid a visit that morning and prescribed cocaine to ease James's pain. Before they left, Michael told them that he suspected that James was being poisoned. The doctors agreed to check this and Dr Humphreys took away samples of urine and faeces, together with samples of the brandy and invalid food that was in the bedroom. Michael suspected the brandy had been tampered with and he replaced it with a fresh bottle.

Tests performed on the urine and faeces by Dr Humphrey proved negative, but he later admitted to having performed a Reinsch test on them in a rather perfunctory manner. Dr Carter, who was the local expert on arsenical poisoning, tested the brandy and invalid food, neither of which Florence had poisoned, so he found nothing suspicious. Although these gentlemen were unable to prove that James was suffering from the effects of arsenic, Florence herself provided the proof that very night.

At 11am nurse Gore went off duty and another nurse came on for a 12 hours day shift. During this period James was fed only food which the nurse had prepared in the sick room. Apart from the invalid food and chicken broth, James drank only champagne and the new brandy. He complained throughout the day of pains in the stomach and of a burning sensation in his throat. He was sick again that evening at 8.15pm.

Nurse Gore returned to duty at 11pm and her first job was to give James some Valentine's meat juice from a fresh bottle which she herself opened.

Shortly after midnight, Florence surreptitiously palmed this bottle and took it to her bedroom. This sleight of hand did not escape the notice of the nurse, and she also watched Florence try secretly to replace the bottle on her return only a minute or so later. Florence had been caught in the act. This bottle was passed to Michael and the following afternoon he handed it to Dr Carter for analysis. It was found to contain 38 mg of arsenic when analysed by a professional analyst. The meat juice had been diluted with a solution of arsenic and water.

Florence had already prepared another bottle of poisoned Valentine's meat juice and hidden it in a hatbox. She exchanged this for the new bottle, and quickly returned to the sick room hoping to replace it before the bottle was missed. Only her clumsy movements gave her away, although at the time she probably thought she had managed the switch undetected since nurse Gore said nothing and left the bottle where it was.

This explanation of the incident is supported by the fact that an unpoisoned bottle of Valentine's meat juice was found in the hatbox. Moreover, it would also explain the swiftness with which the whole operation was carried out – and which was essential if it were to succeed. So damning were the implications of this incident that Florence felt compelled to make a statement at her trial explaining it. According to her version of the story, James begged her to give him one of his powders and this she agreed to do by adding it to the meat juice. As she was doing this she spilled some of the meat juice. She made it up to the mark again with water.

What the poisoned Valentine's meat juice also showed was that Florence had no clear idea of what constituted a fatal dose of arsenic. It seems reasonable to assume that she had been poisoning James with similar small doses. Even if he had consumed a quarter of a bottle of the poisoned beef extract per day he would have taken only about 10 mg, certainly not a large enough dose to kill him. As it turned out, it was certainly the Friday lunch of Du Barry's invalid food laced with flypaper extract and sherry that had delivered the fatal dose, although the small doses that Florence administered after that could only have made his condition worse.

Friday 10 May: Nurse Gore was relieved at 11am after seeing James through an almost sleepless night; he vomited on two occasions. Before she went off duty she warned the day nurse not to give James any of the suspect meat juice. Before she retired to bed at 1.30pm (she was sleeping at Battlecrease

House) she told Michael what she had seen Florence do. Michael went at once to retrieve the meat juice, which he handed to Dr Carter when he called to see his patient.

James's strength failed as the day progressed. His pulse became weaker and one of his hands turned white. The doctor described his tongue as 'simply filthy'. Yet another mixture of medicines was prescribed: sulphonal as a sleeping draught, nitroglycerine for his hands, more cocaine to kill the pain, and phosphoric acid as a mouthwash.

During the day Michael stumbled upon Florence as she was pouring medicine from one bottle to another and he confiscated both of them. No arsenic was found in either. Later that same day, James was overheard to say to Florence: 'you have given me the wrong medicine again' which Florence denied. At 4.30pm a new day nurse took over and at 6pm she heard James admonish Florence with the words: 'Oh Bunny, Bunny, how could you do it? I did not think it of you.' He repeated this three times, and these were his final coherent words before he lapsed into delirium.

By 10.30pm that evening his pulse was so weak as to be almost uncountable and the doctors gave up hope of saving him. Florence, now certain of James's imminent demise, and as always playing the part of the solicitous wife, asked Michael to call his London doctor, Dr Fuller. Michael told her it was too late for that. James continued to sink and in the middle of the night Florence woke Michael as it seemed James was about to breathe his last. Mrs Briggs was also sent for.

Saturday 11 May: The doctor called at 8.30am and confirmed that the end was approaching. At 5pm, young James and Gladys were brought to their father so that he could take his final leave of them. At 8.30pm that evening he died.* Florence fainted shortly afterwards and the nurses put her to bed in the spare room. Whether she passed out from exhaustion, remorse, or shock we do not know, but she was suddenly aware that everyone at Battlecrease House believed her to have caused James's death.

Two hours after James's death, the nurse and nanny Yapp conducted a

---

* In 1911, Florence's son James poisoned himself with potassium cyanide solution. It may have been an accident since he was working at a gold mine in British Columbia where such solutions were used. He drank the cyanide while eating his lunch and although he tried to telephone for help he died before it arrived.

search for poison and found the chocolate box containing the packet of arsenic trioxide with 'for cats' written on the side. The solicitor living next door was immediately sent for by Michael so he could witness the sealing of this incriminating piece of evidence. On that fateful night, this must have seemed to all present as proof positive of Florence's guilt.

Florence remained in a coma until well into the following Sunday morning, and when eventually she did appear, Michael insisted that she went into Liverpool with her brother-in-law Thomas, who had just arrived from Manchester, to arrange the funeral, but this was merely a pretext to get her out of the house. As soon as she had left, Michael, Edwin, and Mrs Briggs searched her room and there they discovered the bottles of arsenic solution, together with meat juice, in one hatbox. In a second hatbox they found the remains of some milk in a glass which also later turned out to contain arsenic. These objects and other suspicious bottles were collected and handed over to the police who were called to the house.

When Florence returned to Battlecrease House she was confronted with a situation beyond her control and took to her bed. Meanwhile, a post-mortem examination of James's body was performed the next day and the acute inflammation of the throat and bowels, indicative of gastroenteritis, was noted. His stomach contents and liver were removed and taken for analysis. The police cautioned Florence on the Tuesday. By now she was completely isolated, even by her supposed friend, Mrs Briggs, who played rather a cruel trick on her. She persuaded Florence to write an urgent note to Alfred Brierley asking him to help her in her plight. This letter was, of course, immediately confiscated by the police when Mrs Briggs told them about it. On the Friday of that week, and six days after James had died, Florence's mother the Baroness arrived from Paris. The day after, Saturday 18 May, her daughter was transferred to Walton jail.

The trial of Florence Maybrick

The coroner's inquest was held on 28 May after an initial adjournment on 14 May. The body of James was exhumed on 30 May and further organs removed for analysis. More arsenic was found in them. The trial of Florence began on 31 July with Judge Stephens presiding. It was the sort of show piece trial beloved of the Victorians and the newspapers of the day were well represented in court.

Florence's defence counsel was headed by Sir Charles Russell QC, MP, then at the height of his fame. He was not unfamiliar with cases of arsenic poisoning, having successfully prosecuted Mary Ann Cotton at Durham in 1873. He knew that it was easy to belittle and trivialize forensic evidence, and he was experienced in exploiting the tensions between the older medical men, who set great store by subjective experience, and the younger doctors with their more scientific approach. In his defence for Florence he tried to offer an alternative explanation for the various bits of evidence produced against her. He hoped to rely on the Styrian Defence mentioned in Chapters 6 and 7: James regularly took arsenic as a tonic, and Florence used it as a cosmetic.

The main line of argument that Russell used was that the amount of arsenic found in James's body was too little to account for his death. His second line of defence was that even supposing James did die of arsenical poisoning, there was no proof that it was given to him by Florence. According-ing to Russell, the arsenic was there as a result of James's habit of self-medication. Russell failed in his defence of Florence, but he was always of the opinion that he had shown that under the law as it then stood, the jury would have had good grounds upon which to return a verdict of not guilty, had they wanted to. That they did not want to, was the fault of the Judge. Russell was upset by Judge Stephens's summing-up, which ended in a vilification of the prisoner, damning her on her sexual and moral failings.

Had a Court of Appeal then existed, Russell would have been able to question the verdict on the grounds of misdirection of the jury by the judge, and very likely his plea would have been upheld. He could have quoted several legal arguments that today would be very telling, the most damning of which would be that the judge had introduced facts detrimental to the prisoner in his summing-up that were not produced in evidence, and so could not be challenged by the defence.

The prosecution was handled by Mr John Addison, QC, MP, who did his best for the defence by calling his witnesses in an order that was without any underlying logic. This left the courts somewhat confused as to the times, dates, and sequence of events. The judge commented on this presentation of the Crown's evidence and it certainly befuddled him, but perhaps that might not have been difficult in any case. Reading the trial report today still leaves one unclear about the chronological order of events, and it was little wonder that the judge sometimes got his dates wrong. Nevertheless the court heard that James's liver was found to contain 20 mg of arsenic, which is about what one would expect as the residue of a much larger quantity. His intestines contained 6 mg and his kidneys gave a positive test but only 0.5 mg was judged to be present. His hair and nails were not analysed since it was not realized at the time that arsenic collected in these. It would be intriguing to have a hair of his head and to subject this to modern neutron activation analysis.

All in all, the prosecution failed to make its points and often pulled its punches. The judge in his summing up had to do their work for them. The prosecution muddled the scientific evidence and there was confusion over the analyses of the samples removed from Battlecrease House. Much of the trial was taken up with an argument about the symptoms of arsenical poisoning, with doctors Fuller, Humphreys, Carter, and Stevenson, the Home Office expert, all supporting arsenic as the cause of death, while doctors Drysdale, Tidy, McNamara, and Paul, finding nothing in the reports of the post-mortem, or in their knowledge of arsenic toxicity, to support death from this cause.

The medical evidence of the defence experts is not without its gems of wisdom. For example, Dr MacNamara of the Dublin Lock Hospital for the treatment of venereal disease used to 'saturate' patients with arsenic in an attempt to cure them. He pointed out that James must have been suffering from gastroenteritis and not arsenical poisoning because blistering of the stomach would stop the vomiting of the former but not the latter. Professor Paul for the defence found that boiling acid could release arsenic from the glaze of pans, and that the fatal dose for a man of James's build would have been three grains (250 mg). Under cross-examination he admitted that he had never been involved in an arsenic poisoning case before this one. So it went on, with the defence really scraping the bottom of the barrel when it called witnesses such as John Thompson who was a wholesale druggist. He had been visited by James a few years before in connection with the employment of James's cousin, now deceased. The implication was that James had once had relatively easy access to arsenic, although there was no evidence that he had ever bought any from Thompson.

The defence also found a chemist from Bangor, North Wales, who appeared in the witness box to attest that people sometimes came to his shop to buy flypapers 'when there were no flies about'. Finally a James Bioletti, a women's hairdresser, took the stand and revealed that he used arsenic compounds to remove unwanted hair. Moreover on a few, admittedly rare, occasions he had been asked by his lady customers to use arsenic as a cosmetic. This evidence was called in support of Florence's contention that she bought arsenical flypapers for cosmetic purposes. However, she had never been one of Mr Bioletti's clients. He also told the court that he had once read in a newspaper that arsenic made hair grow.

Despite rubbishy evidence along these lines, Russell had been tunnelling away under the prosecution case and to some effect. His medical witnesses countered those of the prosecution's quite effectively. But then his tunnel caved in when Florence delivered her voluntary statement to the court. This had to be made without legal advice. Under the law as it then existed she could not be brought to the witness stand, but she could make a voluntary statement, and this she did. In it she admitted putting arsenic in the meat juice, although she said she had done this at James's request.

In his summing-up, Russell alluded to James's mistress, but all he says about her is 'another person, a woman, concerned in the matter'. He hinted at difficulties in getting defence witnesses. (The judge answered this in his summing-up by asking the very relevant question of why Florence's mother, or any of her friends who knew of her supposed habit of using arsenic as a cosmetic, did not appear to give evidence for her.) Russell said that if Florence were guilty why did she not get rid of the arsenic in her room? Then he made a legal *faux pas*. He uttered the astonishing remark that but for nanny Yapp's unethical opening of Florence's letter none of this would have come to light, and there would have been no trial. In other words, it was just hard luck that she had been found out.

Judge Stephens began his summing-up on the sixth day of the trial and completed it on the seventh. In the first half of his address to the jury he made a few good points but also a few blunders over dates. He openly admitted that the terms used in the scientific evidence meant little to him. Yet he did explain to the jury that arsenic may kill even though it is found in only small amounts after death.

The last day of the trial began with a continuation of the judge's summing-up, but the tone was entirely different. Whereas the day previously, Judge Stephens was fair and impartial towards Florence, the intervening night saw a marked change in his attitude towards her. To begin with, he treated her as on the day before, but as the temperature of that hot August day increased so did the heat in his words. He painted Florence as a scarlet lady. He assumed that the money-lender she met at Flatman's Hotel was also her lover. He referred to other letters that had been found in her room, which he obviously had read, and said that 'any woman having the least regards for her character and reputation would have burnt them'. He read out a letter which had not been produced in evidence, and he used this to show Florence to be a liar.

The judge then dealt with the fateful letter she wrote to Alfred Brierley, and showed the jury what a scheming, lying, adulteress she really was. (As indeed she was, but then James was also an adulterer, and violent as well, although that was never dwelt on at the trial.) The judge went on to read Florence's statement from his collection of newspaper cuttings of the trial, and he ended his summing-up in a blaze of vituperation which ran thus: 'For a person to go on deliberately administering poison to a poor, helpless, sick man upon whom she had already inflicted a dreadful injury – an injury fatal to married life – the person who could do such a thing as that must indeed be destitute of the least trace of human feeling. But I need not say more about it, and if I were to say much more it would be very easy to say more than that would be decent to say, and I should be engaged in an odious task.'

A little later he followed this with a further choice phrase: 'You must remember the intrigue which she carried on with this man Brierley, and the feelings – it seems horrible to comparatively ordinary innocent people – a horrible and incredible thought that a woman should be plotting the death of her husband in order that she might be left to follow her own degrading vices.'

After this it took the jury, which consisted mainly of skilled working class and lower middle class Lancashire men – who clearly knew a wicked woman when they saw one – only 38 minutes to reach their verdict of guilty – and of course their verdict was the right one. The judge passed sentence of death. Such was the public furore in the weeks following, that the sentence was commuted to life imprisonment. (Ironically Florence almost killed her-self 3 years later in Woking prison, to which she had been transferred. She nearly bled to death one night when she severed an artery in her vagina while obtaining blood that she was using to prove that she was suffering from TB.)

Despite requests made on her behalf at different times by three US Presidents, she served the full life-sentence of 15 years. After her release in 1904, she returned to America where she made a living writing and lecturing about her experiences. This phase lasted a few years but eventually public interest waned and she retired to the small hamlet of South Kent, Connecticut, in 1917. There she became a recluse, surrounded by a colony of inbred cats and finally died, dirty and neglected in 1941 at the age of 79.

It seems incredible that people should have thought her not only unjustly treated, but actually innocent. Five major facts told against Florence for which she could offer no satisfactory explanation:

1. The quantity of arsenic she had purchased and which was found in her bedroom.
2. The poisoned Valentine's meat juice which James would certainly have consumed had nurse Gore not been sufficiently alert.
3. The traces of arsenic in the jug and other utensils used for the last meal which James ate at his office.
4. The arsenic found in Dr Fuller's medicine which James drank.
5. Why Florence bought a second lot of 24 flypapers and what became of them.

It is true that the total amount of arsenic discovered in the organs removed from James's body was quite small but this is hardly surprising in view of the fact that the first large dose was taken 14 days before he died. This would have been eliminated in that time. James survived for three days after the final dose on that fateful Wednesday and it was chiefly the residue of that which accounted for the arsenic found in his liver. The body expels arsenic rapidly and after three days we would expect the arsenic to be less than a fatal dose and what was in the body was diffused throughout the tissues and no longer just concentrated in the liver.

Antimony

Antimony the great cure-all

For more technical information about the element antimony, consult the Glossary.

GRAM-FOR-GRAM, antimony is about as toxic as arsenic but on a dose-for-dose basis it is less life-threatening simply because antimony salts rapidly cause violent vomiting which expels most of the toxin from the body before it can be absorbed. This curious ability of antimony to trigger the muscles of the stomach to expel its contents generally prevented antimony's misuse as a murder weapon, but occasionally a large single dose did lead to death, as happened to Charles Bravo – see Chapter 10. The fatal dose can be as little as 120 mg, so long as the body retains it. Alternatively, it was possible to kill someone by giving them many small doses, which was George Chapman's way, as we shall discover in Chapter 11.

Antimony is not as widespread in nature as arsenic. It occurs to the extent of only 0.3 ppm of the Earth's crust, and only 0.3 ppb in sea water, values which are a tenth of those of arsenic. Consequently, the amount of antimony that gets into the food chain is correspondingly less. The amount of antimony released to the atmosphere each year is about 1,600 tonnes, most coming from the burning of coal, which contains 3 ppm on average, with metal smelters and municipal incinerators also releasing significant amounts.

Over the centuries the amount of antimony in the environment has increased, mainly in line with lead and copper production whose ores often contain it as an impurity. While the release of this element is of some concern the impact of antimony may have been underestimated. Professor William Shotyk of the University of Heidelberg, Germany, is an authority on antimony and his researches on peat samples taken from Swiss and Scottish bogs show that the level of antimony today is up to a thousand times higher than it was 5,000 years ago. Like lead, antimony has no biological role and indeed it is ten times more toxic, and like lead it is a cumulative poison. In Chapter 5, we saw how arsenic can pass through the gut wall into the blood stream and it can substitute for phosphate in metabolic processes, but this is not possible for antimony despite the fact that it resembles arsenic chemically. But while it may be harder for antimony to breach the body's defences, it tends to be more difficult to remove once it has.

Antimony in the human body

The background level of antimony in bodily fluids used to be almost undetectable but techniques have improved in recent years so that it is now measurable at parts per trillion. The level in urine is normally around 250 p.p.t. (which is equivalent to a mere 0.000000025%). Some organs of the body have relatively high amounts, such as the brain with 0.1 ppm, the hair with 0.7 ppm, the liver with 0.2 ppm, and the kidneys with 0.2 ppm, yet none of this poses a threat to health. In all these organs there is usually more antimony present than arsenic for the simple reason that arsenic is rapidly excreted from the body whereas antimony tends to linger. Tests on patients who were given antimony-based drugs to treat parasitic infections were found to retain the element for longer than expected. A typical total dose, given by a succession of injections over several days, would amount to around 500 mg of antimony and even after six months the level in urine was relatively high at 1 ppm, and after a year it was still registering 0.25 ppm.

The average daily intake of antimony in our food and drink is around 0.5 mg and it comes mainly from the traces of antimony in vegetables. The total body burden of antimony in the average person is around 2 mg and it is associated with the sulfur atoms that are present in proteins. Provided these sulfur atoms are not those located at the active sites of an enzyme then the antimony will not seriously interfere with the body's metabolism.

A far more deadly form of antimony is the gas stibine (antimony hydride, SbH_3) but this is rarely encountered outside certain industries, although in the 1990s it was thought to be responsible for cot death, as we shall see. Stibine can be formed when certain antimony alloys come in contact with strong acids. It is the most toxic form of antimony and causes headaches, nausea, and vomiting, and it can block the *anticholinesterase* enzymes which ensure the heart beats properly. Another source of industrial poisoning has been the solution of antimony trichloride in hydrochloric acid known as bronzing fluid, a name which describes it usefulness in imparting a bronze-like finish to other metals such as cast iron. The liquid is sometimes used in furniture polishes and in patent leather treatment. Deaths from drinking bronzing fluid have been reported but this has never been a suitable agent for perpetrating a murder because the liquid is corrosive and it would be impossible to administer it without the intended victim being aware of it.

When antimony enters the blood stream it first accumulates in the liver from where it transfers to other parts of the body. It is slowly excreted by the kidneys but this is not helped by the drop in urine production, which is a symptom of antimony poisoning. If a victim of antimony poisoning survives for 48 hours, then the prognosis is complete recovery, given moderate care and treatment. Post-mortem findings may reveal no sign of antimony poisoning beyond inflammation of the gastrointestinal tract.

The best test for antimony in Victorian times was the Marsh test which gave a metallic mirror, just as it did with arsenic. The antimony mirror was blacker and was deposited on the tube walls nearer the flame used to decompose the stibine gas that was evolved from the solution being analysed. Moreover, when the antimony deposit was heated in a stream of air, it was not converted to a volatile oxide as happens with arsenic, and this extension of the Marsh test served to distinguish between the two metals. Confirmation that it was antimony was obtained by passing hydrogen sulfide gas over the heated mirror which converted it to an orange-red deposit of antimony sulfide; the same treatment of an arsenic mirror gives a canary yellow deposit of arsenic sulfide. In the 20th century, analytical tests became much more sensitive as described in the Glossary.

Once antimony poisoning has been diagnosed, then treatment can begin and today there is little likelihood of a victim dying because there are chelating drugs – see Glossary – that will scavenge the antimony from the blood and organs and transport it from the body. Before the introduction of these drugs, the standard treatment had been to pump out the contents of the stomach, wash several times with water, and then encourage the patient to drink lots of fluids. This treatment too would often save a patient's life, and there are cases on record in which patients with blood antimony levels of 0.1 ppm – which is more than 60 times the normal level and a sign of severe poisoning – being successfully treated without resorting to chelating drugs.

Antimony in medicine

Antimony in one form or another has been used in the treatment of disease for more than 3,000 years. Egyptian papyri reveal that the naturally occurring mineral stibnite (black antimony sulfide) was given to patients to treat fevers and skin conditions. The Roman physician Dioscorides of Anazarba (Southern Turkey), who lived in the second half of the first century AD, was also familiar with stibnite, which he called *stibi* which he recommended for skin complaints, ulcers, and burns, and that it was to be applied as an ointment mixed with wax. He also noted that stibnite could be reduced to the metal by heating on coals and that it melted like lead. Another advocate of antimony sulfide was Kalaf ibn-Abbas al-Zahrawi, who was known in Europe as Abulcasis, and who died in 1013 AD. He was a distinguished doctor of Arabic Spain and was aware that antimony's use was approved of by the Prophet himself, who recommended it for the treatment of diseases of the eye such as ophthalmia.

Antimony metal appeared in two guises in the Middle Ages: emetic cups and perpetual pills, the former to cure a hangover and overindulgence, the latter to cure constipation. At the end of an evening of eating and drinking to excess, some wine was left in a special goblet made of antimony to be drunk the following day, whereupon it would soon provoke vomiting and empty the stomach. Perpetual pills were small balls of antimony which were swallowed. These would irritate the gut thereby promoting it to action to eject the irritant. The pill was then retrieved from the expelled excrement, washed, and stored for further use. There are reports that such pills were highly effective and passed from generation to generation.

Antimony compounds really came into vogue in the 1500s when they were used to treat all kinds of ailments, but not all doctors believed in them. The so-called Antimony War broke out in the 1600s between empiricists, who followed the teaching of Paracelsus which recommended their use, and the traditional Galenists, who clung to the older ways advocated by the ancient Roman physician Galen. Paracelsus praised the use of antimonals and his favourite remedy was *lilium*, which consisted of an alcoholic solution of the nitrates and tartrates of antimony (4 parts), tin (1 part), and copper (1 part). This he advocated as a treatment for many complaints.

As new antimony compounds were prepared, they were experimented with by the growing number of empiricists. Nevertheless, the Faculty of

Physicians of Paris condemned the use of antimony treatments and succeeded in getting them severely restricted by a law of 1566, an Act that remained on the statute books of France for a century. Some states in Germany were equally convinced antimony should be banned and from 1580 to 1655 all the graduates of the medical faculty of Heidelberg University were required to take an oath that they would never administer antimony (or mercury) to a patient.

Slowly the empiricists gained the upper hand and antimony medicaments were eventually being advocated by eminent doctors in the 1600s. They were supported by the writings of Oswald Croll (1580–1609) who wrote *Basilica Chymica* [Basilica of Chemistry], and this contained 23 recipes that used antimony. He was particularly impressed with antimony trichloride ($SbCl_3$) which has a butter-like texture and indeed was called butter of antimony.* In water it reacts to form insoluble antimony oxide chloride ($SbOCl$) and this was how it was prescribed, otherwise it would have been too dangerous to use, even externally. The virtues of antimony were also popularized in a book *The Triumphal Chariot of Antimony*, published in 1604. While many of its recipes are obscure, it is possible to deduce the chemicals being talking about, such as glass of antimony, which is antimony oxide sulfide, and fixed antimony, which is antimony nitrate.

By the end of the 1600s there were more than a hundred remedies based on antimony compounds. Some physicians who produced antimony-based powders gave their names to the product. Victor Algarotti, a physician of Verona, Italy, who died in 1603, produced one that came to be known in England as *Powder of Algaroth*, and this was prescribed as an emetic. It was antimony oxychloride and taking it produced instant vomiting, which was seen as a way of expelling bad 'humours' from the body. Other patent medicines became even more popular such as the Earl of Warwick's powder, which appeared in 1620. It was concocted by Robert Dudley (1574–1649), who was the son of the famous Earl of Leicester, Queen Elizabeth I's favourite. (The Earl of Leicester married Dudley's mother only two days before his birth and later abandoned her, even denying that he had married her.) On

---

* Antimony trichloride is a solid crystalline material at low temperatures, but at ordinary temperatures it becomes a paste-like mass resembling butter; it melts at 73°C to form a yellow liquid that fumes in air. Until recently, vets used a solution of antimony trichloride dissolved in a solvent like alcohol or chloroform to de-horn calves, although by law it had to be applied within a week of their being born.

the death of the Earl, Dudley attempted to establish his legitimacy and so claim the titles and estate of his father, and when he failed to do this he emigrated to Florence where he converted to Roman Catholicism. There King Ferdinand II created him Earl of Warwick and the Pope ratified his title. Dudley died in 1649.

The Earl of Warwick's powder became internationally famous when it was used with astounding success to cure the 19-year-old King of France, Louis XIV, who fell ill with typhoid fever at Calais in 1657. From that time onwards it never looked back. Its use may have been banned by the Faculty of Physicians of Paris but they could hold out no longer and the law against antimony was repealed in 1666. Even so, it was fraught with danger and brought the medical profession a certain amount of censure. The French playwright Molière was convinced that the death of his only son was due to its careless administration and this is probably the reason why he never missed a chance to poke fun at doctors in his plays. His work *La Malade Imaginaire* [The Imaginary Illness], about a hypochondriac and the efforts to cure him, was a natural vehicle for comedy at the expense of the medical profession. Ridiculed it might be, but the Earl of Warwick's powder continued to be popular. The London Pharmacopoeia of 1721 gave as its ingredients: 2 ounces of scammony (a gum resin that will act as a purgative), one ounce of diaphoretic antimony (probably antimony oxide*), and one ounce of cream of tartar (potassium hydrogen tartrate). This recipe would make more than a thousand doses of the powder.

Johann Glauber (1604–1670) did not reveal how he made a new antimony preparation that he called *kermes mineral*, and which he invented about 1651. It too was reputed to cure all kinds of illnesses and became particularly popular in France, where it was known as *poudre de Chartres* and used to treat not only fevers, but also more serious conditions like smallpox, dropsy, and syphilis. Glauber kept the process a secret during his lifetime, but it passed to a Dr de Chastenay when he died, and he in his turn confided it to a surgeon, known as La Ligerie, who in his turn entrusted it to a monk, Brother Simon, who used it to treat fellow monks with great success. La Ligerie eventually sold the formula for *kermes mineral* to King Louis XIV for an undisclosed but considerable sum. The great chemist Berzelius analysed

---

* Antimony has a common oxide, antimony trioxide (Sb_2O_3), and a rarer antimony tetroxide (Sb_2O_4), which can be formed when the ordinary oxide is heated strongly in air.

it many years later and deduced that it was about 40% antimony oxide and 60% antimony sulfide with small amounts of sodium sulfide depending on its method of preparation. It was still to be found in the US pharmacopoeia as late as 1910.

Glauber made his *kermes mineral* from the black ore stibnite. This was boiled with potash (potassium carbonate) to produce a red material, which he assumed to be a new substance, although we know that it was simply another form of antimony sulfide. It was not until a hundred years later that a chemist called Rose was to show that this chemical could exist in two forms with distinctive black and red colours. When it was precipitated from solution it was red, but would revert to the black form on heating. Glauber had found a way to convert stibnite to the red form, although in the process of doing so he had oxidized some of it.

Eventually these various medicines were eclipsed by a much better, and safer, medicament: James's Powder. This was patented in 1747 by Dr Robert James (1705–1776), a medic and lifelong friend of Dr Samuel Johnson, who produced the first dictionary of the English language in 1755. James's Powder was also known as fever powder and it was given to produce copious sweating and was to be part of the medical pharmacopoeia for 150 years. A dose of the powder would deliver 5–10 mg of antimony and this would have the desired effect. James's Powder was even prescribed for King George III. Samples of it were also recovered from the medical kit of a lost expedition to the Arctic, which set out in the early 1800s and a few of whose abandoned belongings were eventually discovered.

According to the patent, the powder was made by heating antimony metal to form the oxide and then treating this with animal oil, salt, and molten nitre (potassium nitrate). To safeguard his secret formula, James had in fact submitted a false patent specification. Chemical analysis later showed that his compound was a mixture of antimony oxide and calcium phosphate, and while the former may have been produced as the patent said, the latter was almost certainly added as calcined (burnt) bone, this being the only source of phosphate in the 1700s. An officially approved substitute, which appeared in pharmacopoeias, consisted of antimony sulfide or oxide and powdered hartshorn, but there were many who considered that this was a poor alternative to Dr James's original formulation and his powder continued to be manufactured in London well into the 1800s. The only active ingredient in James's Powder was the antimony but there had to be only a little of it to produce the desired effect – hence the need to dilute it with some inert material like calcium phosphate.

The antimony drug that outlasted all these various proprietary medicines was tartar emetic (potassium antimony tartrate), which was first described in 1631 by Adrian Mynsicht, physician to the Duke of Mecklenburg, although it may well have been used for many years under various guises before this. Glauber published a method of making it in 1648, and his recipe involved boiling three parts of argentine flowers of antimony (a form of antimony oxide) with four parts of cream of tartar (potassium hydrogen tartrate) for 1 hour, then filtering the solution and evaporating off most of the water. Crystals of antimony potassium tartrate would grow on cooling.

The efficacy of this drug was reaffirmed in the mid-1700s by two noted English medics: John Fothergill (1712–1780), a Yorkshireman, who became a rich and successful London practitioner, and John Huxham (1692–1768), who was a Fellow of the Royal Society. Huxham came from Devon and he won the prestigious Copley Medal of the Royal Society for his essay on antimony in 1755. In his treatise, he particularly recommended Antimony Wine which was obtained by infusing an ounce of the antimony oxide sulfide in 24 ounces of Madeira wine for 10 to 12 days. The resulting solution was recommended as an alterant (general tonic), attenuant (slimming aid), and diaphoretic (sweating agent), depending on the dose.

Other patent remedies containing antimony potassium tartrate were available under a variety of names, such as Dr J. Johnson's pills, Hind's sweating ball (used by vets), Mitchell's pills, and there were yet more products given French and German names, either because they came from there or were branded as such because of the known efficacy of antimonals from those countries.

Antimony *sodium* tartrate was preferred to the potassium salt because the latter tended to effloresce in dry air, in other words it lost its water component, and if it became completely dry then it was more difficult to dissolve it. The sodium salt did not suffer in this way and it also had another advantage in that it was less of an irritant to the gut. It never succeeded in ousting the potassium salt which was easy to manufacture from readily available ingredients and which crystallized well from solution. Tartar emetic came to be regarded almost as a cure-all, but was seen as particularly good for treating fevers. Nevertheless its use declined in favour of arsenical drugs, such as Dr Fowler's solution (see Chapter 5) and when aspirin became widely available in the 1890s, it became the drug of choice for reducing the high temperatures associated with fevers.

Generally the older antimony remedies fell out of favour in the late 1800s, no doubt due in part to the publicity surrounding the murders that were committed with it. Nevertheless, there was a revival of medical interest in antimony salts in the 1900s after it was discovered, in 1915, to be effective against parasitic infections such as schistosomiasis and trichinosis. Antimony potassium tartrate solution was given by injection in doses that would kill these organisms yet not poison the host body. Treatment with antimony potassium tartrate was generally successful, although patients often experienced the side effects normally associated with antimony poisoning. Occasionally some patients died because of their sensitivity to antimony and there is even a case on record of one patient dying within minutes of being given his first injection.

Drug research into safer antiparasitic antimony compounds came up with antimony sodium bis(pyrocatechol-2,4-disulfonate), which was given the generic name stibophen and marketed under various names such as Fouadin and Trimon. This drug, injected in 100 mg doses, was found to be effective against schistosomiasis, trichinosis, and trypanosomiasis. Side effects were much less likely than with antimony potassium tartrate and only one patient, out of 2,041 people treated with the drug, died as a result of antimony poisoning. Another commonly used drug was anthiolimine (antimony lithium thiomalate), which was injected in doses of 1 ml at intervals of 2 to 3 days, each injection delivering 10 mg of antimony. Slowly the dosage was increased to 4 ml, the object being eventually to administer between 40 and 60 mg of antimony at a time, the amount depending on the bodyweight of the patient.

Schistosomiasis, which is also known as bilharziasis, is the disease caused by small flat flukes of *Schistosoma* that infect the veins of various organs of the body. Some colonize the bladder, while others colonize the bowel, lungs, or the liver. Humans become infected by coming in contact with the eggs of the parasite and these are released from the intermediate hosts that they infect, namely aquatic or amphibious snails. One successful treatment was to inject stibophen in which antimony is in its lower oxidation state, i.e. antimony(III). Today the disease is treated with the drug praziquantel, which does not contain antimony and which requires only a single dose to be effective.

Drugs in which antimony is in its higher oxidation state, antimony(V), are given to treat another parasitic infection, leishmaniasis, in which the protozoa *Leishmania* invades the skin forming pimples that grow large and suppurate. Two million cases of this infection are reported every year, mainly of children, and indeed in some countries it is seen as a childhood disease. Antimony-based drugs are used to treat cases of particularly resistant infections, and to check the deadly form of leishmaniasis in which the liver, spleen, and lymph nodes are invaded, and swell alarmingly. Sand flies spread the disease and they often live in houses. These insects also infect animals and wildlife, making them reservoirs of the parasites, thereby making it impossible to break the cycle of human infection by pest control alone.

Antimony compounds in oxidation state III were originally used to treat patients with leishmaniasis but in the 1950s the preference was to use compounds in which the element was in oxidation state V, such as antimony sodium gluconate (pentostam) or meglumine antimonite (glucantim). However, the disease appears to be becoming resistant to these antimony treatments and the reason lies with the oxidation states. In the parasite cells, antimony(V) is reduced to antimony(III) and it is this form which kills them, but some *Leishmania* cannot reduce antimony(V) to antimony(III) and they thereby protect themselves.

As these uses show, conventional medicine still has need of antimony and, indeed, Professor Nina Ulrich of the University of Hanover, Germany, has recently forecast that its application is likely to increase in the years ahead. If this really does happen then the antimony will also have to be carefully monitored, because its effects really can be unpredictable. Antimosan, the potassium salt of stibophen, was introduced for the treatment of multiple sclerosis, but was not effective. In 1926, a 24-year-old woman suffering from multiple sclerosis was given a series of antimosan injections, and after 17 of these, spread over a period of two months, she appeared to be improving. But with the 18th injection she reacted badly, started vomiting and coughing up blood, then went into a coma from which she never recovered. Some antimony treatments will always be perfectly safe, however. It is still widely used in homoeopathic medicines but such is the dilution of the antimony solution that these are perfectly safe because almost no atoms of the element enter the patient's body.

Antimony has its uses

Antimony is not quite a true metal – chemists refer to it as a metalloid or semi-metal, meaning it has both metallic and non-metallic properties. In its metallic form it resembles lead with which it can be alloyed, and these alloys account for most of its modern usage, being the electrode plates in motor vehicle batteries. Antimony imparts strength and hardness and it is also added in small amounts to the brass used to cast bells. When lead-based pewter fell out of favour, it was replaced by an alloy of 89% tin, 7% antimony, 2% copper, and 2% bismuth. Other products made of lead such as bullets, lead shot, and cable sheathing, are still hardened with a few percent of antimony. The antimony in bullets can be used to identify them, as we shall see in the next chapter.

Babbitt metal was patented by Isaac Babbitt in 1839, although the term is now applied to a range of silver-white alloys with remarkable anti-friction properties. All contain antimony. They consist of relatively hard alloy crystals embedded in a matrix of softer metal and are ideal for machine bearings, so they are used in gas turbines, electric motors, and pumps. Some are lead-based and typically contain 15% antimony, plus small amounts of tin and arsenic, while others are tin-based and have around 7% antimony with small amounts of copper and lead. The alloy is cast on to the metal surface, or it can even be sprayed on by means of a flame arc gun. Babbitt metal is used when there is likely to be incomplete separation between a stationary surface and a moving part. The Babbitt metal quickly adapts its shape to the moving part so that there is less friction between the surfaces and the lubricating oil spreads evenly between the two metals. This leads to fewer hot spots and less likelihood of enough heat being generated to cause the metals to weld together.

Antimony compounds were once used in industry more widely than they are now, for example antimony potassium tartrate was employed in the tanning and textile industries as a mordant, which is the fixing agent needed for dyeing leather and fabrics. The antimony binds chemically to the surface of fibres and the dye binds chemically to the antimony. Nowadays aluminium in the form of aluminium potassium sulfate has replaced antimony as a mordant. Other antimony compounds are still used in certain types of glass and ceramics, pigments, and semiconductors such as gallium arsenide antimonide. Antimony sulfide is needed for medical scintigraphy equipment, which uses radioisotopes and a scintillation counter to get an image of a body organ.

Antimony is an important metal in the world economy. Annual production in 2003 was 140,000 tonnes, and 90% of this came from China, whose reserves are in excess of a million tonnes. The chief ores are stibnite and tetrahedrite (a copper, iron, antimony sulfide mineral) which yields antimony as a by-product. On the world market antimony costs about $1 per pound (which is around $2,400 per tonne). The US gets most of the antimony it needs from recycled lead acid batteries and it has a reserve of 80,000 tonnes of the metal. There is no mine production of antimony in the USA but there are deposits of antimony-rich minerals in Alaska, Idaho, Montana, and Nevada that could be used if needed.

Antimony oxide, Sb_2O_3, is the form in which antimony is added to plastics to act as a flame retardant, especially in car components, televisions, and cot mattresses. This use accounts for about two-thirds of all consumption. Antimony oxide quenches a fire by reacting chemically with the burning plastic to form a viscous layer that smothers the flame. And while this might serve to increase safety within the home, it came to be regarded as the agent that could bring sudden death.

Cot death and antimony

From the 1950s onwards most babies slept on foam mattresses with water-proof PVC covers. In 1988 it became a legal requirement to use additives to reduce the risk of fire from household furniture, and this meant that mattresses had to contain fire-retardants. In most cases this meant adding antimony oxide to the PVC covers. From the 1950s onwards, the incidence of cot death started to increase, and by the late 1980s about 23 babies in every 10,000 were dying for no apparent reason and from no detectable cause. Was there a connection between the introduction of PVC mattresses and the appearance of cot death? Two Englishmen thought there was, and put forward an hypothesis as to why this might be so: stibine gas was being emitted by the mattresses and poisoning the babies. They publicized their theory on radio and in a letter to a leading medical magazine – and so began a campaign by pressure groups that was to last almost 7 years, and which at various times had thousands of young parents discarding cot mattresses and phoning helplines in near panic.

The phrase 'cot death' – which has the emotive power of an oxymoron – was first used in 1954 by a pathologist, Dr A.M. Barrett, to describe the unexpected death of an apparently healthy infant. In 1969 a paediatrician in the USA, Dr J.B. Beckwith, gave cot death the more formal name of sudden infant death syndrome, which is often abbreviated to SIDS. Whatever it was called, it appeared to be becoming more common, and not only in the UK and the USA, but throughout Western Europe, Australia, and New Zealand. Strangely, there were almost no cases in other parts of the world, such as China, India, Africa, and Japan, although babies of Japanese parents in the USA were just as likely to die this way as those of other ethnic groups. It was assumed that cot death was linked to infant care practices or to something in the home environment.

In the 1970s and 1980s cases of SIDS became more and more common, until by the late 1980s it was accounting for around a third of all the deaths among babies under the age of 12 months. In England there were about 20 such cases a week, and these were deaths identified as SIDS rather than being attributed to natural causes, such as an unidentified infectious disease, or accidental suffocation. Something clearly must be causing them, but what? Various suggestions were put forward to explain this type of death, such as an extreme allergic reaction to cow's milk. Another was that babies were more at risk when the parents smoked and that this was somehow responsible. A few commentators with a mystical bent even thought that nearby power lines caused cot death.

In 1990 the hypothesis that the deaths were due to antimony poisoning by stibine hit the headlines in a big way, especially in the UK. According to proponents, the gas was being given off by the fungus *Scopulariopsis brevicaulis* breeding on the antimony oxide fire-retardant in the PVC of cot mattresses. This microbe flourishes in damp conditions as we saw in Chapter 6, and what *Scopulariopsis brevicaulis* could do in making arsenic volatile, it might also do with antimony. What we were witnessing was the antimony version of Gosio's disease. The fungus appeared to be breeding on urine-damped mattresses and giving off the equally deadly stibine gas.

The two men who came up with this hypothesis in 1988 were Barry Richardson, an independent materials consultant and director of Penarth Research International, in Guernsey, and a friend Peter Mitchell, who lived in Winchester, in the south of England. At first they believed that *arsenical* fire-retardants were being used to protect the PVC used to cover cot mattresses and that it was arsine that was causing SIDS. They obtained several such mattresses from various police forces around the country and began to test them to see if this gas was being emitted from them, but none was detected. Instead they thought they had discovered that stibine was being released. They detected and identified it by using dampened silver nitrate test papers, which turn black when exposed to this gas.

Other aspects of cot death seemed to fit the theory. The apparent increase in cot deaths coincided with the introduction of PVC-covered cot mattresses, and its occurrence was only likely to occur in those countries where these were used. In Japan, for example, babies were put to bed on cotton futons that were treated with borate fire-retardants, and in that country cot death was not a recognized problem. What really convinced Richardson and Mitchell that they had stumbled on the cause of cot death was the discovery that cot death babies had higher levels of antimony in their blood. They wrote about their hypothesis and their supporting scientific evidence in a paper which they sent to the *British Medical Journal*, but the editor refused to publish it because they had breached one of the journal's guidelines, that there should be no publicity prior to publication.

What Richardson had done was to contact a reporter on the BBC Radio 4 programme *You and Yours*, which went out at midday, and its producers had alerted the editors of the early morning programme *Today*, who were happy to carry the story and to interview Richardson. So it was on the morning of June 1989 that the British public learned about Richardson's belief that cot death was due to the chemical antimony oxide in cot mattresses. Richardson was finally able to publish his hypothesis in the *Lancet* in 1990 [vol. 335, page 670] in which he claimed that on all the baby mattresses he examined he could detect *Scopulariopsis brevicaulis* and he assumed the gas which this microbe was generating was stibine.

Newspapers took up the story and such was the concern they generated that the UK Government felt obliged to take action, which it did first by commissioning the Laboratory of the Government Chemist to confirm Richardson's findings. This they were unable to do, but that piece of negative evidence counted for naught in the face of media clamour for something to be done. So, on 9 March 1990, the Government announced an official inquiry which would be carried out by a group of experts headed by the late Professor Paul Turner of the prestigious St Bartholomew's Hospital in London. He was the chairman of the Government's advisory committee on toxicity.

The committee reported 14 months later, in May 1991, and concluded that there was no scientific evidence to support Richardson's hypothesis, although the Government did take on board one of Richardson's suggestions, which was that parents should not place their baby on its stomach when laying it down to sleep. He thought this would reduce the amount of stibine from a damp and infested mattress being inhaled by the baby. The Department of Health launched a 'Back to Sleep' campaign in December 1991 which advised parents to lay their babies on their backs. As a result of this advice the number of cot deaths, which were already declining from a peak of almost 1500 in 1988, began to fall even faster and by 1993 were down to around 420 cases. The issue of cot death still had a very high national profile, and became major news when Sebastian, the 4-month-old son of popular television presenter, Anne Diamond, was found dead in his cot one morning.

One of the investigators approached by the Turner committee was Joan Kelley who had been asked by the Department of Health to examine cot mattresses for microbes. She had inspected 50 such mattresses, on 19 of which a cot death had occurred. Altogether she found many kinds of microbes, including the much more dangerous *Aspergillus fumigatus* fungi, but only on three mattresses could she detect *Scopulariopsis brevicaulis*. She concluded that there was no evidence that this microfungus could be linked to SIDS.

Richardson and Mitchell were fast losing credibility by this time, but then on 17 November 1994, *The Cook Report*, a popular and influential television programme, devoted itself entirely to their hypothesis. A new media scare erupted and soon there was widespread alarm. The report was called 'The Cot Death Poisoning' and it vindicated Richardson's theory. A helpline set up by the TV company that broadcast the programme received 50,000 calls from worried parents, who began throwing away their cot mattresses. It did not seem to matter that Richardson's theory had been shown to be wrong and that leading experts were criticizing the programme not only for its alarmist tone but also for its lack of scientific content.

Nevertheless, post-mortem analyses were showing that cot death babies had higher than expected levels of antimony and these analyses were scientifically sound, having been carried out by respected scientist Dr Andrew Taylor, of the Robens Institute of Industrial and Environmental Health and Safety at the University of Surrey. He had analysed the blood of 37 SIDS babies and found antimony in 20 of them with an average level of 0.07 ppm. Among 15 babies who had died from other cases he found antimony in only one and this had less than 0.0005 ppm. This evidence appeared to indicate a significant link between antimony and risk of cot death.

A second edition of *The Cook Report* was broadcast 2 weeks later on 1 December, and this gave the new evidence that cot death victims had higher levels of antimony than other babies who had died of other illnesses. In the eyes of many it seemed to confirm that stibine really was responsible for cot death, and this had to be coming from the PVC covering of cot mattresses. Suspect mattresses were dumped in their thousands.

In response to *The Cook Report*, shops that sold such mattresses quickly removed them from their shelves despite the fact that several manufacturers of cot mattresses had already taken action in response to the first stibine scare in 1991, and removing all arsenic and antimony-based fire-retardants from their products. Some had even gone over to using polyurethane foam that contained no phosphate fire-retardant either. The UK Government's response to *The Cook Report* was to set up another investigation: the UK Expert Group on Cot Death Theories. This time the investigation would be even more thorough, and it was chaired by no less a person than Lady Sylvia Limerick, vice-chairperson of the Foundation for the Study of Infant Deaths. She would guarantee that all sides of the issue would be addressed and experimental research would be commissioned and funded via the Department of Health.

Other researchers were directed to look more closely at the stibine theory, and at London University's Birkbeck College, a microbiologist, Dr Jane Nicklin, and a chemist, Dr Mike Thompson, carried out a series of experiments. Nicklin cut up SIDS mattresses and placed samples in conditions that would allow *Scopulariopsis brevicaulis* to flourish and Thompson analysed the gases given off to see if stibine was evolved: it wasn't. Meanwhile the Scottish Cot Death Trust commissioned an investigation by researchers at the Royal Hospital for Sick Children in Glasgow and they found that the livers of SIDS babies had *lower* levels of antimony than those of infants who had died of other causes. At Liverpool University, Professor Dick Van Velsen analysed hair samples from cot death babies and found that the hair that had grown on babies while they were still in the womb had *more* antimony than that which grew after birth, which again seemed to disprove the theory that the antimony was coming from mattresses. An epidemiological survey carried out by Professor Peter Fleming of the Institute of Child Health at the Royal Hospital for Sick Children in Bristol, came to the conclusion that PVC covered mattresses were *less* likely to be associated with cot death.

In its 9 December 1995 edition, the *Lancet* carried a long article from a group headed by Dr David Warnock at the Public Health Laboratory in Bristol. The group had been asked by the Limerick committee to test PVC mattress covers, and of the 23 they tests they found most contained antimony oxide (generally between 0.7% and 1.5%, although one had 3%) although some contained only trace amounts. Richardson had willingly collaborated with this group and they had carefully repeated his analyses of PVC mattress covers, but they came up with very different results: the microbes that grew on the incubated PVC were not *Scopulariopsis brevicaulis* but a species of *Bacillus*, a common environmental bacteria, and the gas they gave off was not stibine but a sulfur compound, although it too turned the silver nitrate test papers black, just as Richardson had observed.

The *Lancet* in which these results were reported acknowledged that Richardson had tested his theory within the limits of his resources, but in 1989 he had been misled and misinterpreted what he had observed. The same issue also carried a letter from Mike Thompson who had carried out tests on nine mattress covers that contained antimony, and while he too could get fungi to grow on them, including a little *Scopulariopsis brevicaulis*, there was no indication that any stibine was emitted, and he had looked for this using advanced analytical techniques (inductively coupled plasma mass-spectroscopy) that would have detected minute amounts had it been present. In 1995, a BBC TV programme, QED, sponsored a repeat of the Richardson experiments and those of Thompson, and again no stibine could be detected.

The stibine issue was debated in the letter columns of the *Lancet*, with Richardson making claims that there was a link between SIDS and mattresses, while others said that this was not so. The media bandwagon rolled on, with some newspaper columnists firmly supporting the Richardson theory, and belittling those who criticized it. It was even unfairly suggested that the new committee set up by the Government, headed by Lady Limerick, had rejected the Richardson theory in advance and so could not be expected to deliver an independent verdict.

The issue refused to go away and newspapers continued to carry articles saying that stibine was the main cause of cot death. The book *The Cot Death*

Cover Up by Jim Sprott, a consultant chemist and forensic scientist, was published in New Zealand, although it was not a scientific work as such because it gave no references to the sources on which its author made his allegations. Sprott was also part of the Campaign Against Cot Death in that country, where BabeSafe mattresses and mattress covers were being sold that were free of all antimony, arsenic, and phosphorus compounds, and which he endorsed. His newsletter *Cot Life 2000* claimed that his warnings and advice about the dangers of stibine had led to a striking decrease in the incidence of cot death in New Zealand.

Researchers in the UK at De Montfort University, Leicester, led by Professor Peter Craig, were commissioned by the Limerick committee to carry out intensive research into *Scopulariopsis brevicaulis* and its ability to form volatile compounds with inorganic antimony. For the first time ever, they had found that this could generate a volatile product of antimony, but that it was trimethylstibine, not stibine. (Trimethylstibine has three methyl groups (CH_3) attached to an antimony atom.) However, when the antimony oxide was incorporated into the PVC of a cot mattress on which the microbe grew then not even this volatile compound was formed.

The mystery of where the antimony in babies was coming from was finally revealed in March 1997. The *Lancet* carried news of the research by Mike Thompson, done jointly with Ian Thornton, of Imperial College London, and which had been published in the *Environmental Technology* [vol. 18, p. 117]. It reported his analysis of samples of household dust collected at random from 100 homes in Birmingham (an industrial city), Brighton (a seaside town), Richmond (an outer suburb of London), and Westminster (inner London and the seat of Government). All showed unexpectedly high levels of antimony of between 10 and 20 parts per million, compared to the background level of antimony in the Earth's crust of only 0.2 ppm. (In a few homes, the levels of antimony in dust were in excess of 100 ppm and in one home in Birmingham it was an incredible 1800 ppm.) Thompson and Thornton pointed out that the level of antimony in the livers of cot death children was only 0.005 ppm, and that this amount could easily have come from the ingestion of household dust. The amount of dust a baby ingests is thought to be around 100 mg a day, which for most babies would give them around a microgram of antimony. The antimony in a home was correlated to the amount of lead in the dust, suggesting that this was where it was originating.

The news was carried by several newspapers, who questioned its reassuring message about antimony because this still appeared to conflict with other data which were given in the book by Sprott. Other results, however, were now appearing that tended to clear antimony of the charge of causing cot deaths. The analysis of the urine of 148 young babies at the Great Ormond Street Hospital for Children in London showed that antimony was even present in babies tested within 24 hours of birth when clearly cot mattresses could not be to blame. The antimony was almost unmeasurable but nevertheless it was detectable.

Then, in May 1998, the final report of the Limerick committee, which went under the name of *The Expert Group to Investigate Cot Death Theories: Toxic Gas Hypothesis*, was published. It refuted all the arguments that had been put forward regarding stibine and SIDS. It pointed out that cot mattresses were rarely contaminated with *Scopulariopsis brevicaulis*; that there was no evidence that stibine was generated by cot mattresses and that hundreds of experiments had been carried out to test whether this could occur. (This report did say that under laboratory conditions it was possible to convert antimony oxide to trimethylstibine.) It pointed out that there was no clinical evidence that any SIDS baby had died of stibine poisoning; that antimony was present in the majority of infants and was no higher in cot death babies than other babies; that the introduction of antimony oxide fire-retardants in mattress covers in 1988 had not led to an increase in cot deaths, nor had its removal in 1994 led to a decrease. All of these conclusions were backed up by experts in the field. The Limerick study had cost £500,000 and had taken 3 years to complete, but this was felt to be justified if it could show one way or the other whether stibine might be the cause of cot death.

Media coverage of the Limerick report was far less than that given to the earlier scare stories. Some newspapers interviewed those who had been on the firing line as a result of *The Cook Report*. They were scathing in their condemnation of the programme, calling it irresponsible and pointing out the unnecessary distress it had caused parents of babies and young children. One popular newspaper the *Sunday People* even suggested that Carlton Television, which produced it, should have its licence removed. Joyce Epstein, Secretary General of the Foundation for the Study of Infant Deaths, referring to the Limerick report, summed up the years of needless worry and misdirected effort by approving the findings of the Limerick report, saying that it brought to a close 'a ghastly episode in public health scare mongering'.

Despite the reassurance of the Limerick report, researchers continued to check whether SIDS babies really had raised levels of antimony in their bodies but the evidence was that they did not. A particularly intensive study of Irish cot death babies was organized by Professor T.G. Matthews at the Children's Hospital, Dublin. He analysed antimony levels in liver, brain, blood, and urine in all 52 cot death babies that died in Ireland in 1999 and compared these levels with those in babies who died of other causes. In all the babies the level of antimony in tissue was less than 0.01 ppm, and the same in both SIDS and non-SIDS babies. Antimony levels in blood were around 0.3 ppm, with SIDS babies being fractionally higher, and in urine levels were around 3 ppm, but with no difference between SIDS and non-SIDS babies. The conclusion was that there was no evidence to support the theory that SIDS was caused by antimony. Indeed there is still no simple explanation for many of the cases of cot death.

It is likely that the cot death scare will be the last example of public concern regarding the use of antimony. Although this element has a valuable contribution to make to certain materials, it need not be an important part of the domestic environment, and where it might be used it is unlikely to affect human health. Insofar as much of the antimony that is used is added to strengthen lead, then the limits now imposed on the use of this metal will likewise impose a similar limit on the use of antimony. If there is a future for antimony then it is likely to be in the form of new drugs to treat old diseases, but even in the fight against human parasites such are the safety requirements which all new drugs must undergo that it is unlikely that any that are introduced will cause harmful side effects.

Antimony will forever be a part of the world around us but at the end of the day it is a toxic element and nothing can ever alter that. What antimony can sometimes do is provide forensic evidence that would otherwise be unobtainable and thereby help solve a crime. It provided valuable insight into the murder of the most powerful man in the world, as we shall see in the next chapter. It even solved the mystery of how Brazilians were drinking much more Scotch whisky than the Scots were exporting to them.

Fake Scotch whisky

Whisky is very popular in Brazil and in the early 1970s the premium brand, Johnnie Walker, was selling well. Indeed it appeared to be selling well in excess of the amount being imported from Scotland and the obvious implication was that someone was taking a local brew and passing it off as the real thing. The police were informed and their enquiries eventually led to a warehouse where they found a large quantity of whisky which appeared to be genuine Johnnie Walker. What they also found was a suspicious quantity of screw tops, cork rings, and foil caps. Some of these, along with bottles of the suspect whisky were sent to the Radiochemistry Division of the Atomic Institute in Sao Paulo where they were handed to the head of the Institute Dr Fausto Lima.

Whisky contains many chemicals in tiny amounts, and these vary so widely from one batch to another, that it was going to be impossible to prove a whisky was authentic Scotch whisky by analysing for these ingredients. However, it was possible to compare the authenticity of whisky by analysing its minute trace of metals, and the ones that were most abundant in authentic Johnnie Walker were found to be lead, antimony, cobalt, and copper. Comparison of the suspect whisky with that of the authentic showed little difference between them: the authentic whisky had 1,000 parts per billion (ppb) of lead but so had the suspect whisky; the authentic whisky had 330 ppb of antimony, and the suspect whisky 340 ppb; the authentic had 320 ppb cobalt and the suspect had 300 ppb; and the authentic had 200 ppb copper while the suspect had 130 ppb. In other words there was little to choose between them. Perhaps the whisky was genuine after all, as its owner maintained.

The investigating forensic scientists then turned their attention to the foil caps used to seal the bottles and it was this that proved that suspect whisky was counterfeit. Whereas the bottle caps on genuine Johnnie Walker contained only 0.056% antimony, those on the suspect bottles contained 0.237% and the unused foil caps also seized in the raid on the whisky store also had this amount of antimony. The case was solved. The fake whisky was impounded and destroyed.

Requiem for antimony

NTIMONY in a corpse persists indefinitely, and unless a body was cremated, which in former times it rarely was, a murderer using antimony could never be certain that he or she would not one day be brought to account. However, that was a small risk to set aside the potential benefits, which could be large. And there were other benefits in choosing antimony as the murder weapon, not least that it was itself widely used in medical treatment.

Toxic antimony

Poisoners choosing antimony invariably selected tartar emetic (antimony potassium tartrate), and indeed its faint yellow crystals had two advantages. Firstly, they are very soluble in water and, while the solution has a faint metallic taste, this is easily masked by the presence of other flavours. Secondly, the compound was readily available, and all pharmacists stocked it and rarely queried its sale because it was widely used to treat sick animals. Moreover, tartar emetic was cheap; an ounce cost only 2 pence in 1897 (around 50p or $1 today). Pharmacists ordered it by the pound, which gives some indication of the demand for it.

In small doses of about 5 mg, antimony potassium tartrate acts as a diaphoretic, in other words it promotes sweating and will thereby lower the body's temperature. In larger doses of around 50 mg it acts as an emetic. Vomiting begins within 15 minutes and most of the stomach contents are expelled. Thus the poison acts as its own antidote to a certain extent, witness the man who recovered from a dose of around 25 grams (25,000 mg), corresponding to two teaspoonfuls of the crystals, which were accidentally taken in mistake for sodium bicarbonate. On the other hand, some have died after swallowing as little as 120 mg, although such sensitivity to the poison is extremely rare and it would normally require a dose of twice this amount to cause death, assuming it was retained by the body long enough for it to be absorbed. Some individuals are particularly sensitive to antimony, as the Balham Mystery will show, and this sensitivity may well explain the puzzling death of Mozart.

A large single dose of 500 mg or more of antimony potassium tartrate has been known to kill within a few hours, but individual reaction to antimony is so variable that it is impossible to say what constitutes a fatal dose. About 250 mg is thought to be the minimum amount to put life in jeopardy and a dose of 500–1,000 mg would certainly put a person's life at risk if vomiting were delayed for any length of time. Antimony in the body threatens to block the enzymes that are needed for the workings of the liver and kidneys, and particularly the heart muscle.

It is the variable sensitivity to antimony that makes its toxicity unpredictable. In 1854, a 16-year-old girl took 50 grains of tartar emetic (3,200 mg) after which she vomited copiously for 3 hours. She appeared to have recovered by the following morning although in the afternoon she suffered a relapse, soon became delirious, and died later that same night. In 1966, a 43-year-old man took a quarter of an ounce (7,000 mg) of tartar emetic by mistake, realized what he had done, walked a mile to see his doctor, but showed no symptoms of antimony poisoning. However, within the hour he began to vomit and was sent home to bed. The next morning he appeared to have recovered but the following day saw him in a state of collapse and he was given brandy, which seemed to revive him. A day later he died, although conscious to the last.

In the hot summer of 1928, lemonade that had been prepared in an enamel pail affected 70 employees of a Newcastle-upon-Tyne company, who experienced severe antimony poisoning after refreshing themselves with a drink that had been prepared the previous day from so-called 'fruit crystals'. These were mainly citric and tartaric acids and they dissolved antimony from the enamel of the pail, which contained about 3% of antimony oxide. A half pint glass of the drink was later shown to contain as much as 60 mg of antimony, a quantity capable of producing the symptoms from which the workers suffered: nausea, vomiting, stomach cramps, and fainting, yet all but two rapidly recovered and those two were only kept in hospital overnight for observation.

In the next chapter we will see how George Chapman's wives died as a result of being given repeated sub-lethal doses. The victim of several small doses begins to lose weight and waste away through loss of appetite and an inability to retain food. Great thirst is experienced. Death eventually results from a general collapse. A feature of this type of antimony murder was the prolonged and painful struggle for life which the victims, most of them women, put up as they reeled beneath repeated blows of the poison, fed to them in their food, drink, and medicines, often given to them by the very person they trusted most, and who daily played the part of being particularly solicitous of their welfare.

There is one early death which for two centuries has remained a mystery and which now seems likely to have been due to antimony. Whether it was accidental or deliberate, it nevertheless robbed the world of one of its greatest composers at the peak of his career.

The discordant death of Mozart

In the summer of 1791, Wolfgang Amadeus Mozart was approached by a stranger who offered him an extremely generous commission to write a requiem mass. It was this piece of music that Mozart was working on when he died a few months later, by which time he was obsessed by the idea that he was in fact composing it for his own death. As it turned out, he was. The unfinished requiem was completed by his protégé Süssmayr and performed in January 1793, although the motive for completing it was to enable Mozart's widow Constanze to collect payment for it. The man who had been employing Mozart to write the requiem was Count Franz von Walsegg-Stuppach, who commissioned it in memory of his wife and who intended to pass the work off as his own.

Mozart died of miliary fever at his home in Vienna at one o'clock in the morning of 5 December 1791. Miliary fever is no longer recognized as a medical condition; indeed it simply described a condition in which the patient had a raging fever and a skin rash. So what did kill the 35-year-old composer, then at the height of his powers? Over the years this question has generated answers ranging from murder by poison, to natural causes. A recent suggestion is that it was a result of his hypochondria and his predilection to taking patent medicines containing antimony. In his final days this was compounded by further prescriptions of antimony to relieve the fever he was clearly suffering from.

Other theories have been advanced to explain his death. The most recent one, suggesting he could simply have died of food poisoning, was published in 2001 in the *Archives of Internal Medicine* by Dr Gan V. Hirschmann of the Puget Sound Medical Center and the University of Washington in Seattle. His theory was that Mozart might well have died of trichinosis, the infection that occurs when people eat parasite-contaminated pork that has been undercooked. We know Mozart liked pork chops, and about 6 weeks before he became ill he wrote about such a meal in a letter to his wife. Trichinosis generally kills within 3 weeks and Mozart's symptoms of swelling, vomiting, fever, and rashes are consistent with this disease.

In 1966 a Swiss physician, Dr Carl Bar, suggested that Mozart died of the long-term effects of the rheumatic fever that he had contracted as a youth and that the treatments for this, which included bleeding, purging, and sweating, were responsible for his early death.

An Australian physician, Peter J. Davies, published several articles on Mozart's illnesses and death in the 1980s, and showed that it could have been one of many diseases, such as rheumatic fever, blood poisoning, pneumonia, smallpox, hepatitis, and a strep infection. He even suggested that his final illness was contracted during a performance on 18 November, when he conducted a specially written cantata at his Masonic Lodge. It was at that event that he picked up a kidney and urinary infection and from this he died.

Another theory traces his condition to a blow to the head caused when he had fallen over drunk the previous year, that this was responsible for the swelling on his head that Mozart's doctor observed and that his death was due to a blood clot on the brain. The Mozart museum in Salzburg claims to have Mozart's skull among its collection and this appears to confirm the fall and subsequent swelling.

Mozart became ill on 20 November 1791 and took to his bed, remaining there until he died 15 days later. He was seen by Dr Thomas Franz Closset and it was he who diagnosed a swelling of the brain, and also by his colleague Dr Matthias von Sallaba, whom he consulted when he began to fear for his patient's life. However, on Saturday 3 December, Mozart appeared to improve but this was to be a short remission of his symptoms and the following day they returned with a vengeance. Indeed Mozart knew that his end was approaching and on that Sunday he told his sister that death was already knocking on his door. That evening he took a turn

for the worse. Dr Closset was called and came at 11pm when he prescribed cold poultices for Mozart's raging fever, but to no avail, and
2 hours later the composer died. The following day the two doctors gave the
cause of death as miliary fever and this was entered in the official Register of
Deaths, and in the Book of the Dead at the nearby St Stephan's Cathedral
in Vienna.

Mozart was buried the next day and laid to rest in St Mark's cemetery.
Viennese law did not allow a body to be buried in a coffin, the object being to
save space in already overcrowded graveyards, so the composer was laid to
rest in an ordinary grave alongside other recently deceased corpses. The law
was designed to do away with pompous and costly burials and to curtail the
influence of the Church – graveside ceremonies were also banned – and by
burying bodies in groups it would speed their decay. Contrary to popular
opinion, Mozart was not thrown into a pauper's grave and the cost of burial
was not paid for by the local authorities. He was interred according to the law
and his funeral costs were met by his family and friends.* In April 1792 a
memorial service was held at the Freemasonry Lodge to which Mozart
belonged.

Within a few days of Mozart's death, however, others began to suspect
that he had been poisoned, and the fact that his corpse was reported to have
swelled up immediately he died was taken as sure evidence that this was so
and reported in the Prague newsletter *Musikalisches Wochenblatt* [Music
Weekly]. Coincidentally, the hero of Mozart's last opera, *La Clemenza di
Tito* [The Clemency of Titus] is the Roman Emperor Titus, who was also
poisoned.

At an international music festival in Brighton, England, in 1983 a committee was set up to investigate the death of Mozart and its members concluded
that he had in fact been poisoned. Various potential murderers were suggested, including Mozart's student Franz Xaver Süssmayr, who has been
suspected of having an affair with Mozart's wife Constanze, and who some
have suggested was the biological father of Mozart's son, Franz Xaver.

* I am indebted to Professor Dr Hans Gross of Berlin for bringing these facts to my
attention. He also points out that Mozart was far from being poor and that his annual
salary was the equivalent of around £30,000 today and that the publication of his works
he netted something of the order of £150,000 today. That he was often short of money,
and the fact that little of this wealth remained on his death, was due to his profligate
spending, heavy drinking, and gambling.

Another possible murderer was Franz Hofdemel, a court official and a brother Mason, his motive being that Mozart had already betrayed Masonic secrets in his opera, *Zauberflote* [The Magic Flute] and had to be silenced to prevent him betraying yet more lodge rituals. Significantly, and consistent with this theory, was the fact that Hofdemel committed suicide by cutting his throat the day after Mozart died. Indeed it has even been suggested that Mozart had had an affair with Hofdemel's wife, who was a pupil of his, and who was five months pregnant. She had been brutally attacked by Hofdemel before he slit his throat.

Another potential poisoner was Antonio Salieri, who was the Court Composer in Vienna, and who was known to be jealous of his younger rival Mozart. He had every reason to be; his work was soon to be eclipsed by that of Mozart. There was even a report in a German newspaper that during a performance of Mozart's *Don Juan*, Salieri began whistling aloud and ostentatiously walked out of the theatre. Yet there is other evidence that during Mozart's lifetime the two men were on good terms, even when they competed against one another. Indeed Mozart invited Salieri and his son Karl Thomas to attend a performance of *Zauberflote* on 13 October 1791 – and they did.

Salieri had a nervous breakdown in the autumn of 1823 and was committed to the Vienna General Hospital, where he confessed that he had murdered Mozart. At a performance of Beethoven's Ninth Symphony on 23 May 1824 a leaflet was handed to concertgoers containing a poem in which Salieri was depicted as Mozart's poisoner. Salieri even tried to commit suicide by slitting his throat, but was found in time and survived. He died a year later on 7 May 1825. At the time his confession was taken seriously, but few today believe that Salieri poisoned Mozart and put down his claim to the ramblings of senile dementia. Yet there may be more to Salieri's claim than this because he told his father confessor that he had murdered Mozart, and that he then wrote to his bishop to confirm that it was so. Those who claim to have seen this document say that it shows he was not babbling incoherently and that it is a sincere confession.

And what of the poisons that might have killed Mozart? If it was arsenic, as most people assumed, then it may have been given in the form of *Aqua Toffana*, which was reputed to be still available in Salieri's native land of Italy. Mozart himself suspected that he had been poisoned with this, and a month before he was taken ill he said as much to his wife Constanze. That was on 20 October 1791. Accidental poisoning by mercury is another theory, advanced in 1956 by two German doctors, Dieter Kerner and Gunter Duda, who reasoned that Mozart was taking this dangerous cure for syphilis. This seems unlikely because Mozart exhibited few of the symptoms of mercury poisoning, and there is no sign of the telltale spidery handwriting that is associated with this metal in the parts of the score of the *Requiem* that he was known to have composed in his final days.

If Mozart was poisoned then it was accidental and due to his lifestyle and his final days of medical treatment, or so Dr Ian James, a clinical pharmacologist at London's Royal Free Hospital, believes. He put forward this theory in 1991 on the 200th anniversary of the composer's death. When Mozart's final illness began, his doctor first prescribed an antimony powder to treat what he diagnosed as melancholia and which we would see as severe depression. There were good reasons why Mozart would have been depressed. He had lost a lot of money gambling, he was overworking, and he had seen *La Clemenza di Tito* mauled by the critics when it was premiered at Prague that autumn (although the public enjoyed it and it became popular). His debts also included large bills for medicines from local pharmacies in Vienna, although we do not know what he bought, but he might well have been taking antimony-based medicine because this was prescribed for melancholia in those days. There were signs that he suffered a depressive illness, and indeed he was a hypochondriac, regularly dosing himself with a variety of medicines, so much so that he had run up a drugs bill with the apothecaries of Vienna to the tune of £2,000 ($3,000) in today's money.

Ian James believes that Mozart's illness was in fact antimony poisoning since the symptoms fit those that we know he suffered in his final days and hours: violent vomiting, and swelling of his hands, feet, and stomach. He gave off an offensive body and mouth odour. It may be that miliary fever really was a form of antimony poisoning. It started with stiffness of the limbs and a general weakness of the muscles. The patient ran a high fever and was acutely depressed. After a few days a rash appeared on the neck and breast and consisted of raised red spots that eventually turned yellow to the extent that it could be confused with scarlet fever, although doctors were aware of the difference. In some cases of antimony poisoning such a pustular eruption of the skin was observed. Miliary fever generally lasted 7 and 14 days and was not always fatal.

When Dr James consulted the official Pharmacopoeia issued by the Viennese authorities he discovered that the standard treatment for miliary fever involved both mercury and antimony medicaments. Antimony was given to reduce fevers and mercury to act as a laxative. He concludes that Mozart's symptoms were consistent with antimony poisoning, beginning with his being treated for depression and then being exacerbated by more antimony being given to treat his miliary fever. His symptoms of fatigue, renal damage (leading to swelling of face, hands, and feet), his final pneumonia-like illness, and even his pustular rash were consistent with antimony poisoning. It may well be that Mozart was one of those individuals susceptible to antimony poisoning, in which case small doses would have disproportionately severe effects.

Mozart may not have been murdered with antimony but others certainly were. Murder by antimony was so risky a venture that few attempted it, and those who did were generally medical men. The two poisoners who hit the headlines: were Dr Palmer (1855) and Dr Pritchard (1865). As such they ran the least risk of discovery and were sufficiently knowledgeable to control the doses of antimony potassium tartrate to simulate natural illness. Even George Chapman, whose murders we shall examine in more detail in the next chapter, had some medical training.

The successful antimony poisoner needs to persist with frequent doses of the poison until the victim has absorbed enough to kill him or her. Like arsenic, the first symptoms of too much antimony are those of gastroenteritis, and easily mistaken for food poisoning or a viral infection. The poisoner relies on this resemblance. However, the victim needs to be plied with further amounts of the poison and each time the symptoms recur, they are passed off as a relapse. Soon the victim begins to weaken physically and waste away. Even this may pass unrecognized by other doctors as antimony poisoning, and one of Chapman's victims was even registered as dying of tuberculosis, so emaciated had she become.

Dr William Palmer (1824–1856)

As a result of William Palmer's murderous schemes, an Act of Parliament was passed that was commonly referred to as Palmer's Act. It made it unlawful to take out an insurance policy on someone else's life, unless the death of that person would cause the insurer financial loss. In other words it was no longer legal to do what Palmer had done, which was to insure third parties, some of whom were unaware they were being insured, and then poison them in order to collect the insurance. What was unusual about Palmer was that he employed a variety of poisons, but the one that finally brought him to his public execution was antimony.

William Palmer was born on 6 August 1824, the son of a very rich timber merchant and his wife who lived in the town of Rugeley in the English Midlands. Palmer's father was keen for his son to enter a profession and apprenticed him to a pharmacist in Liverpool, but there he stole money from his master and was sent back to Rugeley after only a year or so, during which time his father had died leaving a fortune of £70,000, making him the equivalent of a multi-millionaire today. His mother now apprenticed William to a local doctor, Dr Tylecote but that situation did not last and he was dismissed because of his sexual improprieties. His mother next enrolled him as a student at the Stafford Infirmary but that ended in disgrace when he challenged a 27-year-old plumber, George Abley, to a brandy-drinking contest which resulted in the man's death from alcoholic poisoning.

Palmer went back to Rugeley but was not there for long before his mother persuaded him to register as a student at the famous St Bartholomew's Hospital in London. For 2 years he kept out of trouble and ended with a medical diploma which qualified him to practise as a doctor, and this is what he did when he returned to Rugeley in 1846. Soon he was courting Annie Brookes, the illegitimate daughter of a wealthy ex-officer of the Indian Army whose late father had provided generously for her in a will that left a considerable amount of money in trust, money that had previously been looked after by nominees, but was now handed over to Palmer after they were married. Most of it was to be spent in drinking and gambling. Yet despite his dissolute life-style Palmer was not without friends and he regularly went to church with his wife and took Holy Communion. Today he would no doubt the thought of as a likeable rogue. And he was not unattractive to, as well as being attracted by, the opposite sex.

It is still a matter of debate how many people Palmer poisoned, and there are even apologists who think that he has been much maligned. He almost certainly disposed of ten of the illegitimate children that he fathered by various women across the county of Staffordshire. In some cases he invited the child to come and stay with him in Rugeley whence the child would invariably be taken ill and die. What poison he used we will never know but he kept buying various toxic compounds from his local pharmacy, particularly strychnine and potassium antimony tartrate. It was the latter of these with which he disposed of his mother-in-law.

Palmer believed that his father-in-law Colonel Brookes had left a great deal of money in trust to his wife and that this would come to Annie should her mother die. Nor would the cantankerous old lady be missed; she was often drunk and lived in a house overrun with in-bred cats. On 6 January 1849 she was found unconscious due to too much gin and Palmer had her brought to his home where she died 12 days later, aged 50, after a brief illness in which she displayed all the symptoms of antimony poisoning. Her death certificate was signed by an 80-year-old local doctor, Dr Bamford, whom Palmer often turned to for help in such matters and who was grateful for the small gratuity that Palmer paid him for performing this service. Unluckily for Palmer his gamble failed and he discovered that Colonel Brookes's will was under the control of the Chancery Court and that the members of that legal body decided that ownership of the nine houses that formed the bulk of the colonel's estate should pass to another of the colonel's heirs who was legitimate. Annie got nothing.

Palmer next poisoned 45-year-old Leonard Bladon to whom he had lost several hundred pounds while they were at the races. He too was staying with Palmer when he conveniently took ill and suddenly died on 10 May 1850, ostensibly of an abscess of the pelvis from which he had been suffering. The notebook in which he had recorded Palmer's debts was never found. Despite this piece of good fortune, Palmer's debts with other bookmakers continued to mount reaching many thousands of pounds, but then he devised a plan that would not only pay them all off but leave a little over. In April 1854 he insured his wife with the Prince of Wales Insurance Company for £13,000 (equivalent to £250,000 today), paying a premium of £760. Six months later he poisoned her with antimony and collected the insurance.

Annie caught a cold while attending a concert at Liverpool's St George's Hall on 18 September, and when she returned to Rugeley she took to her bed, but soon started vomiting. Dr Bamford was called to attend her, diagnosed English cholera and when she died on 29 September, aged only 27, he signed the death certificate to this effect. There are those who think she might have committed suicide, depressed by the fact that four of her five children were already dead. There are others who think she was poisoned by Palmer and that he was also responsible for her babies dying. Baby Elizabeth died in January 1851 aged 10 weeks, baby Henry died exactly a year later aged 4, baby Frank died after only 4 hours, and baby John died after four days, and all died of 'convulsions'. (Popular lore had it that Palmer dipped his finger into poisoned honey and allowed the baby to suck it.) Although the insurance company suggested that an official inquiry into Annie's death should be held, they did not press the point and paid up. Not that it solved anything for Palmer because he was soon in debt again as he continued gambling recklessly.

If £13,000 had been a nice windfall, £82,000 would be even nicer – which is what Palmer next tried to insure his brother Walter for. No insurance company was willing to issue such a policy and the most he was able to insure his brother's life for was £13,000 and then only with the help of one of the men to whom he was deeply in debt, who would benefit should the ailing Walter die, and indeed there was every likelihood that might happen because Walter was an alcoholic. A policy was taken out in January 1865 and in August that year Walter dutifully died, but the insurance company refused to pay up. Palmer was outraged but before he could press the matter through the courts he was himself arrested and charged with the murder of someone else.

The man he murdered was 28-year-old John Parsons Cook, a racehorse owner whose mount, Polestar, won him around £2,000 at the Shrewsbury Races on 13 November 1865. On the following Monday, 19 November, Cook collected his winnings and that same night he was taken ill at the Raven Hotel where he and his new found friend Palmer were staying. A woman guest at the hotel had seen Palmer with a glass of brandy and water in the hotel foyer and he appeared to be dissolving something in it. That something was potassium antimony tartrate because the effect of the drink was to make Cook vomit copiously. The next day he felt well enough to move to the Talbot Arms in Rugeley, a pub opposite Palmer's home. There Palmer was able to doctor a cup of coffee which again made Cook very sick. Over the next few days Palmer visited Cook at the pub seemingly attending to his friend and on one occasion he send him some broth which not only made Cook ill, but also a maid who ate some of it. Cook finally died a day or so later, probably of strychnine poisoning, whereupon his money and little black book of betting transactions and debts had mysteriously disappeared.

Cook's step-father was soon on the scene and, his suspicions aroused, he demanded a post-mortem. Palmer was delighted when no strychnine was detected but there was now enough doubt about the deaths of others that the bodies of his wife and mother-in-law were exhumed and indeed antimony was detected in their remains. His brother Walter's body yielded nothing, suggesting that he died of strychnine which does not survive for long in the body.

Palmer was arrested and so strong was local feeling against him that his trial was transferred to London. Not that it saved him. The jury found him guilty and he was returned to Stafford for execution and hanged there at 8am on Saturday 14 June 1856 before a crowd of 30,000. Such was Palmer's notoriety that his waxwork stood in the Chamber of Horrors at Madame Tussaud's in London for 127 years.

Dr Edward Pritchard (1825–1865)

Edward William Pritchard was born in Southsea in Hampshire in 1825, the son of a captain in the Royal Navy. After an apprenticeship with two surgeons in Portsmouth he was admitted to the Royal College of Surgeons in 1846 and obtained a post as an assistant surgeon, also in the Royal Navy. In 1860 he fell in love with a Mary Jane Taylor and they were married that year in Edinburgh. Although Pritchard liked his life in the navy, his wife wanted him to settle down on-shore and she persuaded her parents to purchase a doctor's practice for her husband. This they did in the little town of Hunmanby near the seaside resort of Bridlington in Yorkshire. Pritchard soon became well-known there, partly through his articles in local newspapers, but he began to live beyond his means and was soon in debt, so much so that after 6 years he was forced to sell his practice to pay off his debts.

Pritchard then got a job as the private medical attendant to a wealthy old man who wanted someone to accompany him on his lengthy travels to foreign parts. Meanwhile Mrs Pritchard went to live with her parents, who were eventually persuaded to find the money for another practice for their son-in-law, this time in Glasgow at 11 Berkeley Terrace. Again he made himself popular, giving public lectures, joining the Freemasons and the Glasgow Atheneum Club of which he eventually became a director. His fellow doctors in that city shunned his company and disapproved of his publicity seeking, which manifested itself as lectures about his travels and adventures. An advert for one of them boasted: 'I have plucked the eaglets from their eyries in the deserts of North Arabia and hunted the Nubian lion in the prairies of North America'. Perhaps in those innocent days it was not thought impossible for eaglets to be snatched from their nests in Arabia and lions, Nubian, or otherwise, to be hunted in America. And they may even have been impressed by his walking stick which was engraved 'To William Edward Pritchard from his friend General Garibaldi', although they were not to know that he had never met the great Italian leader.

As was normal in the homes of professional people in Victorian times, the Pritchards employed a live-in maid and she was an attractive young woman, Elizabeth McGirn. Whether she had illicit sexual relations with the doctor can never be proved but she died in mysterious circumstances on the night of 6 May 1863. She was burned to death in her bed on the attic floor. A passing policeman knocked on the door of 11 Berkeley Terrace to tell the occupants that he could see flames and smoke coming from an attic window and when Pritchard answered he said that he was aware that there was a fire but he had been unable to arouse Elizabeth even though he tried to get into her room. The policeman was also beaten back by the flames and sent for the fire brigade. They soon had the fire out and then it was discovered that Elizabeth was in her bed, her body partly burnt. The fire had started near the bed head and it seemed that a candle had set fire to the bedding and that she was so overcome by smoke, that she had been unable to save herself. It is more than likely that she had been drugged unconscious and may even have been pregnant. Pritchard submitted an exaggerated fire insurance claim, saying valuable jewellery had been lost in the fire, but he was only compensated for fire damage to the room.

Soon after the fire, the Pritchard family moved first to rented accommodation in Royal Crescent and then to a house in Sauchiehall Street, which he bought with the help of a £400 deposit given by his mother-in-law plus a £1,600 mortgage. The new housemaid was young Mary McLeod who also became Dr Pritchard's mistress. Mrs Pritchard even caught them kissing one day, although she was persuaded that it was merely a passing impulse of the doctor's and so let Mary stay on. Pritchard, however, had already told Mary that were his wife to die then he would marry her. To bring this about he bought an ounce of antimony potassium tartrate at a local pharmacy on 16 November 1864 and this he fed in small doses to Mrs Pritchard over the next three months.

Her first bouts of vomiting were said by Pritchard to be a chill on the liver, but he hinted that some deeper underlying cause might really be responsible. There were brief periods when she seemed to improve but these never lasted long and throughout that winter she got progressively worse. She had a particularly severe attack of vomiting with pains in the stomach and cramps in her legs in the early evening of 1 February 1865. Pritchard feigned puzzlement at his wife's condition and called in a Dr J. M. Cowan of Edinburgh for a second opinion. He naturally failed to diagnose antimony poisoning and suggested a stomach upset as the cause of her illness, prescribing rest and iced champagne as the best cure. He even suggested that Mrs Pritchard's mother be asked to come and look after her daughter. On the day of Cowan's visit Pritchard bought another ounce of potassium antimony tartrate. When Dr Cowan's treatment seemed not to work, Mrs Pritchard asked to see another doctor and a Dr Gairdner who lived nearby, was summoned. He blamed the continuing attacks of vomiting, stomach pains, and muscle cramps on the iced champagne and said that no stimulants should be given to Mrs Pritchard. A light diet and rest was all that she needed.

On 10 February Pritchard's mother-in-law came to look after her daughter, but within two days of her arrival her daughter's condition became worse. Soon Mrs Taylor also began to be afflicted with the same symptoms of vomiting and stomach pains and yet another doctor, Dr Paterson, was called to attend her on 4 March. He did not go along with Pritchard's diagnosis that the old lady was suffering as a consequence of her long addiction to drink and opium, but he did opine that she was dying and indeed she did die the following day. At a later date some of the medicine that Dr Paterson had prescribed was analysed and found to contain antimony.

Paterson also had noted that Pritchard's wife seemed very ill and he even suspected this was being caused by an irritant poison like antimony, although he said nothing to Pritchard himself, who had also asked the new doctor to attend to his ailing wife. Pritchard had now decided that his long-suffering wife could be sent to join her mother and on 15 March he gave her an egg-flip (raw egg, sugar, and whisky) that made her seriously ill. She died at one o'clock in the morning of 18 March 1865, despite the administrations of Dr Paterson. Her death might well have been attributed to natural causes had not an anonymous letter been sent to the Procurator Fiscal, the head of the legal system in Scotland. The writer, who signed themselves *Amor*

Justitae (lover of justice) said that both Mrs Taylor and Mrs Pritchard had died under suspicious circumstance.

A post-mortem examination of the latter's body revealed large quantities of antimony, and when the former lady's body had been exhumed her remains likewise were found to contain antimony. Pritchard was arrested and tried for murder, found guilty, and condemned to be hanged. (He confessed to the murders while awaiting execution.) A crowd estimated to be around 100,000 turned out to watch. His death was not instantaneous and the hangman had to pull on his legs before all signs of life were extinguished. Although buried in an unmarked grave, Pritchard's skeleton was unearthed many years later when the foundations for a new Glasgow Court Building were being excavated. The clue to the skeleton's original identity was a pair of elastic sided boots that he was known to have worn on the day he was executed and which were still in recognizable condition.

Pritchard's motive for murdering his wife and mother-in-law is not clear and, while he was in debt to the bank, he was not being pressed for payment. Maybe he really was in love with Mary McLeod.

Florence Bravo (1846–1878)

Antimony potassium tartrate was also responsible for the death of 30-year-old barrister Charles Bravo of Balham in 1876, and he appears to be a rare example of someone dying as a result of a single dose. There was an earlier example of such a death when, in 1856, a Mrs McMullen had given her husband a 'quietness' powder of potassium antimony tartrate, ostensibly to treat his liking for drink but more likely to quell his sexual demands. He died and she was convicted of manslaughter. The Bravo affair may have had its origins in a similar motive.

Charles lived with his 29-year-old wife Florence in an imposing house, The Priory, on Bedford Hill Road, Balham, then on the fringes of south London. They had been married only five months when Charles was poisoned. He was Florence's second husband; her first had been Captain Alexander Ricardo of the prestigious Grenadier Guards regiment and he had died of alcoholism in 1871 leaving his 25-year-old wife with a fortune of £40,000 (equivalent to several million pounds today). During the course of that marriage, Florence had taken to heavy drinking and had even been on the water cure at Malvern in 1870 where she had been personally treated by the famous doctor James Manby Gully (1808–1883) who was 62.

Gully had set up his water cure institution in 1842 with his partner James Wilson, and it had been particularly successful in treating the rich for nervous disorders. While she was at Malvern, Florence fell in love with the doctor and following the death of her husband she became his mistress. Such was the strength of their attachment that he eventually moved to Balham to live near her at a house only 5 minutes' walk away. The two of them spent several holidays together on the continent and it was even possible that she became pregnant by him and that he performed an abortion on her. This affair with Gully caused Florence to be rejected by her father and it was in order to be accepted back into the family circle, that Florence married the eminently acceptable Charles Bravo. He was respectable even though he had a mistress and she had borne a child by him – such was the accepted morality of the times. Florence and Charles were frank enough to tell each other of these relationships and Florence agreed never to see Dr Gully again while Charles agreed to give up his mistress.

The marriage on 3 December 1875 of Florence to Charles Bravo united two wealthy families, the Bravos of Kensington Place and the Campbells* of Buscot Park, Berkshire. It brought together an odd couple, marrying for the wrong reasons, she to enable her to move respectably in society, he for her money. He admitted this to an acquaintance who congratulated him on his engagement to Florence the previous year: 'damn your congratulations. I only want the money!' was how this was reported at the inquiry into his death. Soon both of them regretted their decision to marry. He sometimes resorted to violence and to have engaged in 'unnatural practices' with Florence, which appear to have been anal intercourse.

On the evening of Tuesday 18 April 1876, Charles, Florence, and Florence's long-standing companion Mrs Jane Cox ate dinner together. Mrs Cox was 43-years-old when she was taken on as a paid companion to Florence after the death of Captain Ricardo and by a curious coincidence she already knew Charles Bravo's father whom she had met in Jamaica many years earlier. That evening Florence drank herself into a stupor on sherry and marsala wine, both of high alcoholic content. Indeed the two women managed to consume a bottle of each that evening, and they went to bed early. At

* Florence was born Florence Campbell.

9.15 pm Charles retired to his own bed in a separate room – Florence was recovering from a miscarriage – but before getting into bed he took a large drink from the carafe of water that was always placed on his bedside table. That water had been poisoned with antimony potassium tartrate and as a result Charles died a few days later. The 'Balham Mystery', as it was called and indeed as it has remained, centred on the question of who poisoned the carafe. Was it his wife Florence? Or could it have been the faithful Mrs Cox?

By 9.30pm that fateful evening Charles was already vomiting and writhing in agony. His shouts failed to waken Florence but Mrs Cox heard him and found him retching out of the window on to the roof below (from where samples of vomit were later collected for analysis). She sent one of the domestic staff to get a local doctor who came straight away. He found Charles unconscious and with his heartbeat almost undetectable. A Dr George Johnson of King's College Hospital was also sent for and he arrived to find Charles again vomiting violently and bringing up blood. When he asked him what he had taken, the only answer he got was that he'd earlier had some laudanum mainly to ease his pain of a fall he'd had earlier in the day, and the bottle of laudanum was there in the bedroom. Over the next two days the doctors proved powerless to prevent Charles's condition deteriorating and he died on Friday 21 April, 55 hours after drinking the water from the carafe.

It was estimated that Charles had taken a dose of between 20 and 40 grains of potassium antimony tartrate, which is around 1–2 grams, or about a small teaspoonful. At the inquest, which was obligingly held at The Priory, Mrs Cox tried to persuade the jury that it was a case of suicide, but they returned an open verdict, of death by tartar emetic poisoning. Forensic analysis had established this as the cause of death and although none of the water in the carafe was available for analysis, the investigators were able to collect and analyse a sample of vomit from the roofing outside the bedroom window and this clearly revealed antimony to be present. The coroner's verdict was far from being the end of the matter. It was clearly not a verdict to satisfy Charles's parents and in the weeks that followed there was so much rumour and speculation that the Lord Chief Justice quashed the verdict of the inquest jury and ordered another inquest. This one was held at the Bedford Hotel, Balham on 11 July and was in all but name the trial of Florence on the charge of murdering her husband.

Charles's death could have been an accident or even suicide, although that seems unlikely. It came to light later that Charles had bought a patent cure for alcoholism for Florence and that this consisted of powders, each containing 35 mg of tartar emetic, enough that when added to a drink would produce the desire effect of causing the drinker to vomit. The idea behind the cure was what later became to be known as aversion therapy. After a few such doctored drinks, the body would associate alcohol with sickness so that the mere thought of taking a drink would provoke a feeling of nausea. The cure did not work for Florence, indeed she probably never tried any of the powders, but she eventually died of alcoholism in 1878.

If the anti-drink powders were the source of the poison, then someone must have added the contents of at least 30 such powders to Charles's bedside carafe that evening. That was not really feasible. What was more likely was that the tartar emetic came from the stables, where their coachman, Griffiths, who had previously worked for Gully, occasionally used it to de-worm the horses. He was away in Kent at the time of the murder, but he had previously been heard to say when drinking in Balham that Mr Bravo 'would be dead in a few months' although he did not say why. Either Florence, or her devoted companion Mrs Cox, could easily have taken tartar emetic from the stables and put a teaspoonful of this into the carafe of water in Charles's bedroom. Perhaps it was intended simply to 'quieten' him.

In their five months of marriage, Florence had twice become pregnant and twice miscarried, the last time being only 2 weeks before Charles died. Unhappy, depressed by her condition, and finding release in alcohol, Florence no doubt thought that all her troubles stemmed from her hasty marriage to Charles. She appeared to have a motive for murder, but her faithful companion tried to deflect suspicion away from her. At the first inquest Mrs Cox testified that the dying man had confessed to having deliberately taken the poison saying: 'Mrs Cox, I have taken poison . . . don't tell Florence.' At the second inquest she consented to fill in the missing '. . .' and now said that Charles had really spoken the words: 'Mrs Cox, I have taken poison for Dr Gully. Don't tell Florence.'

The first inquest had found that Charles had died of antimony poisoning but that there was insufficient evidence to show how it came to be in his body. The second inquest jury were to learn a lot more about the people involved from various witnesses, including Dr Gully, but they still could come to no definite conclusion. They rejected Mrs Cox's evidence that Charles had committed suicide, and they returned a verdict that he had been 'wilfully murdered but there was insufficient evidence to fix the guilt on any person or persons'. And that was that as far as the law was concerned, although throughout the summer of 1876 the newspapers delighted their readers with details of the 'Balham Mystery'. As far as true crime writers were concerned it continued to fascinate them for the next century – and it continues to do so. Murder in high society is not something that is encountered every day and there are enough combinations of suspects and motives to ensure many theories as to what happened that night.

The most likely explanation is that Charles had taken laudanum earlier that day to relieve the pain from a riding accident that he had had, when his horse had bolted. He had returned to The Priory that afternoon dishevelled and limping. Maybe the discomfort was such that he drank too heavily that evening. It appears to have been his custom to drink a large glass of water before he went to sleep, this being the traditional way of avoiding a hang-over.* Another possibility was that he planned to take an emetic to empty his stomach before going to sleep, and had added too much tartar emetic to the water, but this was not a realistic explanation and was disproved by the fact that no tartar emetic was found in his bedroom after his death.

What happened to the protagonists after the inquest? Florence died at Southsea in 1878 at the young age of 32, neglected and disowned by her family. Mrs Cox, who was dismissed by Florence immediately after the inquest, returned to Jamaica where she lived for many more years, inheriting substantial properties there. Dr Gully, now a social outcast, lived for another 9 years and devoted much of his time to spiritualism.

Death of a President

Bullets are made from lead and they are generally made in batches of 50,000 or more at a time. The lead from which they are made contains traces of other metals, such as copper and silver, but it is the amount of antimony that is present which varies from batch to batch because this metal is deliberately added to harden the lead. There may be as little as 0.4% antimony or as

---

* Drinking alcohol causes the body to lose water and the dehydration this accounts for most of the symptoms of a hangover. Drinking a pint of water before going to bed is always advisable, assuming you are not too drunk to remember to do it. Putting the water by your bed before you go out on the razzle is one way to ensure you do remember.

much as 4% if a particularly hard bullet is required. The amount of antimony will vary slightly from bullet to bullet because the alloy from which they are made is not perfectly homogeneous, but they will not vary by more than a small amount. A bullet, or even a fragment from a bullet, can generally be identified as coming from a particular batch by neutron activation analysis (see Glossary) and this has proved important in many criminal investigations, because it may be possible to link the bullet found at the scene of the crime with other bullets in the possession of the person who committed it, even when they have sought to evade detection by disposing of the gun. (Bullets can be linked to a particular gun by the marks that it makes on the bullet as it is fired.)

President John F. Kennedy was assassinated in Dallas, Texas, on 22 November 1963. The chief suspect, Lee Harvey Oswald, was arrested soon afterwards although never brought to trial because he himself was killed by Jack Ruby while in police custody. The results of the forensic investigation of the bullets and fragments that were collected after the assassination were withheld for many years, and it appears they had been carried out in a very haphazard manner. A paraffin wax cast of Oswald's right hand and right cheek, where minute fragments from the bullets might have become lodged, was sent for neutron activation analysis to Oak Ridge National Laboratory but the samples were carelessly handled and nothing useful came of the analysis.

Bullet fragments from the President's limousine, from his body, and from that of Texas Governor John Connally who was also wounded in the incident, were analysed by conventional techniques in the FBI laboratories and all they could report was that they came from the same brand of ammunition. Or so it was claimed at the time. We now know that they too were sent to John F. Gallagher, forensic scientist at the FBI and he carried out a neutron activation analysis of them in May 1964. It was only in 1975 that these results were released and then only after a long legal battle. The battle was fought by Professor Vincent Guinn and Dr John Nichols of the Kansas Medical School. Guinn was by now a leading expert in analysing the lead of bullets and their fragments and what he found when reading the FBI report was such a wide variation in the amount of antimony they detected that these results were different from those of any bullet he had analysed. He asked permission to re-examine the evidence and submit it to further neutron activation analysis, which had become a much more sensitive technique in the intervening years.

In 1977 he got that permission from the House of Representative Select Committee on Assassinations, and the tests were carried out using the nuclear reactor of the University of California, Irvine, where Guinn now worked. His results showed the amount of antimony in the lead of the bullets was extremely low; in other words the bullets were very soft which is why they fragmented so easily. In the pieces taken from the President's brain there was 621 ppm of antimony (in other words only 0.0621%), and 602 and 642 ppm on fragments taken from the limousine. A fragment taken from Connally's wrist had 797 ppm, while the mysterious bullet that was found on the stretcher on which Connally was taken to hospital, and which was believed to have come from his thigh wound, had 883 ppm antimony (0.00883%). While these results vary somewhat, they vary within the range expected for a particular batch of molten lead.

Guinn concluded that Oswald had fired only two bullets into the limousine, one that wounded the President and the Governor before finally coming to rest in the latter's leg, and the one which killed the President and broke into fragments inside his skull. What the antimony analysis revealed was that there was no second gunman firing at the President, as was popularly believed.

Severin Klosowski *alias* George Chapman

S EVERIN Klosowski was born on the morning of 14 December 1865 in the village of Nagornak near Kolo in part of Russian-occupied Poland. He died 38 years later, as George Chapman, on the morning of 7 April 1903 in London, hanged for poisoning three of his partners with antimony in a way that was long and painful but which made it appear they were dying of natural causes. What is rather unusual about these murders were the many witnesses to the way that he carried them out.

Severin Klosowski

Severin was the son of 30-year-old Antonio Klosowski and his wife, Emilie, 29. They were Roman Catholics, and Antonio was the village carpenter. When he was 7 years old, on 17 October 1873, Severin started primary school, which he attended for the next 7 years, leaving on 13 June 1880, with a good final report. Later that year, on 1 December, he was apprenticed to the barber-surgeon, Moshko Rappaport, in Zwolen, 90 km south of Warsaw, training to be a *feldscher*, an occupation combining the roles of barber and minor surgeon. This qualification would allow him to perform small operations by himself, or to assist major surgery carried out by a fully qualified surgeon.

In the summer of 1885, when he was 19, Severin left Zwolen and armed with a good reference from both his employer and a local doctor, he set off for Warsaw with the idea of becoming a fully qualified surgeon. To finance himself through his studies he took a job as an assistant to a barber-surgeon in the suburb of Praga, and that October he enrolled for a three-month course in practical surgery at the Hospital of Infant Jesus nearby. In January 1886 he took a job as an assistant surgeon to a D. Moshkovski and continued working thus until 15 November that year. The following month he came of age and that allowed him to apply for a passport and he was also allowed to sit the entrance examination for the degree of Junior Surgeon at the Imperial University.*

In February 1887 Severin paid one month's study fees to the Warsaw Society of Assistant Surgeons, but then dropped out of college. What he did for the rest of that year and in the early months of 1888 is unclear, but he might well have started living with a partner, and he appears to have fathered a child, with another one on the way, which may be the reason why he could not pursue his studies. In any event he emigrated to London. A woman with two children followed him there but he would have nothing to do with them, and eventually they returned to Poland.

We know from the address written inside a Polish book of Severin's entitled *500 Prescriptions for Diseases and Complaints* that when he first came to London he lived at 54 Cranbrook Street, Bethnal Green, and indeed he may have lodged there for quite some time. For five months he worked as an assistant to a Mr Abraham Radin who ran a hairdressing shop in West India Dock Road, Poplar. He showed Mrs Radin his Polish documents and told her he had been medically trained and on the strength of this he helped nurse the Radin's young son when he was ill. His next job was with a barber whose shop was in the basement under the White Hart public house at 89 Whitechapel High Street. There were two independent witnesses who remembered him working there in the autumn of 1888. One was a fellow Pole, Wolf Levisohn, a travelling salesman in hairdressers' requisites, and who knew him as Ludwig Zagowski, an alias that he had adopted in London.

The second man who remembered seeing Klosowski in the basement shop was none other than Inspector Abberline who was in charge of the hunt for Jack-the-Ripper. In the autumn of 1888, the Whitechapel area of East

---

* In order to apply to the university he needed to provide evidence of his birth, education, and training as a feldscher, together with character references. This collection of certificates were still in his possession when he was arrested in 1902, and it is from these that his early life can be reconstructed, even to knowing that while in Praga he lived at 16 Muranovskaja Street.

London saw a series of murders of prostitutes in which their bodies were badly mutilated. Just how many murders can be attributed to Jack-the-Ripper is debatable but in the autumn of that year six women met their deaths. The first murder actually occurred on a landing of George Yard Buildings, a stone's throw from the shop where Klosowski worked. The woman, 35-year-old Martha Turner, was stabbed 39 times with two knives in the early hours of the morning following Bank Holiday Monday, 6 August 1888. Whether Klosowski was that infamous murderer is doubtful but there were good grounds for suspecting him and he invariably comes into the frame when the Ripper murders are discussed, mainly on the grounds that Inspector Abberline was convinced he was the man. This speculation does not really add to the story of him as an antimony poisoner, however.

Klosowski may eventually have taken over the hairdressing business but by June of 1889 he had his own barber's shop in Cable Street near the docks and had reverted to his real name of Severin Klosowski. He was becoming more sociable and while attending the Polish Club in St John's Square in Clerkenwell he met Lucy Baderski, a woman who came from the German-occupied part of Poland. Klosowski had little trouble getting Lucy to live with him and by the August Bank Holiday of that year she was telling her brother that they were married and she moved into the flat above the shop in Cable Street. In fact Klosowski and Lucy were not married until 23 October, when Lucy had become pregnant. As soon as Lucy and the baby were fit to travel, they set off for the USA in May 1890 and there he opened a barber's shop in New Jersey.

Perhaps this time the shop was a success, but that was certainly not true of his family life; rows between Lucy and Klosowski led to violence. By February of the following year Lucy had had enough and although she was then six months pregnant was prepared to brave the rough seas of a winter crossing of the Atlantic to return to London with their baby son. She went to live with her sister Mary at Scarborough Street, where on 12 May 1891 she gave birth to a baby girl.

Two weeks later her husband returned from the USA, and they were reconciled. Mary was persuaded to find alternative accommodation and moved out so that the family could be together, but the reconciliation did not last long and soon Lucy and Klosowski agreed to part permanently with the children staying with their mother. Although they never lived together again, they never got divorced.

From May 1891 to the autumn of 1893 there is no record of Klosowski's whereabouts and he may well have returned to the USA because among his possessions discovered after his arrest was a sheet of paper on which he had written: 'came from America in 1893 independent' and 'deposits £100 when from America I had £1,000.' What this last comment means is impossible to say, because it appears to imply that he returned from the USA with the latter sum, which was quite considerable for its time and equivalent to 10 years' salary for an ordinary working man.

On his return to London, Klosowski became an assistant in a hairdressing salon at 5 West Green Road, Tottenham, owned by a Mr Haddin, and it was while he was living and working in Tottenham that he met Annie Chapman. He asked her to become his housekeeper which she did and moving in with him in November 1893. Soon they were calling themselves Mr and Mrs Klosowski, although they were not married, but their partnership was not to last and she left him. However, towards the end of January, Annie Chapman called to tell him she was pregnant and that he was the father. Klosowski denied it was his and refused to have anything to do with her. Stung by this, she went to see a solicitor but he told her he could do nothing for her because she was not his wife and so could have no legal claim on him. The last time she was to see Klosowski was in February, when he cycled round to her home to tell her that he was leaving the district. The only thing he was taking of hers, although she was not to know it, was her name. Henceforth, Klosowski called himself George Chapman, and from now on we shall do the same because it was the name under which he gained his notoriety. He never again answered to his real name and even insisted he was George Chapman when he was about to be hanged.

After he left Tottenham, Chapman may have returned to the East End, possibly working in Shoreditch, until he applied for the job of a hairdresser advertised by a Mr Wenzel of 7 Church Lane, Leytonstone. He said he was George Chapman, and he got the job. Perhaps he thought this choice of name would also be a good smokescreen because he still lived under the threat that when Annie Chapman's baby was born she might serve a paternity order on him – but she would surely never come looking for him under the name of Chapman. He sometimes posed as an American who was orphaned when young. He was always nattily dressed and liked to be thought of as a gentleman but his attitude to women was far from gentlemanly. With his handsome looks and masterful manner he had little difficulty in attracting them and persuading them to live with him, but they could never be married to him because he never divorced Lucy. What was unusual about Chapman was that he never simply walked away from these partnerships or even threw his partners out: he disposed of them by slowly poisoning them.

The Murder of Mary Spink

At Leytonstone, Chapman met 39-year-old Mary Isabella Spink, the estranged wife of a railway porter. During their marriage they had had two children, Shadrach, the eldest being taken by his father, and William, born after they separated, living with his mother. When Mary met Chapman, William was about 5 years old. Mary told Chapman that there was a furnished room to let in the house where she lodged in Forest Road, and he went to see it and decided to take it.

It was not long before her relationship with Chapman became serious and one day the landlord's wife found them kissing on the stairs. She told Chapman in no uncertain words that was to be no such 'carrying on' in her house, to which he replied that he and Mary were soon to be married. Indeed, on Sunday, 27 October 1895 the couple donned their Sunday best clothes and left the house early, returning at 10am to say that they had been to the Roman Catholic Church in the City of London there to be married. Needless to say no marriage took place.

Mary was not without her faults; she was a heavy drinker, which was the reason her real marriage had ended on the rocks, and she had a child in tow. But she had one asset that more than compensated for these: she had inherited about £600 from her late grandfather. Some of this she had spent while living with Mr Spink, but she still had £500 which she was willing to invest in a new business. So it was in May of that year they moved to the holiday resort of Hastings on the south coast and there they went to see a local solicitor to enquire about using some of Mary's inheritance, which had been left in trust, to purchase a hairdressing business in the town. They had seen a shop in George Street that was for sale for £195 and wanted to draw money from the trust to purchase it. The solicitor agreed that it was in order to do this and on 11 June 1895 the sale went ahead. At first business was disappointing, but then Chapman had a brainwave and installed a piano in the shop which Mary played to entertain the customers while they were waiting.

Mr and Mrs Chapman and her son appeared to be settling down in Hastings. Business improved to such an extent that Chapman bought a small boat, *The Mosquito*, and some nautical attire. On several Sundays they went on trips in it until one day the boat capsized and they had to be rescued by a fisherman. Chapman was also a keen photographer and cyclist. Mary on the other hand had started drinking heavily again and seemed to be in a permanent daze. She was also neglecting to look after little Willie who was thin and weakly, but she still had £300 in her trust fund.

One of the regulars at the Chapman hairdressing parlour was Mr William Davidson who ran a pharmacy in the High Street. On 3 April 1897 Chapman purchased an ounce of antimony potassium tartrate, costing two pence. We know the exact date because Chapman signed the poison register in the shop and even wrote down the reason for the purchase, but this was illegible. In any event it hardly mattered because the purchaser was personally known to the pharmacist.

Chapman's roving eye was already seeking out new territory; he wanted to enter a new line of business and take a new soul mate. For a while he pursued a domestic servant Alice Penfold, even to the extent of telling her he was looking for a pub and suggesting that she might like to move into it with him. They even went to nearby St Leonards-on-Sea to inspect one that was for sale. Nothing came of it, mainly because Chapman was tiring of living by the sea and longed for the excitement of London. He eventually decided on a beer house called the Prince of Wales which was situated in Bartholomew Square in Finsbury, London. Again the solicitors agreed to advance money from Mary's trust fund, and indeed to wind up the fund, and on 31 August 1897 a cheque for £250 was handed to Mary who deposited it at Lloyds Bank in Hastings. Within a week it had been withdrawn and given to Chapman and they departed for London where the money was deposited in his account at a London bank.

Although she had been sick twice that year, in May and again in August, Mary arrived at the Prince of Wales beer house quite well. She had a tendency to drink too much and Chapman ordered her out of the bar on one occasion when she was obviously drunk. As far as he was concerned, Mary was now more of a liability than a help so he proceeded to poison her with antimony. Separation was out of the question because of the money she had lent him, which meant she had to die.

It seems likely that Chapman began to poison Mary in November because by 12 December her condition was so bad that a regular at the pub, Martha Doubleday, was asked to come in and stay with Mary during the night, while a Mrs Jane Mumford came in to look after her during the day.

About two weeks before Christmas, Martha told Chapman that he should call in a doctor to look at Mary, and a Dr Rogers, who lived nearby, was brought to see her. He diagnosed tuberculosis and prescribed medicines for her, but of course to no avail. Whatever Mary ate or drank seemed to make her vomit, and Martha noticed that this was particularly true when Chapman gave her some brandy. Brandy was a traditional treatment for an upset stomach and he put a bottle of the very best brandy by Mary's bedside with strict instructions that it was reserved solely for her. This not only made his concern for his wife seem genuine but also ensured that only Mary drank it.

As Christmas approached, Mary grew ever weaker. She had diarrhoea as well as vomiting almost constantly and suffered greatly from thirst. Her vomit was green with bile, and she had violent stomach pains. A Mrs Elizabeth Waymark, a part-time nurse, was brought in to look after her. She found her patient in a terrible condition and Chapman said it was *delirium tremens*, in other words she was an alcoholic, but now it seemed that the doctor's diagnosis of TB was correct and that it was 'galloping consumption' in other words it was progressing rapidly. There was no let up in the disease and as she neared her end she even had a haemorrhage from her womb. On Christmas morning Mary had a particularly violent attack of vomiting and died at 1 o'clock in the afternoon.

According to the women who attended to Mary in her illness Chapman displayed a rather ambivalent attitude towards his wife. On the one hand he insisted on giving her medicine himself and sent Martha out of the room while he did so, and he also pressed her to take brandy, although whenever she had either the medicine or the brandy the vomiting began soon afterwards. Clearly both were poisoned. On the day she died he shed a few tears, but to the amazement of the two women who had been nursing Mary, he went downstairs and opened the pub as usual. Mary was buried on Thursday 30 December 1897 at St Patrick's Cemetery, Leytonstone, in an elmwood coffin at the bottom of an 18-feet deep grave. She did not go to her final rest unmourned. Several regulars of the Prince of Wales followed the coffin and members of the local Whist Club, who regularly met at the pub, even collected for a wreath.

For 5 years Mary's body lay undisturbed, during which time seven other coffins were interred on top of it. When the grave was opened in December 1902 the stench from these was quite appalling but on opening Mary's coffin those present were astounded to find that her body was virtually unchanged. The nurse, Mrs Waymark, was able to identify her and it was noted that her eyes were still intact. An autopsy revealed that the spleen, kidneys, bladder, heart, and lungs were all normal and there was no sign of consumption. The cause of death was acute gastroenteritis caused by a chemical agent and that agent was antimony. Nor could this have percolated from the surrounding soil, which was tested for its antimony content, and none was found.

Mary's body was saturated with antimony and this was chiefly responsible for the remarkable state of preservation of her corpse because of the severe dehydration it caused. The organs analysed yielded antimony as follows: kidneys 4 mg; stomach 2 mg; bowels 27 mg; liver 57 mg. This last amount revealed that she had been given a large dose not long before she died. Altogether there was 90 mg of antimony found in the parts analysed.

Once Mary was out of the way, Chapman then had to do something about her orphaned son Willie. He did not want a 7-year-old-boy to look after and on 30 January 1898 he took him along to Dr Barnardo's Home to see if they would take him in. This the Home were reluctant to do and said that they must check that the boy had no living relatives who might look after him. Chapman gave them various addresses which he said were of Mary's family living in Leytonstone, but it turned out that these were all false and they refused to take Willie in. No doubt Willie was capable of doing odd jobs about the pub because Chapman made no further attempts to off-load him until the following year. Then on 20 May 1899, and 15 months after his mother's death, Chapman finally sent him to the Shoreditch workhouse and never saw him again.

The murder of Bessie Taylor

A few weeks after Mary Spink's death, Chapman advertised for an assistant to work at the Prince of Wales pub, and from the applicants he chose 32-year-old Miss Bessie Taylor. She was the daughter of a Cheshire farmer and cattle dealer. Bessie was an independent minded young woman who left home at 21 when she came of age and moved to London. There she worked in various restaurants over a period of 10 years. Her photograph shows her to be somewhat overweight but she clearly attracted Chapman. Not only did they work together but she was soon sharing his bed, again posing to the outside world as man and wife. Chapman must have confided to Bessie that he could not marry because he was already married and unable to obtain a divorce because of his religion, but not long after she moved to St Bartholomew Square they announced that they were married. What date this was deemed to have taken place is not known, but we do know that on 18 July 1898 a cheque for £50 arrived from Bessie's father as a wedding gift and this was deposited in Chapman's bank account where it joined the £250 that he had 'inherited' from Mary.

In their spare time that summer he and Bessie went cycling together, and they even joined the local Police Cycling Club. Maybe they cycled as far as Bishops Stortford because in August 1898 he negotiated to buy The Grapes public house there and on the 23rd of that month he drew out all the money from his bank account, a little over £369, and they moved to that country town on the border between Hertfordshire and Essex. The venture appears to have been doomed from the start. Unfortunately Bessie was troubled with her teeth and had to go into hospital for an operation when she developed an abscess in her mouth. This was just before Christmas 1898 and it coincided with a visit of an old friend of hers, Elizabeth Painter, who had been invited to stay at The Grapes for two weeks over the Christmas season. The relationship between Bessie and Chapman was going through a bad patch at this time and Elizabeth witnessed several rows. On one occasion Chapman even threatened Bessie with a revolver.

Life in the country was not really to Chapman's liking and early in the New Year of 1899 he sold The Grapes and they moved back to London. In May, he secured another pub, The Monument, in Union Street, Southwark, and for a while things between him and Bessie improved. She was quite a character and became a popular figure in the pub and the surrounding district. She attended All Hallows Church in Pepper Street, just off Union Street, and gave generously to its charitable organizations. She kept up with her cycling and her parents came to visit her on several occasions and stayed at the pub. Indeed her mother thought a lot of her supposed son-in-law and complemented Bessie on her choice of husband. And so things continued through the rest of that year and most of the next.

Bessie began to fall victim to antimony potassium tartrate in December

1900. About 2 weeks before Christmas she went down with sickness and constipation and on the advice of Martha Stevens, a regular customer at The Monument, she consulted Dr Stoker of the New Kent Road about half a mile away. Whatever he gave her, it did no good and the bouts of vomiting continued. Martha was a nurse by occupation and about ten days before Christmas, Chapman asked her to come and look after Bessie, and this she began to do. At first she came only during the day, but as Bessie's condition deteriorated she stayed through the night as well.

When Bessie's brother William, who lived in London, called to see her, he found his sister ill and shrunken, and looking very much older than her 34 years. She complained of violent internal pains and was sick while he was there. Another person to visit Bessie that December was her old friend Elizabeth, who found her fading away. Chapman occasionally made a pass at Elizabeth, and on one of her visits he gave her a kiss.

Bessie was so ill by Tuesday 1 January 1901 that Dr Stoker was called and thereafter he visited The Monument every day until Bessie died 6 weeks later, on Wednesday 13 February. (Queen Victoria had died on Tuesday 22 January.) The doctor found her suffering from regular vomiting, diarrhoea, and stomach pains. Sometimes she would improve for a few days but invariably she would suffer a relapse.

During January, Dr Stoker called in expert medical opinion in an effort to discover the cause of Bessie's illness. Dr Sunderland, a gynaecologist came to see her and thought she had something wrong with her womb. When examination proved it not to be so, a Dr Thorpe, who lived nearby, examined her and he concluded that her symptoms were psychosomatic, although the term he used in those pre-Freudian days was 'hysteria'. In that Bessie's symptoms were not the product of any recognizable organic disease, he was right. Nevertheless, her symptoms were real enough and finally a Dr Cotter was consulted. He thought that Bessie might have cancer of the stomach or intestines and acting on his suggestion a sample of vomit was sent to the Clinical Research Association for inspection but no cancerous cells were observed. In early February, Bessie's mother was sent for and when she came down to stay she made herself responsible for preparing all her daughter's meals. Slowly Bessie recovered and by Sunday 10 February she felt fit enough to get up and when Dr Stoker called he found her playing the piano.

Chapman was not finding it so easy to dispose of Bessie as he had of Mary Spink. Presumably under the watchful eye of her mother, he could not press upon her the small regular doses of antimony that he had been giving her before her mother arrived, and so he must have decided that the only solution was to give her a single large dose, which he did in the evening of Tuesday 12 February. She died that night at 1.30am. The doctor who was called could do nothing to save her and on the death certificate he wrote that death was due to intestinal obstruction, vomiting, and exhaustion. Bessie was buried in Lymm churchyard in her native county of Cheshire. Chapman pleaded poverty and Bessie's brother William paid the funeral expenses.

The body of Bessie lay undisturbed for 21 months, long enough for it to decompose. However, when it was lifted from its grave on 22 November 1902 it was observed to exude no odour of putrefaction and with the exception of a layer of mould it was in a remarkable state of preservation. The corpse was dissected and all the major organs found to be free of disease with the exception of the stomach which showed the expected signs of gastritis. The inner surface of the bowel was found to be coated with yellow antimony sulfide, indicating that the fatal dose may well have been introduced into the body in the form of an enema. This compound would be produced from antimony reacting with the hydrogen sulfide that forms as protein decomposes.

The chemical analyses showed the stomach to contain 8 mg of antimony, the kidneys 20 mg, the liver 107 mg, and the intestines 548 mg. This last amount being the largest ever recorded in the victim of antimony poison. The total amount of antimony was 693 mg. The soil around the grave was tested but no antimony was found. The evidence all pointed to a massive single dose given to Bessie on the Tuesday evening before she died.

The murder of Maud Marsh

Chapman now needed a barmaid and in August 1901 he saw an advertisement placed in a newspaper by an 18-year-old Maud Marsh, who was seeking such a position. Chapman replied to the advert and a few days later Maud, accompanied by her mother, came to be interviewed. In reply to Mrs Marsh's questions about his status, Chapman told her that he was a widower, and when he said that Maud would be expected to live on the premises, Mrs Marsh wanted to know who else was resident in the building. She was reassured to learn that Chapman had let off some of the upstairs rooms to a family, and with her mother's approval Maud then took the job. At 15, she had gone into service as a housemaid, but after working for two families, she worked as a barmaid at a pub in Croydon. It was when she lost her job there that she advertised for a similar position and ended up at The Monument.

Chapman immediately took a fancy to Maud and even delayed her coming to The Monument until he could evict the family living upstairs. From the day that she arrived at the pub, Maud and Chapman were the only residents in the building. From the start Chapman pressed his attentions on her and there is every reason to believe that these were not unwelcome to Maud, but remembering her mother's advice she kept him at arms length. However, when he proposed, she accepted and they became engaged. What this meant to Chapman was, he hoped, easy access to Maud's favours without the necessity of having to bother with yet another charade of marriage, but to Maud it meant only that they were engaged and that sharing the same bed would have to wait until after the ceremony.

In the middle of September she wrote a desperate letter home in which she said: 'George says if I do not let him have what he wants, he will give me £35 and send me home . . . but I am still engaged . . . and if he does not marry me I can have a breach of promise, can't I?' Her mother replied straight away, telling Maud to come home, but she received another letter from her daughter in reply saying that everything was all right again and that she and Chapman would cycle out to Croydon on Sunday to visit them. This they did, and to prove his intentions were honourable, Chapman showed Maud's mother and father a draft will that he had drawn up in which she would inherit £400 when he died.

On Sunday 13 October, Maud and Chapman put on their finery and leaving Maud's younger sister in charge of the pub, they set off to get married at the Roman Catholic Church in Bishopsgate Street. At 1pm they returned and broke the news that they were now Mr and Mrs Chapman. Maud's mother arrived from Croydon and, not unexpectedly, was non-plussed at the news and asked Maud to show her the marriage certificate. Maud told her that Chapman had put it away with his papers, and her mother did not press the point. In any case she accepted that they were married and stayed for a meal. Whatever reservations she had were gradually allayed over the next few weeks as it became apparent from Maud's letters that her daughter was very happy. As time went by, the Marsh family came to accept Chapman and made frequent visits to The Monument. Maud's father sometimes helped in the bar.

Chapman meanwhile had thought up a new scheme to make some quick money. He decided to insure The Monument and burn it down. The lease had almost expired, which did not give him much time to plan a proper job. He made it seem such an obvious case of arson that in the end the insurance company refused to pay up. In fact he had completely emptied the premises of furniture and valuables before he lit a fire in the basement of the building. The *Morning Advertiser* reported the incident rather too libellously, with the result that they were issued with a writ by Chapman. This was withdrawn when the police started to investigate the fire but no official action followed and the incident blew over.

Just before Christmas, Chapman secured the tenure of a much more important public house, The Crown, at 213 High Street, Borough. This was a larger pub much frequented by medical students from the nearby Guy's Hospital; it even had a billiard room. Chapman must have felt he was finally having some success in life. Then Maud became pregnant, but in April he persuaded her to let him carry out an abortion by syringing her womb with a dilute solution of the disinfectant phenol (then known as carbolic acid).

Around this time Chapman found himself in the newspapers once again, although this time he was the victim of a crime rather than its perpetrator. A commercial traveller, Alfred Clark, had come into The Crown accompanied by Matilda Gilmor. They were a pair of confidence tricksters and their scam was to sell stolen shares in the Caledonian Gold Mining Company. These shares were virtually worthless and so far they had not found a buyer for them. Chapman knew Matilda by sight and very soon he was in conversation with the couple. Impressed by what they told him, he agreed to pay £700 for the shares and secured the deal with a deposit of £7. He then withdrew the remainder of his money from his bank, but before he handed it over he wisely made some enquiries about the Caledonian Gold Mining Company and discovered its shares were worthless.

Chapman went to the police and reported the fraudsters. He also told the police that he had paid them £700 for the worthless shares. The couple were arrested when they went to hand over the shares to Chapman and they appeared in court at the Newington Session in June 1902. Chapman went into the witness box for the prosecution and Maud supported his story. Although Clark's defence counsel tried to break Chapman's version of events, they failed and the jury found Clark guilty and the judge sentenced him to 3 years in jail. Matilda was acquitted. As it happened, Chapman was able to supply the police with a list of the numbers of the notes of the money he had withdrawn from the bank. When he was arrested later in 1902, some of the notes were found in The Crown. Alfred Clark received a free pardon in December of that year.

Although Maud had stood by her 'husband' while he perjured himself in court, her misplaced loyalty was not to stand her in good stead. Her position at The Crown was already being undermined by a new barmaid who was taken on in June 1902. Florence Rayner had been a mid-day regular at the pub for some time when one day Maud asked her whether she would like a job there as a barmaid earning 5 shillings per week and full board included. She agreed and relieved Maud of working over the lunch period, enabling her to do her domestic chores and prepare a meal, so that she and Chapman could eat together when the pub closed during the afternoon. Within a fortnight Chapman was on very friendly terms with Florence although his advances towards her had not gone beyond the odd furtive kiss.

When Maud discovered what was going on there was the inevitable row between them, in which Maud threatened to leave him, and to which Chapman replied that if she went out that day then she could stay out for ever. Unfortunately for her she stayed. Shortly after the row, Maud went down with a bout of sickness and diarrhoea, but not before she dismissed Florence from their employ.

By thwarting Chapman in his attempt to replace Maud as his partner, Maud had in fact sealed her own fate. By the middle of July, Chapman had begun to poison her with antimony potassium tartrate. After several bouts of vomiting and diarrhoea, alternating with constipation, she was persuaded by her sister to seek medical advice. Unlike his two previous partners, Maud had very close family ties, and especially with her sister Alice and mother. In the end it was their concern that proved to be Chapman's downfall, although it did not save her life. In July, Alice's action earned Maud a temporary reprieve. Despite Chapman's protestations that he did not want Maud to go to hospital, because of what the doctors might do to her, Alice insisted that she must see someone about her illness. To placate Chapman, they promised to go first to a doctor round the corner from the pub, but when they did not find him at his surgery, Alice insisted on taking her sister to Guy's Hospital, and there she left Maud waiting to see a doctor.

Maud was so ill that she fainted in the hospital waiting room but the doctor who examined her diagnosed nothing serious enough to warrant admitting her and sent her home. Another dose of poison was administered shortly after her return, which sent her hurrying back to the hospital and this time she convinced them of the seriousness of her condition and she was admitted to Guy's Hospital where she remained for the next 4 weeks. Maud's symptoms at the beginning of her stay in hospital were a high temperature, rapid pulse, diarrhoea, vomiting, and abdominal pains which came with great severity at times. For 2 weeks she showed little change but on 10 August she began noticeably to improve; her temperature began to fall and after another ten days she was pronounced fit enough to leave. The doctor in charge, Dr Targett, thought she had had peritonitis.

Maud returned to The Crown, and for a month she was fit and well until Thursday 7 October. The new routine at The Crown was for Maud to serve in the bar at midday and to have a late lunch by herself. On that day some potatoes were kept aside for her from the food eaten by Chapman and two other employees. When Maud came to eat them later that afternoon they made her sick and she felt so bad that she went to bed. The next day her sister Louisa called but found her vomiting and having diarrhoea. The final attack on Maud had begun. The next 2 weeks of her life are much better documented that the last days of Chapman's other victims. From what happened at The Crown we learn how he carried out his poisoning, and he knew that the poison he was using would enable him to evade detection. About this time he boasted to one of his customers that he could give a person 'a little bit like that [indicating a pinch between his thumb and forefinger] and 50 doctors would not find out'. The fact that he had got away with murder twice, the second time under the noses of four medical men, three of them specialists, may only have served to convince him that antimony was the perfect poison.

The rest of that week saw Maud no better and on Friday Chapman went round to Dr Stoker's surgery asking for some medicine to stop her vomiting and diarrhoea. It was on this visit that he confided in the doctor that he and Maud were not really married. The doctor gave him a stomach mixture consisting of chalk, bismuth, and opium. He put her on a light diet, and called to see her the following day, Saturday 11 October, and found her still very ill. Maud asked him if it was peritonitis again, but was reassured that it wasn't. The following day the doctor found her much improved and for Sunday lunch she had pork, potatoes, greens, bread, and ginger beer. Chapman waited until teatime to strike again, and by 5.30pm she was vomiting copiously. Chapman had opened a bottle of champagne during the course of the evening and gave some to Maud, but when Louisa tasted it she found it tasted odd and said so, whereupon Chapman emptied the bottle into the chamber pot.

Monday 13 October, was the first anniversary of Maud and Chapman's 'wedding' but Maud was too ill to want to celebrate. From that day forward, she became progressively worse. The vomiting and diarrhoea continued and were again accompanied by stomach pains and leg cramps. By Wednesday she was unable to keep any food down and the doctor said that she must now be fed anally. Chapman asked one of his regulars, a Mrs Toon, if she would nurse Maud and feed her. The nature of the latter task made her decline the offer. Chapman found others equally reluctant to do the job so he has to do it himself. Mrs Toon did agree to nurse Maud each evening until the pub closed its doors and she went home at 1am.

Mrs Toon found Maud very thirsty but the drinks she was given brought no relief – she vomited after most of them. Nor was the anal feeding successful – it too was quickly expelled. Maud refused to have Dr Stoker's medicine, saying that it only served to make her worse. Chapman made himself responsible for all Maud's needs and was never out of her sight for more than a few minutes. He insisted that all that she ate and drank must only be given by him. Maud's father called to see her on Saturday evening and found this behaviour of Chapman rather suspicious, saying that he would call in a second opinion if she showed no improvement by Monday. Maud, how-ever, showed signs of improving as she had the previous Saturday, suggesting that over the weekend Chapman was too busy in the bar to devote any time to poisoning her.

Sunday found Maud well enough to have some rabbit for lunch, a meal that she believed was responsible for the recurrence of her symptoms later that day. A servant who ate some of the dish was also sick. The Marsh family organized themselves to attend to Maud and her mother arrived at The Crown on Monday morning to find her daughter very ill and complaining of excessive thirst and pain in the lower abdomen. She had been given brandy and soda to drink by Chapman but each time she had some she vomited it back. Martell's three star brandy no less.

Dr Stoker called and finding that Maud was much weaker he took her mother to one side and confided that her daughter was probably dying. He discussed the wording of her death certificate with Mrs Maud, and possibly confided in her that Maud was not married to Chapman. At nine o'clock the next morning, Tuesday 21 October, Maud's father went to see his own doctor Dr Grapel, and asked him to go and see his daughter that day. This he did, and arrived about 3.30 in the afternoon. Dr Stoker, who had already called about mid-day, was sent for and together the two doctors examined the sick woman. Dr Grapel found her pulse quick but her heartbeat strong. Her appearance was jaundiced, her skin sallow, and her breathing shallow. She was now in a semi-coma. The doctors agreed that it was probably a case of food poisoning and arranged to meet again the following evening.

Later in the day, Maud began to improve and when her father called to see her after work, he found her feeling better and more cheerful. He told Chapman that he thought she would pull through, to which Chapman replied ominously: 'She will never get up no more.'

Maud was not much better by the evening and her mother stayed the night again. The nurse left at 1am and Chapman brought up a large glass of brandy for Maud and put it by her bed in case she needed a drink during the night. For a while Maud slept but at 3am she awoke and her mother gave her some of the brandy with soda. Shortly afterwards she vomited it back. At 5am, Mrs Marsh, worn out after two days of nursing Maud, decided that she too was in need of something to sustain her and had a drink of the brandy, adding some ice and water to it. Within minutes she too was vomiting and during the next 2 hours she needed to go to the lavatory six times.

Maud's mother put two and two together and decided that the brandy was suspect. She asked Mrs Toon to taste it, and this she did but found it too strong and said it burned her mouth. The two women did not have time to take any action, however, because Maud had a fit shortly after Mrs Toon's arrival. One of her arms had turned a deep red colour and her lips had gone a dark grey. A severe uterine discharge occurred. She was now sinking fast but about noon she regained sufficient consciousness to realize that her end was near. 'I am going,' she whispered to Chapman and the two women, to which he replied: 'Where?' Then with a last gasp she bade him goodbye. It was 12.30pm. Overcome with emotion, Chapman burst into tears, but then found enough inner strength to pull himself together and half an hour later he opened The Crown as usual. Dr Stoker called at 3pm and finding Maud dead, he refused to sign a death certificate, saying that he intended to perform a private post-mortem. Maud's mother had said the previous day that she wanted this to be done if Maud were to die.

The arrest and trial of Chapman

In the afternoon, Mrs Marsh sent a telegram to her husband which he received at 4pm, and immediately took it round to Dr Grapel. He assured Mr Marsh that action would be taken to investigate the real cause of his daughter's death. In fact he confided his suspicions about arsenic poisoning in a telegram to Dr Stoker the following day and as a result of this Dr Stoker removed part of Maud's liver, rectum, and her stomach during the autopsy and sent them round to a Dr Bodner of the Clinical Research Association for analysis. The post-mortem itself revealed no direct cause of death; the liver, kidneys, lungs, and ovaries all appeared to be healthy. The post-mortem was a private one, and slightly irregular because of that very fact, but it was properly conducted and carried out in the presence of two other doctors as witnesses.

On Friday 24 October, Dr Bodmer began his analysis of Maud's remains, looking in particular for arsenic. In fact he did detect traces of arsenic which he reported to Dr Stoker, who immediately informed the police and wrote to the coroner. In doing so he was a little premature because further forensic analysis revealed the presence of much larger quantities of antimony in the fluid from Maud's stomach but Bodmer had not mentioned this because Dr Stoker had specifically asked him to look for arsenic. (He assumed the antimony had been given medicinally.)

Meanwhile, back at The Crown, Chapman began to realize that his perfect method of disposing of unwanted partners had come unstuck. In any event he took steps to cover himself by paying off Mrs Toon and warning her to keep her mouth shut about the events of the past week. He collected and burned all the cloths, towels, and bedding from the sick room which had been contaminated with Maud's vomit and which Mrs Toon had put aside at Maud's mother's request. He disposed of the remainder of his stock of antimony potassium tartrate and the only antimony to be found by the forensic scientists was a trace in the residue of a bottle of Dr Stoker's medicine. Fortunately for the police, Chapman had kept the label from the original bottle of poison he had bought from the pharmacist in Hastings many years ago.

Saturday 25 October, was the day of the Royal Progress through South London, following the Coronation of Edward VII. Like many other shop-keepers along the route, Chapman has festooned his pub with flags and filled his windows with sightseers who were willing to pay for such a vantage point from which to watch the procession go by. Chapman, however, was not destined to benefit from this little earner. Just before noon, two detective inspectors entered The Crown and arrested Chapman, charging him with murdering Maud Marsh. They also searched the premises and removed Chapman's personal papers, £268 in bank notes and gold sovereigns, and a loaded revolver. Then began a series of legal proceedings which today seem bizarre. By writing both to the coroner and the police, Dr Stoker had set in motion two independent inquiries into the death of Maud Marsh. As we have seen, the police moved in almost straight away, arrested Chapman on the Saturday and brought him before the Police Court on the Monday, where he was refused bail. The coroner was also active and opened his inquest on the day after the funeral, and after the first Police Court hearing. This was adjourned for a second post-mortem and analysis of the remains, performed this time by a Dr Stevenson. He found arsenic and antimony, but it was the latter which was present in far greater amounts that normal, and this was the cause of death. His analysis showed the stomach to contain 21 mg, the bowels 390 mg, the liver 46 mg, the kidneys 9 mg, and the brain 11 mg. Altogether 477 mg of antimony was found.

Over the next two months, the Coroner's Court sat four times and heard evidence about the deaths not only of Maud Marsh, but of Mary Spink and Bessie Taylor, whose bodies were exhumed and analysed towards the end of November. Finally on Thursday 18 December 1902, the Coroner's jury returned a verdict of death by wilful murder against George Chapman and he was committed for trial at the Central Criminal Court. The trial itself was a foregone conclusion. Beginning Monday 16 March, it consisted of 3 days of prosecution evidence, the first day establishing that George Chapman was really Severin Klosowski, a Pole, the second day was taken up with the murder of Maud Marsh, and the third day heard the medical and forensic evidence and dealt with the murders of Mary Spink and Bessie Taylor. The fourth day of the trial was in fact the last day.

In his summing up, the judge, Mr Justice Grantham, took a strong anti-Chapman line, but also rebuked the medical men involved in treating Chapman's victims for their inability to diagnose that their patients were being poisoned. It took the jury only 10 minutes to find Chapman guilty and he was executed on 7 April 1903. He never admitted he was Severin Klosowski even at the end of his life, and although his real wife, Lucy, called repeatedly at Wandsworth Prison to see him, he always refused to see her.

Lead

The empire of lead

For more technical information about the element lead, consult the Glossary.

LEAD is useful, surprising, unpredictable, dangerous – and deadly.

Lead is useful. Previous generations found it to be an essential part of civilized living; pipes, pewter, pottery, paints, and even potions were made with it. Toy soldiers were cast from it, port wine protected by it, grey hair was disguised with it, church roofs covered with it, cosmetics contained it, and cans of food were sealed with it.

Lead is surprising. In 1859, Professor Lyon Playfair was taking the 18-year-old Prince of Wales, and future King Edward VII, round the chemistry laboratories of Edinburgh University when they came across a pot of molten lead. Playfair then carried out a remarkable demonstration: he poured the molten metal over the fingers of his assistant and to the Prince's amazement the young man's hand was unharmed. The Prince too wanted to try the same test, and after rinsing his hand in dilute ammonia solution, lead was poured over his fingers as well – again without scalding them.* This demonstration was still being performed before stunned audiences in the 1950s, and the trick is to have the hand wet, so that as the metal hits the film of water on the skin it forms an instant layer of vapour that both protects the skin and causes the lead to bounce off in tiny droplets.

Lead is unpredictable. In Shakespeare's *Merchant of Venice* those who would seek to marry the fair Portia are offered three caskets from which to choose: gold, silver, or lead. The man who chooses the casket containing her

---

* Ammonia had to be used to remove any grease and to ensure his hand was thoroughly wetted.

portrait will win her hand in marriage. Of course the winning choice is lead, the casket with the curiously foreboding inscription: 'Who chooseth me must give and hazard all he hath' and her would-be suitor Bassanio reasons that this must be the right one saying:

> '. . . but thou, thou meagre lead,
> Which rather threat'nest than does promise aught,
> Thy plainness moves me more than eloquence,
> And here choose I; joy be the consequence!'
>
> [*Merchant of Venice*, Act III, Scene II]

All of which suggests that Shakespeare was aware that lead had a dark and threatening side. Not that that stopped it becoming economically far more important that silver or even gold in the centuries ahead – but there was a price to pay in the blighted lives of millions of people exposed to its baleful influence.

Lead is dangerous. The Roman architect and engineer Marcus Vitruvius, who lived in the first century, observed that labourers in lead smelters always had pale complexions. The Greek physician Hippocrates described a severe attack of colic in a patient who was a lead miner. Neither attributed the cause to lead, and nor did most of the physicians who down the centuries treated patients affected by it, although there were times when a few doctors realized how toxic it could be. One such was Tanquerel des Planches of the Charity Hospital in Paris who wrote a definitive report in 1839 on occupational lead poisoning. While this alerted the medical world to its dangers and led to better diagnosis, it had little effect on the extent to which lead was permeating daily life and poisoning almost everyone. It was to take another 150 years before the threat of lead was finally to be lifted from the world.

Lead is deadly . . . and not only as bullets. In the 1940s in the USA and the UK some children of poor families were affected by lead poisoning when they burnt the discarded cases of lead batteries as cheap fuel. In the UK at the village of Canklow, near Rotherham, in Yorkshire, a junk dealer sold these at a shilling a sackful with the results that 25 children were eventually hospitalized, and two of them died. Everywhere that lead was used, ill health was sure to follow, with heavy exposure to lead leading to death. According to one theory it might also have been deadly to empires by killing off those who ruled over them. The upper classes of the Roman Empire may well have been fatally weakened by it, and the rulers of the British Empire were not unaffected either.

The effects of lead on the human body

The body's response to a sudden large dose of lead is to expel the poison by the usual routes of vomiting and diarrhoea, but when it comes in the form of many small doses it can infiltrate the body and be partly assimilated. Then its baleful effects may go unrecognized for years provided the daily dose is small. The best indicator of exposure to lead is to analyse a sample of blood. This can be related to the symptoms it is causing, and those to the metabolic processes it is disrupting. Table 12.1 is a rough guide to lead levels, symptoms and root causes.

Table 12.1 A rough guide to lead levels, symptoms, and root causes

Blood lead level (micrograms per 100 ml)[a]	Likely symptoms	Causes
10	None.	
40	Headaches, indigestion, constipation, irritability, lack of concentration.	Excess ALA[b].
80	As above, but more severe, plus depression, colic, anaemia, lack of energy.	Excess ALA[b] and other metabolites.
100	As above plus insomnia, tingling in hands and feet, blue line on gums[c], infertility in men, abortions in women.	Weakening of peripheral nervous system and disruption of central nervous system.
150 and above	As above plus convulsions, paralysis, blindness, delusions, coma.	Swelling of the brain (lead encephalopathy) leading to permanent damage.

[a] To convert micrograms per 100 ml of blood to ppm divide by 100, so that 10 micrograms per 100 ml becomes 0.1 ppm
[b] Aminolevulinic acid – see text.
[c] This was also indicative of poor dental hygiene.

People respond idiosyncratically to lead. When workers have been tested for their exposure to lead, some individuals with levels in excess of 150 micrograms per 100 ml have shown no symptoms of poisoning at all. Yet others display symptoms of acute poisoning with half this amount in their blood, but in any case a value in excess of 80 micrograms per 100 ml indicates the need for antidote therapy. Some quite remarkably high blood lead levels have been encountered, and one premature baby born to a Sikh woman in Adelaide, Australia, in 2001, had record high levels for a child. The mother had been taking a herbal remedy sent from India whose tablets contained 9% of lead compounds. Her baby's blood was found to have almost 250 micrograms per 100 ml, and was suffering from various symptoms of advanced lead poisoning. Treatment with antidotes eventually saved the baby's life and after three months the blood lead level was down to a less life-threatening 35 micrograms per 100 ml. This penetration of lead from the mother's blood into her foetus shows how insidious it can be.

For much of the 1900s it was assumed that there was a 'threshold' level of lead in blood, below which no effects were to be observed. The most famous lead chemist of his day was Robert Kehoe, an academic researcher who eventually became medical director of the Ethyl Gasoline Corporation, and he set the threshold at 80 micrograms per 100 ml, a value that went unquestioned until the mid-1960s. We now know that this was far too high and it was to be revised downwards several times: to 70 micrograms per 100 ml in the 1960s, then 60 micrograms in 1970s, then 30 in 1975, and finally to 15 in 1990.

Most people of course had nothing like this amount of lead in their blood and in the 1970s the average for US citizens was 15 micrograms per 100 ml, declining to around 10 micrograms in the 1980s, and well below this in the 1990s. Among children, the decline was particularly marked in the 1980s and 1990s when average blood lead levels fell from around 13 micrograms per 100 ml to 3 micrograms per 100 ml. Whereas at one time almost all children in the USA had blood levels in excess to 10 micrograms, by the early 1990s fewer that one child in ten had a blood lead level this high. Yet even in 2000, one child in twenty under the age of 2 still had a blood lead level in excess of 10 micrograms per 100 ml, and these were mainly African-Americans living in inner city areas.

Lead disrupts three key bodily functions: the formation of blood, the workings of the nervous system, and the kidneys. The first of these leads to a build-up of two precursor chemicals in the body – aminolevulinic acid (ALA) and coproporphyrinogen – and it is these which cause the symptoms. Lead blocks the essential enzymes needed to convert ALA to the haem of haemoglobin, which is the key component of blood. Consequently ALA builds up in the body and it affects the stomach, hence the severe pains of colic; it paralyses the intestines, hence the immovable constipation; and it affects the muscles and nerve fibres, hence the general feeling of weakness and numbness of the limbs. The most serious effect of all is on the brain where it weakens the walls of the blood vessels so that they leak and this causes a pressure of water to build up in the brain, and ALA itself can then permeate the brain. These symptoms may range from mild headaches, depression, and sleeplessness in a mild attack, but become hallucinations, insomnia, fits, blindness, and coma in severe cases. The swelling of the brain, known as lead encephalopathy, is the danger most to be feared in children because it can result in permanent damage.

The attack on the nervous system interferes with the ability to transmit messages to the limbs, and the attack on the brain disrupts the action of two messenger molecules: serotonin and dopamine. Serotonin controls the body's sleep pattern, and this leads to insomnia, while lack of dopamine makes a person withdrawn and depressed. Lead destroys brain neurons which are replaced by so-called glial cells which cannot function in the same way. Damage to the kidneys is caused by their attempts to remove the toxic metal from the body which they do by extracting it from the blood and bonding it to proteins. However, the lead is then only removed slowly to the urine, and meanwhile it impairs the activity of these vital organs.

How much lead can the human body accommodate?

The total amount of lead in an average adult today is probably around 100 mg. In the next generation it might be as little as 50 mg, and in subsequent generations even as low as 10 mg. Most lead in the body is to be found in the skeleton, which can have up to 30 ppm of lead. The level to which people in the past were exposed to lead is preserved in their bones and it has been possible to mark the degree of lead exposure by analysing bones from former ages. If the date of death is accurately known, then this can be a useful guide to conditions prevailing at the time. By the late 20th century the level of lead in recently deceased children's bones was down to 2 ppm although for most adults it was still around 5 ppm. The analysis of bones taken from the crypts of Polish churches has proved particularly important. These have been stored in dry conditions and thereby prevented from absorbing lead from their surrounding. Some of these showed levels of lead in excess of 100 ppm. In the Middle Ages levels were around 30 ppm, but began to rise thereafter, so that skeletons from people living in the 1700s had on average around 50 ppm and this rose to 60 ppm for those who died in the 1800s. Clearly their environment was heavily contaminated.

Where does the lead in our body come from today? Most now comes from our diet, but some we breathe in. The average daily diet probably contains more than 200 micrograms of lead of which about 10 micrograms gets into the blood, where it is joined by about 5 micrograms of lead from our lungs (depending on where we live) so that our daily intake probably comes to about 15 micrograms and the body can easily rid itself of such an amount.

Thankfully, only a few per cent of the lead in our food and drink gets into our blood stream, but about half of the lead in the dust we breathe in gets so absorbed. This was shown by research carried out in 1995 by a group at Columbia University, New York, headed by Joseph Graziano. Six volunteers were given a capsule of soil collected from a site which had been an old lead mine and which was contaminated with 3,000 ppm of lead. The isotopic ratio of lead-206/lead-207 (see Glossary) was lower than normal and so the amount of lead entering the bloodstream could be assessed by seeing how this ratio changed. If the capsule was given on an empty stomach the results showed that as much as 25% of the lead was absorbed. Volunteers who were given the capsule after they had eaten a good breakfast absorbed only 3%.

Of the lead that enters the blood stream, either from the lungs or the gut, 5 micrograms are extracted by the kidneys and excreted in the urine, and 10 micrograms is deposited as insoluble lead phosphate in the skeleton. The bone takes up lead in the same way as it does calcium, by forming an insoluble phosphate. Day by day our skeletal lead burden increases, until by the time we are 40 it could well exceed 100 mg, but then as we get older it begins slowly to be leached out again, just as calcium is leached out, and our bones become weaker. Earlier generations whose skeleton had accumulated much more lead than this would then be at risk of suffering a mild but continuous form of lead poisoning later in life.

Lead gets into the food chain because all plants contain some lead, although not very much. For example, sweet corn has 0.02 ppm (fresh weight), while fruits contain hardly any at all, with tomatoes having only 0.002 ppm and apples a mere 0.001 ppm. The amount of lead that plants absorb depends on the lead content of the soil; lettuces grown near a lead processing plant were found to have up to 3 ppm, although even eating lots of this would be unlikely to result in symptoms of lead poisoning. Drinking water, unless it comes through lead pipes, is not likely to contain much lead, and even that which comes through lead pipes will not dissolve much lead if the water is hard, which means that it contains dissolved calcium and magnesium salts. The World Health Organization reduced its recommended upper limit for the level of lead in drinking water from 0.05 ppm to 0.01 ppm in 1995 saying that all countries should aim to achieve this by 2010. The US Environmental Protection Agency lays down a safe limit of 0.015 ppm but in Washington DC it discovered that out of 6,000 homes whose water it tested more than 4,000 of them exceeded that limit. The water taken from one house even had as much as 48 ppm.

Lead in water is present as lead ions, Pb^{2+}, whereas that in food is likely to be in an insoluble form although this may dissolve and release its lead if it comes into contact with solubilizing molecules, such as fruit acids. Lead ions carry the same charge as calcium ions, but they are much larger in size and in that respect they will not easily squeeze through the intercellular junctions of the gut wall to penetrate the channels of the gut membrane, and thereby gain access to the blood stream. Nevertheless, some does get through.

Lead and the decline of empires

The ancient Greek poet and physician Nicander described the symptoms of lead poisoning, including hallucinations and paralysis, and recommended strong laxative treatments to cure it. Yet lead continued to poison unchecked, largely because the link between the metal and its adverse effects on health was not obvious. On the other hand its benefits were obvious, indeed the more lead was used in a society, the higher the standard of living of its citizens. Lead can be an extremely useful metal because it is easy to win it from its ores, it melts at a relatively low temperature, and it makes an ideal solder. Lead is easily worked and can be hammered into sheets, and used to make pipes, pans, roofing, and cisterns, and it is impervious to attack by the oxygen of the air and by water.

Lead has been mined for more than 6,000 years, and it was certainly known to the Ancient Egyptians who used lead pigments as well as casting the metal itself into small figurines. Cosmetics made from lead ores have been found in tombs of the second millennium BC, and these consisted of black galena (lead sulfide), white cerussite (lead carbonate), white laurionite (lead chloride), and brown phosgenite (mixed lead chloride carbonate).

The Egyptians may have got some of their lead from Phoenician traders who were mining lead in Spain about 2000 BC, but it was the ancient Greeks who really began producing lead on a large scale, inadvertently as it happened, because they were really mining for silver. From 650 to 350 BC the Athenians exploited a large deposit at Laurion from which they eventually extracted 7,000 tonnes of silver – and more than 2 million tonnes of lead.

The silver from Laurion underpinned Athens economic power, until the mines became played out in the fourth century BC, after which Athens declined. By then more than 2,000 pits had been dug and 150 km of galleries excavated. The waste lead from these mines was still being exploited hundreds of years later by the Romans, who found more and more uses for the metal and its compounds. Builders, plumbers, painters, cooks, potters, metal-workers, coin-makers, dentists,* vintners, and undertakers all made use of it. (Lead coffins were used extensively throughout the Roman period to bury important individuals.)

The ancients saw lead as a god-given benefit and in Egypt it was associated with the god Osiris, while the Greeks linked it to Chronos, and the Romans to Saturn, which is why lead poisoning is still sometimes called saturnism. In reality lead was really a metal sent from hell. A puzzling feature of the Roman Empire was the surprisingly low birth rate of its ruling classes, and this too has been linked to the high level of lead in the diet. If the fate of a ruling class is what determines the fate of an empire, then the theory that one of the greatest of all empires was destroyed by lead may not be so fanciful as it first sounds. In fact more than just the aristocracy appears to have been less than reproductive. The Empire's population remained stable at around 50 million, despite such social benefits as adequate food supply, high standards of hygiene, and the growth of science, technology, and medicine, all of which should have led to an increase.

Some researchers have put forward the theory that lead was to blame, and we know from the analysis of the bones of its citizens, that their use of this metal was undermining their health. The theory that lead led to the decline of the Roman Empire was first advanced in 1965 in the *Journal of Occupational Medicine* by S.C. Gilfillan of Santa Monica, California, and his

---

* The ancient Romans used lead fillings for teeth.

arguments were subsequently reinforced by Jerome Nriagu of the National Water Research Institute of Canada. Nriagu, writing in the *New England Journal of Medicine* [vol. 308, p. 660, 1983], estimated that a typical aristocrat would be absorbing 250 micrograms per day, while ordinary Roman citizens would get around 35 and slaves only 15, most of which would come from wine in the case of the first two groups. Nriagu has even linked the medical complaints and bizarre behaviour of the Roman emperors to their high lead intake. Many of them suffered from gout as a result. Claudius who reigned from 41 to 54 AD displayed many of the symptoms of lead poisoning, including recurrent attacks of colic. Nriagu expanded on the theory in a scholarly but controversial book, *Lead and Lead Poisoning in Antiquity,* published in 1983.

Lead contaminated the homes of Romans in many different ways. Drinking water was transported along lead-lined aqueducts, through lead pipes, stored in lead cisterns, and maybe drunk from lead pewter vessels. The walls and woodwork of rooms were painted with lead-based paints. But one item in particular must have contributed to the lead in their diet, and that was a sweetening agent known as *sapa*. The famous Roman writer Pliny (23–79 AD) gives the recipe for making *sapa* and specifically mentions that it must be made in lead pans.

Roman cooks had only two sweetening agents that they could use for desserts: honey and *sapa*.* *Sapa* was made by boiling down unwanted wine, or wine which had become sour, in lead pans and we now know that the syrup so produced tasted sweet because it contained a lot of lead acetate. The lead came from the pan in which it was prepared, the acetate came from the wine that was being made sour by the action of enzymes and air which can convert alcohol to acetic acid. The crystals that form from such syrup looked and tasted like the sugar we know today, and were eventually to be known as sugar of lead. Old recipes for making *sapa* have been repeated in recent times, and analysed, showing that the syrup contained around 1,000 ppm of lead (0.1%). A spoonful of *sapa* would deliver a dose of lead that would undoubtedly lead to some of the symptoms of poisoning. Yet the popular Roman book, *The Apician Cookbook*, had *sapa* as an ingredient in 85 of its 450 recipes, and *sapa* was used by vintners as well.

* Sugar was unknown to the Romans. Sugar cane was originally to be found only in Polynesia, and gradually spread westwards reaching Europe about 800 AD.

Sapa was used to preserve wine, and especially Greek wines. These were popular in Rome but had a reputation for causing sterility, miscarriages, constipation, headaches, and insomnia – all of which would be true if they had been doctored with *sapa*. Roman prostitutes were reputed to eat *sapa* by the spoonful because it acted as a contraceptive, gave them attractive pale complexions (due to anaemia), and would also cause abortions.

The Romans mined lead in Greece, Spain, Britain, and Sardinia. At the height of the Empire the British deposits were the main source of supply and the annual rate of lead production was in excess of 100,000 tonnes per year. (In total, the Romans are estimated to have mined and used more than 20 million tonnes of lead.) Originally the Romans left the mining and refining in private hands, but ultimately it was deemed so important that it was all state-controlled. The Romans were not unaware of the risks of lead mining so it was done mainly by slaves, and at the height of the Empire 40,000 slaves worked the mines of Spain.

The collapse of the Western Roman Empire brought an end to economic development in Europe for almost 1,000 years. The causes of the Fall of Rome were a combination of climate change, plague, economic decline, religious dissent, power politics, and outside pressures. Indeed from 250 AD onwards, all these factors came into play. As the Earth's climate became colder, northern peoples began to move southwards and invade. Plague appeared and epidemics ravaged the Empire. Meanwhile internal military and religious disputes raged on. Lead was at most a minor factor in Rome's downfall. Was it also a factor in the decline of the British Empire 1,500 years later?

After the Dark Ages in Europe, which stretched from 500 to 1000 AD, things began slowly to improve, not least of which was the return of a warmer climate, and with this came renewed agricultural and economic activity – and the reintroduction of lead. It now reappeared in all its old guises, and more: lead-glazed pottery was invented in the 12th century, and this produced a finer quality of table and kitchen ware, but one which brought with it a dangerous increase in lead contamination of the diet; lead printing and lead type appeared in the 15th century, and lead bullets were the preferred weapon from the 16th century onwards.

By the time of the British Empire, the exposure of its leading citizens to lead was comparable to that of the Roman aristocracy. The British relied on lead in many aspects of their daily lives and in addition to the uses known to the Romans – including the adulterating of wine – they also poisoned their food with lead glazed pottery and their drink with lead crystal and lead pewter. They took lead-based medicines, used hair dyes and cosmetics made from lead compounds. They ate canned foods which were sealed with lead solder and they covered many things with paints consisting mainly of white lead. Lead was also present in the water collected from leaded roofs, and in pubs, beer was pumped through lead pipes from the cellar to the taps in the bar. In all these ways lead was in their diet and a comparison of the two Empires shows just how exposed to lead their respective citizens were:

Table 12.2 Lead in the Roman and British Empires

	Roman Empire 1–400 AD	British Empire 1700–1960 AD
Drinking water	Lead-lined viaducts	Lead piping
	Lead storage tanks	Lead storage tanks
Tableware	Lead pewter	Lead pewter
		Lead glazed pottery
		Lead crystal decanters and wine glasses
Kitchen ware	Lead-lined cooking vessels	
	Lead solder	
Food	Leaded wines	Leaded wines
	Lead syrup *sapa*	Lead solders on canned foods
Paint	White lead	White lead
	Red lead	Red lead
		Chrome yellow (lead chromate)
Medicinal uses	Lead skin plasters	Lead skin plasters
		Lead medicines
Cosmetics	Black lead oxide eye liner	Lead acetate hair dye
Buildings	Lead roofing	Lead roofing
	Lead solder	Lead solder

Drinking water was often collected from lead roofs and stored in lead cisterns, and as such could often carry a sizeable amount of dissolved lead. Rain water dissolves lead because it is slightly acidic and it may pick up as much as 1 mg per litre. These are the sorts of levels that would also be found in water supplies that come through lead pipes in soft water areas and especially if the water had been standing in the pipe for some time. There have been cases of people suffering lead poisoning from such water.

Clearly lead could have been a factor in the decline of the Roman and British Empires if one assumes that their fate lay in the hands of rulers whose brains were affected by the metal. And while lead may have affected the fertility of the Roman population, it certainly did not affect that of the British who reproduced at such an alarming rate in the 1700s and 1800s that leading thinkers of the time, such as the economist Thomas Malthus (1766–1834), warned that it was likely to increase beyond a point at which there was sufficient food to support it, and that efforts should be made to control human reproduction.

One of the puzzling features of the British Empire was the vigour of its scientists, seamen, inventors, and engineers, who created the wealth on which it was built, and yet it fell apart in the first half of the 1900s, to be replaced by a Commonwealth of Nations, a loose federation of 50 countries, and their dependencies, which still exists. As with the Roman Empire, the lead-laden ruling class of the British Empire must take some of the blame. The decline of their empire was so rapid in the 20th century that it will always puzzle future historians as to its cause. Clearly lead cannot be blamed for this, but one might speculate that some future historian will conclude that some hidden factor undermined it from within and conclude that lead was the reason.

Under leaden skies

Every breath we take adds to our body's burden of lead. This was even true for prehistoric man but it is not something we worry about today, although a generation ago it was a cause for concern because the amount of lead dust in the air of cities was high. It was assumed that such dust would settle near to where it was emitted because lead compounds are heavy solids, and indeed most of it did – but not all. Some travelled thousands of miles and this was revealed by geologists working in Greenland in the 1960s. A paper in *Geochemica et Cosmochimica Acta* in 1969 [vol. 33, p. 1247] reported the finding of Masayo Murozumi, of the Muroran Institute of Technology, in Japan, and Tsaihwa Chow and Clair Patterson, of the California Institute of Technology. They alerted the world to the extent to which lead was entering the atmosphere, coming mainly from the massive use of lead additives in gasoline. They also wanted to discover the amount of lead in the Earth's environment in pre-industrial times and so they took ice core samples from deep within the Greenland and Antarctic snows and analysed for lead. This allowed them to calculate a background level of lead from prehistoric times and which represented the lead from natural sources, such as volcanoes, windborne sea spray, sandstorms, and soil dust. But there was much more lead than expected in the snow of recent times.

In the layers of snow that had fallen during the various Ice Ages there was only a little lead amounting to 0.5 parts per trillion (ppt) on average. During the Roman era the level rose to 2 ppt, although it fell back as the Empire declined, and then revived again in the Middle Ages. However, it was the advent of the Industrial Revolution that really stimulated lead mining and smelting, and from 1750 to 1940 the amount of lead in the snow rose from around 10 ppt to around 80 ppt. Then it increased to more than 200 ppt in the next 25 years and eventually reached a maximum of 300 ppt by the late 1970s. Since then it has fallen due to the phasing out of leaded gasoline and controlling smoke stack emissions. Even so, the level of lead in snow falling today is still higher than it was in the 1800s.

Clair Cameron Patterson (1922–1995) is best remembered as the man who proved that the Earth was 4.5 billion years old, and this he did by means of lead isotopes and which he was able to study by developing sophisticated techniques for analysing them in minute amounts. It was thanks to Patterson that lead isotope geochemistry became an important part of the way that lead in the environment can be tracked back to its source because different lead minerals have different isotopic compositions. Patterson was instrumental in showing the extent to which lead was now contaminating humans and the environment. One result of his work was that leaded petrol was eventually phased out, and indeed many of the measures taken to remove lead from other contaminating sources, such as paint and the lead solder on canned foods, were based on his observations. Patterson was able to show that the natural level of lead in humans, determined from the bones of 1,600-year-old skeletons from Peru, was much less than the levels that were accepted as safe. Patterson was elected to the US National Academy of Sciences in 1987, despite the efforts of the lead industry to discredit him, and he has even been honoured in having an asteroid and a mountain peak in the Queen Maude Mountains named after him.

From his researches, Patterson calculated that the body burden of lead in an adult in prehistoric times would be a mere 2 mg whereas by the mid-1960s in the USA it was more like 200 mg. Patterson also speculated that the effect of lead on the human brain may have had an influence on the course of human history.

There are several environmental archives that preserve a record of lead deposition from the atmosphere, namely polar ice, glaciers, peat bogs, lake and marine sediments, and even trees, although some of these can be contaminated by local pollution. The various archives show a natural background level of lead of between 9,000 and 24,000 tonnes per year. Some of them, such as the ice at the poles, in Greenland, and of high Swiss glaciers, are free of interference and these not only preserve a reliable record but the ratio of lead isotopes (lead-206 compared to lead-207) can tell us where any increase in lead was coming from. Nor is it only ice that preserves a record. The lead analysis of a peat bog on a Swiss mountain has allowed William Shotyk of Bern University to construct an atmospheric lead profile going back 14,500 years. His results show that agriculture started in Europe around 4000 BC with forest clearing and soil tilling. Then about 1000 BC, the isotope ratio changed when serious lead mining began, probably by the Phoenicians, because this increased the level of lead-207 compared to lead-206. It increased even more in the 19th century when Australian lead ores were mined because these had even more lead-207.

Six thousand years ago, the level of lead in the atmosphere began to rise due to human activity but mining as such only really became important with the introduction of silver coinage round 700 BC when concomitant lead production was about 10,000 tonnes per year. When the Romans began mining lead to use as a metal, production eventually reached 100,000 tonnes per year, but declined markedly with the end of their Empire in the 5th century AD. Mining stayed at a relatively low level until the Middle Ages, when it was boosted by the mining of the metal for its silver content in Germany and by the Spanish in the New World, but it really took off with the advent of the Industrial Revolution, when it increased dramatically reaching about 1 million tonnes per year in the early 20th century and quickly rose to 4 million tonnes a year once lead was added to gasoline, which accounted for more than half of this total. Despite the phasing out of this use of lead in most countries, production of the metal has continued to grow and now exceeds 6 million tonnes per year.

A world driven mad by lead

Not all compounds of lead are heavy solids; some are liquids and some of these are slightly volatile. Such a compound was discovered in Germany in 1854. It consisted of a lead atom to which were attached four ethyl groups (CH_3CH_2) and it was known as tetraethyl lead (TEL). It is these groups that give it a relatively low boiling point, for a metallic molecule, of 202°C, which is not high but is high enough for its vapour to be dangerous to breathe. Little interest was paid to this molecule for more than 50 years until the advent of the motor car and the problem of 'knocking'. This is the sound that the internal combustion engine makes when it misfires, generally prematurely, so that not all the fuel in the combustion chamber is burnt. The search for anti-knock additives started in 1912 and it became urgent as the popularity of motoring grew and as the oil refiners struggled to produce enough high-grade gasoline to meet the demand. An additive that solved the knocking problem might solve theirs as well, because it would allow engines to burn lower-grade fuel, more of which could be extracted from each tonne of crude oil.

All kinds of chemicals were tried as anti-knock additives but it was TEL that solved the problem. A few millilitres of TEL per gallon of petrol eliminated engine knock even when the compression was doubled. DuPont began to manufacture TEL at its plant in Wilmington, Delaware, and it went on sale on 2 February 1923 at a service station in Dayton where motorists could purchase gasoline with TEL added by the attendant. Slowly the sales of leaded gasoline began to rise.

The manufacture of TEL and its dangers were suddenly brought to public notice by the death of Ernest Oelgers on 24 October 1924, the day after he had been admitted to hospital suffering severe hallucinations and behaving insanely while working with this gasoline additive. Other workers were also showing curious symptoms. In the factories manufacturing TEL there were soon many cases of serious lead poisoning and many deaths. At least 50 men died at the DuPont plant. The trouble with TEL was that it could be absorbed through the skin.* The TEL crisis worsened during 1924, and newspapers reported more and more deaths. Of the 49 men working with TEL at the Standard Oil Company, five died and 35 ended up in hospital.

---

* It was later revealed that TEL had been tested as a chemical warfare agent by the United State War Department because its vapour was so toxic.

Meanwhile tests on gas station employees showed they too had elevated levels of lead in their blood.

The problem had to be solved and the manufacturing companies instituted a set of regulations that effectively reduced exposure of those working with TEL. Nor were gas station employees to be expected to add the TEL, this was done to tankers of gasoline before they left the refineries. In 1926 the US Surgeon General, Hum Cummings, set up a committee of investigation into TEL and recommended a maximum concentration of 3 ml per gallon, as opposed to the 5 ml per gallon being suggested by the Ethyl Corporation. This recommendation was implemented by the industry, even though it was not mandatory.

The Ethyl Corporation then went on to manufacture almost 7 million tonnes of TEL during the next 60 years, and was still defending its use as late as 1985. It strongly maintained that there was no alternative, but the introduction of TEL probably held back the earlier development of high-efficiency engines and more highly refined petrol grades.

The major manufacturer of TEL outside the USA was a British company Octel, and it too argued that it was an essential ingredient in fuels. Indeed it was so essential that the British Government set up the company in 1938, as it prepared for World War II, because it was worried that supplies might be cut off. Without TEL, the Battle of Britain might well have been lost because the Spitfires of the RAF needed it to fuel their Rolls Royce Merlin engines. Octel claimed that despite the search for other fuel additives, and the testing of almost 1,000 possible alternatives by the oil companies, no substitute for TEL had been found beyond the chemically similar tetramethyl lead (TML) which has four methyl groups (CH_3) rather than ethyl groups. When leaded gasoline burns, the TEL and TML are converted to inorganic lead compounds which are emitted as dust particles from the vehicle's exhaust. While the heavier particles tended to settle nearby, the lighter particles were carried away on the air and, as we have seen, were even deposited in the snow in polar regions.

The polluting of city air with lead did not appear to change much with the introduction of TEL but that was because the pollution from coal burning in cities, which also emits lead into the atmosphere, was declining. In any case in the 1940s and 1950s none of the inhabitants of cities who were tested had a blood lead level in excess of 80 micrograms per 100 ml and it was only when this threshold safety level was lowered to 60 micrograms per 100 ml in the 1960s were some individuals found who were in excess of it (11 out of 2300 tested). Moreover even people in rural areas had lead in their blood, albeit much less.

Also in 1994 another archive of lead levels turned up, this time discovered by Richard Lobinski of the University of Antwerp, Belgium, in the unlikely cellars of a wine producer. He analysed bottles of Châteauneuf du Pape which comes from a vineyard at the junction of the A7 and A9 motorways in the Rhône region of France and found increasing lead levels in the wine from the 1950s onwards. One of the best vintages, that of 1978, had the highest level of lead, and indeed many times higher than more recent vintages; that year represented one of the highest. His findings also reflected the changes in the compound added to petrol. TEL residues in the wine declined from the 1960s onwards, while TML, which replaced it, increased in those years. Together they reached a maximum 0.5 ppb by 1978. The researchers concluded that if the 1978 wine was drunk regularly it could cause mild lead poisoning, but this is most unlikely to worry anyone because this is one of the best vintages, and is quite expensive. Since 1980 the levels of lead in Châteauneuf du Pape have fallen and by the mid-1990s were only a tenth of these earlier years.

In February 2004 Ezra Susser, a psychiatrist at Columbia University New York, reported to annual meeting of the American Association for the Advancement of Science that lead in gasoline and paint might have been responsible for up to a quarter of cases of schizophrenia. His results were based on blood samples taken from pregnant women in the years 1959 to 1966 who were contacted later in life after they had had children. It appears that those whose children were most exposed to lead while in the womb were twice as likely to develop schizophrenia.

Not every scientist believed that leaded gasoline was a threat to the planet. The dangers of TEL and TML were greatly exaggerated according to Zbigniew Jaworowski, of the Central Laboratory for Radiological Protection in Warsaw, who wrote a lengthy paper entitled 'The Posthumous Papers of Leaded Gasoline' in 1994, which was published in *21st Century*. This claimed that the work of Patterson had been misinterpreted and showed conflicting evidence because the lead in snow at some sites had actually fallen during the period 1966 to 1971 when the use of leaded gasoline was rapidly increasing, and that the level of lead in Greenland snows was more strongly correlated to volcanic eruptions. At the end of his funeral oration for lead,

Jaworowski concluded that the decrease in lead levels in humans had been primarily due to its removal from the domestic environment and not by outlawing leaded gasoline.

Painting men, painted ladies, and paint-eating children

White lead is an exceptional material. It is brilliantly white and a small amount will cover a large surface area. It was a key ingredient of paints for thousands of years. Housepainters employed it; artists admired it; and women wore it to be admired. To a greater or lesser degree, all were to suffer because of it, but most affected were those to whom its use had little meaning: babies and young children. They would be poisoned by it, sometimes fatally, but an awareness of the threat came only slowly, and the battle to provide future generations with a lead-free home life was really only won in the last 20 years.

In Roman times, the best white lead came from the island of Rhodes where it was made by laying thin strips of the metal across bowls of vinegar. Over a period of months the lead would be coated with a white layer which was then scraped off, ground to a powder, pressed into cakes and left to dry in the sun. This method of making white lead persisted for hundreds of years, until the Dutch found a way of boosting production in the 1600s by stacking the lead next to pots of vinegar, and surrounding these with heaps of manure. The whole would be left in a sealed room for 90 days after which most of the lead would have been transformed into white lead.

The chemistry of the process occurred in two stages: first the acetic acid from the vinegar would react with the surface of the lead to form lead acetate; this would then react with oxygen, water, and carbon dioxide of the air to form white lead, which is a combination of lead carbonate and lead hydroxide in the ratio of $2:1$, and its chemical formula is $2PbCO_3.Pb(OH)_2$. The Dutch process boosted the yield because the decomposing manure not only provided heat but generated ammonia and carbon dioxide, both of which would increase the chemical attack on the metal.

White lead was widely used in cosmetics in the 1700s and 1800s, when to be white skinned was to be beautiful. In the USA one particular product, called Bloom of Youth, occasionally led to its users dying of lead poisoning. Yet such cosmetic uses of white lead stretched back to ancient times in both

Europe and the Far East; and was essentially for those working in the theatre and for the geishas of Japan.

Artists all over the civilized world regarded white lead as *the* white pigment – nothing came close to matching its brilliance and depth. Although alternative white pigments were available, made from calcined bones, oyster shells, pearls, and common chalk, they did not compare to white lead. Barium sulfate, which became available in the early 1900s was also very white, but had less covering power and it was expensive.

Were famous painters affected by lead? In 1713, the physician Bernardino Ramazzini speculated that Correggio and Raphael might have been victims of lead poisoning. Goya has also been thought to have been affected, and especially Van Gogh who was particularly fond of sucking the paint from his brushes, and this might well have contained some lead. His erratic behaviour and mental state are consistent with the effects of lead poisoning. Nor was white lead the only danger to which he might have been exposed. Some paint pigments were also made of lead such as chrome yellow, which was lead chromate ($PbCrO_4$), and red lead, which was a lead oxide (Pb_3O_4), and both were employed well into the 1900s. (The ancient Romans were particularly fond of covering their walls with red lead paints.) Painters of the Middle Ages used lead stannate yellow ($PbSnO_4$) until it was displaced by the richer hues of chrome yellow.

The trouble with artists using white lead was that it did not stay white when paintings were hung in houses and churches that were heated with coal, whose sulfurous fumes would react with its constituents, reducing them to black lead sulfide. (Barium sulfate offered a way to prevent this happening.) The same effect can be seen in many ancient manuscripts, where faces that were originally painted pale pink are now deep black.

Recognition of the risks associated with white lead were slow in coming, although even in the 1800s occupational lead poisoning was a recognized affliction of those who worked as housepainters, and it was noted that painter's colic was less prevalent among those who worked outside than indoors. Those who worked with white lead were very much at risk, especially young women. The playwright, George Bernard Shaw, even inserted a plea on their behalf in his play *Mrs Warren's Profession* where she questions the benefits of those women who follow a 'respectable' profession as opposed to her own disreputable, but lucrative, business as the madam of a brothel:

Well what did they get for their respectability? I'll tell you. One of them worked in a white lead factory for 12 hours a day for nine shillings a week, until she died of lead poisoning. She only expected to get her hand a little paralysed, but she died. That was worth being respectable for, wasn't it?

Some countries legislated for improved industrial hygiene and in the 1800s a law in Germany forced employers to reduce lead exposure or face the cost. In the UK the Factory and Workshop Act of 1895 made it mandatory to report occupational poisoning by lead, and among white lead manufacturers the number of workers affected fell from 1,058 cases in 1900 to 576 in 1910. France banned white lead, first in indoor paints, then from all paints in 1909. In 1994 the European Union banned the sale of white lead except for a few specialized uses, one of which in the UK is the painting of the outside woodwork of Grade I listed buildings. These have to be renovated using exactly the materials that were employed when they were first constructed.

White lead accounted for as much as a third of lead production well into the 1920s, after which it was in competition with titanium dioxide which is just as brilliantly white and, more important, is non-toxic. But lead did not go quietly and indeed housepainters were said to prefer lead-based paints because the paint wore away slowly and evenly, making it easy to apply future coats.

Children are particularly sensitive to lead, but rarely did poisoning occur *en masse* so it was generally undiagnosed. Just occasionally it could not be ignored, such as in the early 1900s in Queensland, Australia, where there was an outbreak of lead poisoning among children. The cause was traced to their ingesting flakes of lead paint from sun-weathered verandas. Childhood lead poisoning was likely to show itself as vomiting and diarrhoea, but a more insidious effect was its attack on the central nervous system, leading to impaired learning ability, which may well confer a lifelong disadvantage on the child.

The suspicion that lead was causing behavioural difficulties in children was first suggested in the late 1930s by Randolph Byers, a paediatric neurologist, and Elizabeth Lord, a psychologist, working at Boston's Children's Hospital in the USA. They identified 128 children with lead poisoning and monitored their progress up to 1943 when, sadly, Lord died of leukaemia. Their paper, published that same year, showed just what a serious effect the lead had on children.

In the 1930s it was discovered that a form of meningitis could be traced to lead-based face powders used by nursing mothers, and that this may have been harming and even killing children for centuries. Records of the Baltimore City Health Department showed that lead poisoning cases rose steadily from 1936 onwards, when there were 83 cases (32 being children), reaching 493 cases by 1950 (253 children, of whom 9 died). It was pointed out that if this was only in the Baltimore area then the numbers affected all over the USA must have been enormous.

In January 1971 President Nixon signed into law the Lead-Based Paint Poisoning Prevention Act with funding of $30 million to assist lead-paint abatement programmes. These have been responsible for preventing thousands of cases of juvenile lead poisoning and the deaths of hundreds of children. More than 1 million children had been screened by 1975 and more than 4 million by the early 1980s. A quarter of a million at-risk children were identified and 112,000 homes were cleared of lead paint and plaster. Amendments to the Act lowered the permissible lead content of paint from 1% to 0.5% and then to 0.06%. The last reported death of a child from eating chips of lead paint was in Wisconsin in 1990. The paint chips it had consumed were 30% lead.

Renovating older buildings may require painted surfaces to be sanded down to bare wood before being repainted, but this presents problems in that the dust created can be dangerous if inhaled. This fact was not always appreciated, and curiously it tends to affect the upper classes who often live in historical mansions. Indeed, during the Presidency of the first George Bush, in early 1990s, the White House was renovated and such was the amount of lead released, that the family dog Millie nearly succumbed to lead poisoning.

Children are no longer haunted by the spectre of white lead poisoning, but in some cultures black lead can be just as dangerous. The traditional Al Kohl, which was widely used in the Middle East, was powdered galena (lead sulfide, PbS) and it was not only applied round the eyes of women, but also of children and even babies. In some cases children were exposed to the fumes of heated lead in the belief that this would have a calming effect. The high incidence of premature labour in mothers who use Al Kohl is thought to be due to mild lead poisoning, and still the toxic compound continues to be used.

Modern uses and misuses of lead

In the 1900s, lead arsenate was used as a pesticide against leaf-eating insects, although not permitted on crops destined for human or animal consumption. It was regarded as safe when dusted on tobacco crops, but the result was that smokers had more lead in their bodies than non-smokers. That misuse of lead had now ceased and at least smokers need no longer worry about their exposure to these two toxic elements.

More lead today is mined, recycled, and used than in any previous time in history. Production in 2003 was around 6.5 million tonnes of which 60% was recycled lead. The popularity of the motor car and the production of lead went hand in hand throughout most of the 1900s. Lead production doubled and doubled again as it was used in many ways: lead undercoats, lead in tyres, lead batteries, and lead in the gasoline. The last use has since almost ended but not the love affair between car makers and lead producers. Car batteries still account for more than three quarters of all lead consumed in the USA. In lead batteries the anode is spongy lead, the cathode a paste of lead oxide on a lead alloy metal grid, and all of this can be recycled.

What cannot be recycled was the leaded glass used for the cathode ray tubes of television sets and PCs. These contained a lot of lead, maybe up to a kilogram or more. However, though it can't be reclaimed, it poses no threat, not even when it is disposed into a landfill site because the lead is bonded to the glass and does not dissolve when it comes in contact with ground water.

Mined lead is still refined to remove the silver it contains (as much as 1.2 kg per tonne). This is done by adding zinc to the molten metal and allowing it to cool slowly until the zinc settles out as a separate layer on top of the lead carrying the silver with it and this is separated off to reclaim the silver. The molten lead is then heated under vacuum to drive off any remaining zinc, giving a product that is 99.99% pure.

Other uses for lead are in sheeting, cables, solders, lead crystal glassware, ammunition, bearings, and in sports equipment, for weight-lifting and to balance golf clubs. It is used as a stabilizer for PVC although this is to be banned in the EU from 2015 onwards. Lead is very poor at transmitting sound and vibrations, so that it is added to plastic sheeting and tiling designed to block out noise. Lead is still used in architecture as roof cladding and for stained glass windows. It provides protection that will last for centuries, even in industrial and coastal regions, nor does it cause discoloration of surrounding stone or brickwork. In urban or industrial environments the protective surface layer slowly changes over the centuries from lead oxide to lead carbonate, and finally to lead sulfate. In all cases the lead compound that is formed is insoluble and, unlike rust, does not flake from the surface, and it forms a protective coating.

Some uses of lead were no threat to humans, but they have been phased out because of their threat to wildlife. Unfortunately lead shot is still widely used, but lead sinkers are a thing of the past in countries where swans are to be found. Fishermen used lead sinkers and these were often lost to the bottom of rivers and lakes where swans could scoop them up as they fed in the mud along the bottom of rivers. The sinker would then reside in the bird's stomach, slowly weakening it over many months until it died of lead poisoning.

Curiously as one form of lead pollution is ended, there seems always to be another way in which it can be exploited to the detriment of some. In 1994 there was an outbreak of lead poisoning in Hungary, due to red lead being used to colour paprika, the spicy flavouring made from dried red peppers. Eighteen people were arrested but the extent of the damage it caused is hard to assess because Hungarians used the paprika to colour many foods, such as goulash, sausages, and salami. Thankfully, the deceit was not practised for so long that it put lives at risk, although one wonders if this was really a new way to adulterate food, or a rediscovery of an ancient food colouring additive.

Lead is still a useful metal, but only for a few products, and none of which should release its lead so that it could find its way into the human body. At the start of this chapter I quoted the words written on the lead casket in *Merchant of Venice* and commented on how prescient the phrase was: 'Who chooseth me must give and hazard all he hath.' At the end of this chapter I hope you agree. When we expose ourselves, our society, and our environment, to lead then we do indeed hazard all we have. Thankfully that lesson has now been learned.

Lead and dead

Gout was once a common malady that immobilized many of the upper class males of Ancient Rome and Imperial Britain. Both societies blamed it on too much rich food and wine, and they may have been right. The Roman writers, Seneca, Virgil, Juvenal, and Ovid all poked fun at the sufferers of gout, as did the London cartoonists; the popular belief was that it was a just punishment for over-indulgence. Physicians knew of the pain it caused and discovered that it was due to sharp crystals of uric acid between the joints of the bones; but what caused these to form?

Among those affected by gout were Benjamin Franklin, one of the founding fathers of the United States, British Prime Minister William Pitt, Alfred Lord Tennyson the poet, Charles Darwin the biologist, and John Wesley the founder of Methodism. It has been suggested that Alexander the Great, Kubla Khan, Christopher Columbus, Martin Luther, John Milton, and Isaac Newton also suffered its agonies. In the last century, it was found that more than a third of those suffering from gout had high levels of lead in their blood. It now seems likely that earlier generations had exposed themselves to gout by a fondness for port wine, which was invariably tainted with lead, and kept in lead crystal decanters.

At various times in the 1700s, the British were at war with the French and no longer imported their wines or brandy, although quite a lot was smuggled into the country. Instead Englishmen turned to drinking the wines of their most faithful ally Portugal. These, like port and Madeira, contained lead, and they became so popular that by the 1820s more than 20 million litres of port were being imported annually. Bottles of this age have been analysed for their lead content and shown to have in excess of 1 ppm suggesting that the risk of serious lead poisoning from such drink was relatively low, although it would have had an effect. Indeed the lead may have simply served as an irritant to the gut, which is why a glass of port at the end of a meal was reputed to have a laxative effect by the following morning.

Whether lead in drink really did cause gout is still debated because there seems to be no reason why this metal should suddenly cause crystals of uric acid to form between the joints, but it does. These minute, but splinter-sharp fragments, cause intense agony whenever the joint moves and they can form overnight. Two groups particularly at risk of suffering what is known as saturnine gout were those who drank port in the 1800s, and those who drank moonshine liquor in the south-eastern states of the USA in the 1900s. Those in the USA got their lead from alcohol that had been distilled through car radiators used as condensers, the lead solders of which dissolved slightly in the alcohol condensing in them.

Throughout history men have been more prone to gout because the level of uric acid in their blood is higher than that of women. Women have on average 4.3 mg per 100 ml of blood compared to a typical man who has 5.6 mg per 100 ml. Uric acid is the metabolic end product of purines, which are essential constituents of DNA. In other animals, uric acid is further changed to more soluble compounds and so excreted, but humans lack the enzyme that carry this out so we have to excrete it as uric acid which is not very soluble. Too much uric acid and it will crystallize from the blood. Lead appears to interfere with the ability of the kidneys to excrete uric acid so the blood level increases until it reaches a critical level when it suddenly crystallizes. (Gout is not an uncommon complaint even today, but causes other than lead are to blame.) But why should alcoholic drinks in the past have been contaminated with lead?

Leaded delights

The juice crushed from grapes will ferment naturally due to the yeasts from the grape skins, and the resulting wine can contain as much as 13% alcohol. Such wine has been made, traded, and enjoyed since ancient times but there were risks to the vintner in that other yeasts could invade the wine and convert a lot of its alcohol to acetic acid. This sour wine we know by its French name of *vin aigre*, i.e. vinegar, and while it could be sold as a commodity in its own right there was much less demand for it, and it represented a dramatic loss in value compared to the wine from which it had originated.

We saw in the previous chapter how the ancient Greeks and Romans used lead acetate to 'improve' wine by adding *sapa*, but the practice of adding lead to wine did not end with the Roman Empire.

There is no record of who first discovered that wine could be preserved, and even made more palatable, by adding lead oxide to it, but it was a trick of the trade that vintners often used. An anonymous book entitled *Valuable Secrets Concerning the Arts and Trades* was published in London in 1795 and gave the recipe for doctoring wine with lead. The trick was to saturate wine vinegar with litharge (lead oxide, PbO) and then add a pint (about half a litre) of this concoction to each hogshead of wine; a hogshead was a large cask holding 50 imperial gallons (about 225 litres). The resulting wine would then be protected against going off – and it would even taste a little sweeter. The litharge reacted with any acetic acid in the vinegar to form soluble lead acetate. The resulting concentration of lead in the wine might have exceeded 50 ppm and be quite enough for the metal to deactivate the enzymes of rogue yeasts that threatened to destroy its alcoholic content. This amount of lead acetate would not be enough to interfere with the taste.

The adulteration of alcoholic drinks with lead, either deliberately or accidentally, was to bedevil many who drank them in the Middle Ages and later centuries, just as it had in Roman times. Sometimes the amount of lead caused outbreaks of severe lead poisoning which went under various names, often linked to the region where it first appeared, thus there was Picton colic in France in the 1600s, Massachusetts dry gripes in the American colonies in the early 1700s, and Devon colic in southwest England later that century. All involved intense stomach pains, severe constipation, and mental disturbances.

The fortified wines that the British liked may not have had much lead in them but when they were left in lead crystal decanters they could leach more lead. In an article in the *Lancet* in 1991 [vol. 337, p. 142] Joseph Graziano and Conrad Blum of Columbia University, New York, reported their analyses of port stored this way and found that after four months the level of lead could be as high as 5 ppm. (After 5 years, a sample of brandy from such a decanter had 21 ppm.) Crystal glassware can contain as much as 32% lead, this being added to the molten glass as lead oxide PbO. (Unless a glass contained at least 24% lead it could not be called lead crystal, and this is the amount that is used today.) Lead crystal that is produced in Australia has a different ratio of lead isotopes with more lead-206 than is found in lead from the USA. This enabled the researchers to calculate how much of the lead which dissolved from the glass into the wine then passed from the stomach into the blood stream. Using an Australian decanter to store sherry and then giving the contents to Americans to drink it was possible to prove that 70% of the lead that was leached into the drink was absorbed by the body.

Gout was not the only affliction of the middle and upper classes in the 18th and 19th centuries. People often complained of felling unwell and of suffering from ill-defined maladies, which we now suspect was mild lead poisoning, and they were often advised by their doctors to 'take the waters' at one of the fashionable spa towns, of which Bath in England's West Country was the most famous. It was there that those afflicted by 'gout, rheumatics, agues, lethargies, apoplexies, forgetfulness, shakings, and weakness of any member' were taken to be subjected to sitting 3 hours at a time up to their necks in warm water, several times a week for up to six months, while also drinking copious quantities of the local spring water. The cure generally worked and indeed research in the 1980s by Drs J.P. O'Hare and Audrey Heywood, of the Royal Infirmary in Bristol, showed that such a regime would increase the flow of urine and remove significant amounts of lead from the body.

While taking the waters may well have worked for mild lead poisoning, it would have been little help for those suffering severe lead poisoning, as occurred in Europe in the Middle Ages. In various outbreaks, the cause was eventually traced to lead-adulterated drinks but the people of each region had to discover the cause for themselves. In Germany, doctors were able to trace colic to the sweetening of wines with litharge, and the adulterating of wine with lead was even made a capital offence in some states; in Ulm, convicted wine adulterers were in fact executed. Meanwhile in other countries the practice was tolerated.

Picton colic appeared in Poitiers, France, and began in the 1570s; it reached epidemic proportions in 1639. Picton colic was named after an ancient Celtic tribe who inhabited the region around Poitiers. The same disease under the name of 'dry gripes' afflicted the men of the West Indies and the American colonies in the 1600s and 1700s, and was due to rum which had been contaminated with lead from the equipment used to manufacture it. The Massachusetts Bay colony especially was plagued by dry gripes, and the cause of the disease was traced to rum that had been distilled in lead stills and the process was forbidden by law in 1723. The outbreak ceased. However, it persisted in the West Indies until Thomas Cadwalader showed in

1745 that it was caused by Jamaica rum contaminated with lead. Yet even as late as 1788, British soldiers in Jamaica were afflicted by the dry gripes and the cause this time was traced to rum stored in lead-glazed earthenware containers.

The most famous outbreak of this kind of accidental lead poisoning was Devon colic, which affected thousands of people in that English county in the 1700s. The victims were mainly men, who experienced such alarming symptoms as paralysis, madness, blindness, and even death. It was first reported in 1703 and the disease spread year by year. There was an upsurge in the number of cases in 1724, a year that was noted for its abundant apple crop. One investigator, John Huxham, published his findings in 1739 and he blamed the disease on cider, but this explanation was not accepted because the residents of other cider producing areas, such as Herefordshire, were rarely if ever afflicted.

The man who did most to clear up the mystery surrounding Devon colic was the Queen's physician, George Baker. He had corresponded with Benjamin Franklin who told him about an outbreak of lead poisoning that had occurred in Boston when he was a teenager, and that the cause had been traced to lead stills used to distil the rum. A letter of Franklin's still exists, which he wrote to his friend Benjamin Vaughan, about lead poisoning and part of which went as follows:

Philadelphia July 31 1786
Dear Friend,

. . .

The first Thing I remember of this kind, was a general discourse in Boston when I was a Boy, of a Complaint from North Carolina against New England Rum, that it poison'd their People, giving them the Dry Bellyach, with a Loss of the Use of their Limbs. The Distilleries being examin'd on the Occasion, it was found that several of them used leaden Still-heads and Worms, and the Physicians were of the Opinion that the Mischief was occasion'd by that Use of Lead. The Legislature of the Massachusetts thereupon pass'd an Act prohibiting under severe Penalties the Use of such Still-heads & Worms thereafter.

. . .

When I was in Paris with Sir John Pringle in 1767, he visited La Charite, a Hospital particularly famous for the Cure of that Malady, and brought from thence a Pamphlet, containing a List of the Names of Persons, specifying their Professions or Trades, who had been cured there. I had the Curiosity to examine that List, and found that all the Patients were of Trades that some way or other use or work in Lead; such as Plumbers, Glasiers, Painters, &c. excepting only two kinds, Stonecutters and Soldiers. These I could not reconcile to my Notion that Lead was the Cause of that Disorder. But on my mentioning this Difficulty to a Physician of that Hospital, he inform'd me that the Stonecutters are continually using melted Lead to fix the Ends of Iron Balustrades in Stone; and that the Soldiers had been emply'd by Painters as Labourers in Grinding of Colours.

This, my dear friend, is all I can at present recollect on the Subject. You will see by it, that the Opinion of this mischievous Effect from Lead is at least above Sixty Years old; and you will observe with Concern how long a useful Truth may be known, and exist, before it is generally receiv'd and practis'd on.

I am, ever,

Yours most affectionately

B. Franklin

In 1767, George Baker proved that the cause of Devon colic was lead poisoning. He showed conclusively by chemical tests that Devon cider often contained lead, whereas that from other districts was not so contaminated. From a flagon of three gallons (about 14 litres) of Devon cider, Baker extracted a grain of lead (65 mg). We can calculate that such cider contained around 5 ppm, and quite enough to produce the symptoms of mild lead poisoning even if drunk at the rate of only a pint a day (about half a litre). Some Devon farmhands drank a gallon a day (about four litres), in which case the symptoms of severe lead poisoning would soon manifest themselves.

Having discovered the poison, Baker then uncovered the agent responsible for putting the lead in cider. The culprit was either a lead-lined apple press or fermentation vessel, or the lead pipes used to convey the apple juice to the fermentation vat. As cider making was a village industry, the degree of poisoning varied from one locale to the next, and depended very much on the amount of lead incorporated into the cider-making equipment. In some villages where men had died of the disease, a particular lead-lined vessel could be indicted.

George Baker knew that many would disagree with his findings because lead compounds were used in normal medical practice with apparently beneficial results. Despite being vilified from the pulpit of Exeter cathedral as 'a faithless son of Devon', he broadcast his findings at lectures and backed them up by a general attack on lead as a likely cause of other undiagnosed complaints, such as sickly children, who may have chewed lead-painted toys, and men engaged in the trades of plumbing and painting. Even as late as 1916, a survey of painters showed 40% of them to be suffering from lead poisoning.

As a result of Baker's publicity and that of others, some of which showed the dangers of drinking the rainwater collected from lead-covered roofs, the public became aware of the effect of lead on their health in the late 1700s. However, the concerns about lead were not to last, and lead piping and lead glazes were universally used throughout the next century.

Lead glaze

Glazed pottery appeared in the early Middle Ages and became popular from the 1500s onwards. The glazing had a high lead content, which could leach lead from the walls of such a vessel when anything acidic like wine, cider, vinegar, fruit juices, or pickled foods were put in it. Homemade wines were to be a source of lead poisoning especially in the 1800s and 1900s, and sometimes the source of the lead was difficult to trace. A typical puzzling case was the illness of a 52-year-old village butcher in England in 1958 (the medical report does not reveal his name or address). He was found to have a high level of lead in his urine which was in fact extracting 0.4 mg per day of the metal from his body, and when treated with the antidote Edathamil he was recorded as passing as much as 15 mg of lead. All this was coming from his homemade elderberry wine, which was tested and found to have 7 ppm of lead, but where was this lead coming from? In fact it came from a large earthenware vessel that he had inherited from an aunt who had told him that it made wonderful wine, as indeed it did. The investigating team reported the wine had a good bouquet and was a rich red colour although with a slightly metallic aftertaste. The trouble was that the glaze on the vessel was pitted and this allowed the fermenting liquor to extract lead from it.

Cases of lead-poisoned drink still crop up occasionally. In a previous book *Was it something you ate?* my co-author Peter Fell recounted a case that happened to a member of his family, a successful businessman living and working in Madrid. In his mid-thirties he began to lose weight although this was not intentional. It was slow at first, but then over a period of months it accelerated until he had shed almost 25 kg (around 50 lb). Chronic constipation and severe stomach pains had accompanied this loss but the cause was undiagnosed. When he was hospitalized, tests revealed the highest blood lead level the hospital had ever experienced.

The cause of his poisoning was traced to lead in the wine they drank. He had a weekend cottage in the mountains in Spain, and there he had bought local pottery. One favourite piece was a 3-litre jug which fitted neatly in the fridge, and in which he stored his favourite drink, made from red wine, fresh fruit, and lemonade. This was left to chill and would keep for days. He had not been aware that this jug had a lead glaze which had not been properly fired, and as a consequence the lead continually leached into the drink.

Glazed cups or mugs may contain as much as 50 g of lead in the glaze. Even today, lead glazes can kill, and there are cases on record of fatalities caused by faulty glazing of homemade pottery. In a case reported in the USA a man drank Coca-Cola every evening from a cup that his son had made for him. This gave him a daily dose of 3 mg of lead and eventually he died of lead poisoning.

One of the earliest warnings about the dangers of lead-based glazes came in a letter to the editor of the *Scot's Magazine* from a James Lind of Edinburgh, and published in their May issue of 1754. He warned of the dangers of putting lemon juice in earthenware vessels that were so coated, and mentioned an incident in which enough lead was extracted from the glaze that when the lemon juice was concentrated by boiling and then allowed to cool crystals of sugar of lead were deposited.* He also warned against using such vessels for pickling vegetables, saying that if preserved food tasted sweet then it should be discarded. He noted that the popular Delft pottery was particularly prone to leach lead from its surface.

Those who worked in the pottery industry were badly affected by lead. In 1875 the UK Registrar General described this trade as one of the unhealthiest in the country. Every year in the north Staffordshire region, which is where the British pottery industry was based, there were 400 cases of lead poisoning in its advanced state where victims were paralysed, having convulsions, and almost blind. That region was also noteworthy for the number of miscarriages, still births, and newly born babies dying within a few weeks, their short lives blighted by recurrent fits.

---

* These would have been lead citrate, not lead acetate.

The fight to remove lead glazes from pottery was long and hard. The glazing of pottery with lead was already widespread in the 1600s and in the early days it was applied as a powder of galena (lead sulfide mineral, PbS) to the surface of the wet clay before firing. In the following century it was applied as a wet paste to already fired pots, known as biscuit ware, which were then given a second firing. The prejudicial effect of working on the glazing process was known even then and alternatives were sought, and found, but were either not as good or more expensive.

Those who did the dipping of pots into the glaze were noted for their cadaverous looks and were known to be prone to Picton colic. Despite Acts of Parliament to prevent children from working at such employment, little was done until the 1890s when a vigorous campaign led by Gertrude Tuckwell of the Women's Trade Union League brought the issue into the spotlight. She alerted the public to the dangers of the industry and urged them to buy lead-free glazed pottery and earthenware. The laws passed in the 1890s reduced the number of lead poisoning cases which declined from 573, and 22 deaths, in the 5 years from 1899 to 1903, to only three cases 50 years later, and one death, in the 5 years from 1949 to 1953. There were no cases or deaths in the years thereafter. Ultimately her campaign was successful and it led, via various regulations, to an end of lead-glazed pottery under the UK Pottery (Health) Special Regulations 1947 which decreed that as of 7 October 1948 only lead-free glazes could be used.

Medical uses of lead

Despite its poisonous nature, lead was used by doctors for around 2,000 years to treat all kinds of illnesses. The practice started with Tiberius Claudius Menecrates, physician to the Roman Emperor Tiberius (ruled 14–37 AD). His diachylon plasters contained a paste made from lead oxide and they were used to treat skin complaints such as sores, boils, and other infections. The Roman recipe recommended heating litharge until it went gold-coloured then grinding this with linseed or olive oil and marshmallow root. Diachylon plasters were still in use in the late 1800s and it attests to their effectiveness that they were still to be found in the British Pharmacopoeia, and those of other countries, in the mid-1950s. There was little danger from using diachylon plasters because lead is not readily absorbed through the skin. The plasters contained about 33% of lead oxide, and were recommended for treating chilblains, corns, bunions, and chronic leg ulcers. Diachylon paste is still used in a product known as Lestreflex in which it is spread in strips along flesh coloured crepe-like bandages.

Diachylon paste itself found a new and illegal use as a way of procuring an abortion in the days when this was against the law. There was an outbreak of lead poisoning among women in Birmingham in the 1890s and it was due to the paste of plasters which was being scraped off and consumed as a way of terminating an unwanted pregnancy, which it successfully did.

Another ancient remedy was Powder of Saturn, a form of lead carbonate that was precipitated by adding potassium carbonate to lead acetate solution. It was recommended for TB and asthma. Lead pellets were sometimes used in an attempt to re-open twisted bowels, and in 1926 a Columbia University medical professor, Cater Wood, tried injecting a suspension of finely ground lead into cancer patients, claiming that 20% of them benefited from the treatment to some extent. More effectively, lead acetate was used to staunch 'internal' bleeding in the case of women, i.e bleeding from the vagina, and for piles. Lead salts have an astringent reaction and will promote the coagulation of blood by the formation of insoluble lead proteinate.

Lotions made from lead acetate were advocated by an 18th-century surgeon, Thomas Goulard, of Montpellier, France, who wrote a book *The Extract of Saturn* advocating the use of lead in medicine. He made his lotion by boiling golden litharge with wine vinegar. This was to be used externally for bruises, wounds, abscesses, erysipelas, ulcers, skin cancers, whitlows, piles, and the itch. Erysipelas is a streptococcal infection of the skin producing deep red patches; whitlows are inflammations around fingernails and toenails; the itch was the name for scabies, a contagious parasitic infection of the skin, often around the genital area. In the 1930s in the USA, lead acetate solution was used to treat poison ivy dermatitis.

All the above treatments were really external and unlikely to cause lead poisoning in those being treated. Not so the lead medicaments that were to be taken by mouth. Lead acetate mixed with sulfur was prescribed for tuberculosis, and pills of lead acetate and opium were given to cure diarrhoea, which they did. These latter pills contained about 100 mg of lead acetate, which was enough to cause constipation, while the opium deadened the pain of any colic. Lead acetate was sometimes used as a sedative to treat hysteria and convulsive cough.

Nothing with lead as one of its ingredients would be approved for medical

Lead still finds its use in folk medicine
◄═╋═►

In 2000 a physician in Walla Wall, Washington state, was treating a 2-year-old child who had been admitted to hospital, and he realized his young patient was suffering from acute lead poisoning. In fact the lead level in the child's blood was 124 micrograms per litre. The cause was a traditional Mexican folk medicine known as *greta*, and which was a bright orange powder. The parents had used it several times to treat the child whom they thought was suffering from stomach-ache. When *greta* was analysed it proved to be almost pure lead oxide. Other cases of children affected by lead also turned up among the Hispanic communities of the Western States and several of these traditional remedies, known by a variety of names such as *Rueda, Maria Luisa, Coral, Azarcon*, and *Liga*, were found to be mainly lead oxide. Their importation into the United States is now banned. These remedies were traditionally used for all kinds of stomach upsets and especially diarrhoea, which they would in fact control.

use today, but there is one condition for which lead acetate is available, and as an over-the-counter treatment. It is the active ingredient in some hair gels for men, because it will turn grey hair dark brown. Hair, including grey hair, contains a lot of the sulfur-containing amino acids cysteine and methionine and lead will bind strongly to the sulfur atoms in these and the molecular structures that form are thereby permanently stained dark brown. Nothing else cures greying hair like lead.*

The poisonings of Handel and Beethoven

George Frederick Handel (1685–1759) was the first of the great entre-preneurial composers. Born in Halle, Germany, he was eventually appointed musician to the court of the Elector of Hanover, and when the Elector became King George I of England, Handel moved with him and lived the rest of his life in London, producing a stream of operas and oratorios. His most popular work was the *Messiah* which was premiered in 1742. This

---

* Although it is not recorded, it is more than likely that those exposed to high levels of lead may not have had grey hair.

masterpiece may well have been written while Handel was suffering from lead poisoning although the only way this affected him was to give him gout, and clearly it did little to hamper his creative spirit. The lead undoubtedly came from his drink and he was particularly partial to port wine. On one of Handel's manuscripts he jotted a reminder to order 12 gallons of port from his wine merchant, and doubtless this drink was the cause of his gout, but there is no indication that Handel was affected in any other way by lead.

One previously unsuspected sufferer of chronic lead poisoning was Ludwig van Beethoven (1770–1827). We now have proof positive that at least during the last year of his life he was heavily exposed to the metal and there is every indication that the terrible colic that afflicted him throughout his life was due to lead. Beethoven exhibited the symptoms of chronic lead poisoning even as a young man and in 1802 he knew something was seriously wrong with his health. He was already showing signs of deafness. He wrote a letter to his brothers Johann and Caspar which said that after his death they should do all they could to find out what was the cause of his ill health. He never sent this letter to them and it was still in his desk when he died 25 years later, and it is known as the *Heiligenstadt Testament*, named after the village on the Danube where the composer lived.

We have scientific proof that Beethoven was afflicted with lead. It was common in the 19th century to cut locks of hair from those who had died and to put these in lockets. Beethoven died on 26 March 1827, in Vienna, and the following day 15-year-old musician Ferdinand Hiller came to pay his last respects and was allowed to cut a lock of the great man's hair as a memento. Hiller, who went on to become a composer and conductor, put the hair in a locket which he later gave to his son Paul. Although it passed to other members of the family, its provenance was never in doubt, and descendants of Paul's used the locket to buy safe passage out of Nazi occupied Denmark in World War II. Their escape to Sweden was organized by a Danish doctor, Kay Alexander Fremming.

When Fremming died, his daughter sent the locket to Sotheby's in London and there it was auctioned in 1994 and bought by the American Beethoven Society. Six strands of the hair were analysed in 2000 at the US Department of Energy's Argonne National Laboratory where they were bombarded with electrons travelling close to the speed of light in a synchrotron. These create X-rays that excite electrons in the atoms of the hair and these can then be identified and their amounts measured. What this sophisticated analysis revealed was that Beethoven's hair had 60 ppm of lead, a hundred times more than normal. (The amount of mercury in his hair was normal, thus scotching a long-standing rumour that Beethoven's illness was due to mercury, and that this had been prescribed as a cure for syphilis.)

Of course the hair in the locket could only show the level of lead to which Beethoven was exposed during the last months of his life, but there is no reason to suppose that his eating and drinking habits were different then to what they had always been. Nor is it likely that the lead had been applied to the hair as a dye, because the strands cut from his head were grey, white, and brown, showing that he was naturally going grey and not trying to disguise it. There could have been many sources of lead contaminating the food and drink of the great composer. The more likely one would have been water taken from a lead cistern, wine, pewter drinking vessels, or lead-glazed pottery in which acidic drinks or foods like sauerkraut were stored. All these were known to provide high doses of lead at the time, and those afflicted suffered just as Beethoven did with painful colic and constipation, with damage to the nervous systems that could account for his irascible manner, and maybe even his increasing deafness, which became total when he was about 50.

Another famous man whose hair has shown high lead levels was Andrew Jackson (1767–1845), seventh President of the USA (1829–1837). An authentic sample of his hair from 1815 has been found to have 131 ppm, supporting historians who believed that he too suffered from chronic lead poisoning. Its source remains a mystery: did it come from medicines or alcoholic drinks?

The poisoning of George III

During his long reign from 1760 to 1820, King George III had several attacks of an unusual illness. Most of these were mild, but some of them involved mental disturbances that alarmed his family and his Government ministers. The minor illnesses of 1762, 1790, and 1795, were free from such madness, as was the more serious attack of 1765 when he was 26 years old. However, the attacks of 1788, 1801, 1804, and 1810 were all accompanied by alarming mental disturbances.

The illness of 1788 was the most serious because it had political implications that resulted in what was called the Regency Crisis. The Prince of

Wales, his eldest son and eventually successor, favoured the Whig Opposition in Parliament, and he believed his father's madness to be permanent. This would necessitate a regency with himself exercising the royal power. The Tory governing party naturally would have been out in the cold had this happened but they were able to drag their feet over passing the requisite legislation through Parliament. The result of their procrastination was that they remained in office because the King recovered and resumed his role as Head of State. His 'madness' had only been a symptom of his condition. The Royal malady and recovery had important repercussions for the treatment of all lunatics, once it was realized that madness could be a curable condition.

The King's illness of 1788 is worth a closer look because it has been clearly documented. Its chief features were severe constipation, colic, weakness in the limbs, difficulty in swallowing, sleeplessness, with progressive mental disturbances, starting with talkativeness, then irritability, progressing to delirium and coma. These read like a textbook case of acute lead poisoning.

It all began on 11 June 1788 when the King returned to Kew Palace after reviewing the Duke of York's Regiment on Wimbledon Common. Sir George Baker, who by now had been knighted and made the King's Physician, was called in the following day to attend to the King. His Majesty was suffering from colic. For 2 weeks the King was ill but then he left to take the waters at Cheltenham Spa from 9 July to 11 August. There he regained his health, but the recovery was only temporary. A second attack of colic started on Friday 17 October, and again Baker was called to attend the King at Windsor Castle. He arrived to find the King in great pain with stomach pains and cramps in his arms and his legs. For the next 3 weeks the King's health went steadily downhill although on some days he was much worse than on others. He was also constipated, suffered from insomnia, and was weak in the limbs.

Towards the end of October, signs appeared that his brain was being affected. He became talkative and agitated. He rambled, and at times became giddy. On Wednesday 5 November he had a big row at dinner with the Prince of Wales when the conversation turned to the subject of murder. Stung by a remark of the Prince, the King attacked his son physically. The Queen had hysterics and the Prince burst into tears.

What drove the King to such extreme behaviour? It may well have been the suspicion that the Prince would not have been too unhappy to succeed him. Indeed the Prince was in such financial and matrimonial difficulties that only his elevation to the throne could have solved them. Despite a parliamentary grant of £221,000 the year before, plus an increase in his annual income, the Prince had again fallen badly into debt. His riotous living eventually piled up a total of £630,000 in debts over the next 6 years (equivalent to something like £50 million at today's prices). Moreover, the Prince had secretly, and illegally, married his mistress, Mrs Fitzherbert, in 1785 and she was a Roman Catholic. This meant that the marriage was not only unlawful but politically disastrous, if the knowledge of it were to leak out because of the intense anti-Catholic feeling in the country.

The weekend of Sunday 9 November 1788 was a crisis. The King had obviously lost control of his mind and his physical condition deteriorated so rapidly that in London the rumour went round that he had died and by Monday 10 November the King was only semi-conscious – but then he began to recover physically. His mental condition, however, did not improve and to all intents and purposes he was mad. This symptom of his illness dominated all others, since it was so important to the functioning of Government. It has also coloured his popular image ever since. When it became obvious that the King would survive, fresh doctors were called in, doctors who specialized in the treatment of the insane. The King was confined in a straight-jacket and strapped to a heavy chair in order to control his irrational outbursts. He was dosed with mercury chloride and castor oil for his obstinate constipation, and given quinine for his fever.

Despite relapses about Christmas and the second week of January, the King slowly recovered his sanity. In the middle of January the doctor in charge started putting antimony tartrate in his food to induce vomiting. This treatment was kept secret from the King and it greatly distressed him. It was continued for 6 weeks at the end of which time his Majesty was declared to be cured. There was national rejoicing, except among the Prince of Wales and his Whig supporters whose disappointment was ill-concealed. The King suffered similar attacks of colic, constipation, hoarseness, muscular pains, sleeplessness, and delirium in 1801, 1804, 1810, and 1812. The last attack, when he was 73 years old, left him blind and permanently mentally disabled. The Regency was then a necessity, but the Prince of Wales had had to wait 20 years.

The authors of the book *George III and the Mad Business* were the medical historians, mother and son, Ida Macalpine and Richard Hunter. They made an exhaustive study of George III's illness and deduced from the symptoms that his doctors recorded that he suffered from a metabolic disorder which affected the production of the essential body chemical porphyrin. The condition is called porphyria. They were led to this conclusion by the discovery among the doctors' notes of the period that the King's urine was sometimes red in colour. This can be a feature of porphyria; it can also be a feature of lead poisoning. Macalpine and Hunter thought a genetic defect to be the most likely explanation and indeed it is now known that some of George III's descendants were also afflicted with porphyria. They claim even to have traced the original genetic defect back to Mary Queen of Scots (1542–1587), mother of James I of England from when it passed to other royal families of Europe, including members of the House of Hanover of which George III was a member. Anyone with a predisposition of porphyria would be high susceptible to lead. As we saw in Chapter 6, the level of lead in a sample of his hair was 6.5 ppm, more than ten times the normal level. Clearly he was exposed to lead, although not to the same extent as others mentioned in this chapter. However, if the King suffered from the hereditary defect of porphyria then a relatively low intake of lead could have a disproportionately adverse effect.

Lead poisoning and porphyria exhibit the same kinds of symptoms because they stem from malfunction of the same metabolic process in the body. Without modern aids to investigation there is no way of distinguishing the one from the other. Usually lead poisoning is a mild poisoning and it requires a sudden intake of lead, or a sudden release of the lead which has accumulated in the bones, to threaten life. The King's condition was not diagnosed as either lead poisoning or porphyria, which was not to be recognized as a medical condition for another hundred years. It was the symptom of mental disturbance that blinded those treating the King to the underlying cause.

In a mass lead poisoning of 1849, some of the victims suffered in exactly the same way as George III. In that episode, a 30 pound bag (14 kg) of lead acetate was accidentally mixed with 80 sacks of flour and used to bake bread. Five hundred people were affected, some of them seriously. Instead of the typical ashen look, characteristic of lead-induced anaemia, some of the victims had a ruddy complexion (as did King George) and some passed red coloured urine. Another feature of the poisoning was the recurrence of the symptoms several weeks after they ate the contaminated bread, just as had happened with the King. The effects of a massive dose of lead is different to the effect of slow lead poisoning of the Devon colic type, which is perhaps why Baker failed to realize what he was dealing with.

Could George III have been a victim of lead poisoning? Considering the times in which he lived, it is more than likely that he would have been taking in too much lead from his food and drink and there are two particular sources of the poisoning that we can deduce from his diet. The King was very fond of lemonade and sauerkraut. Both these foods are very acidic and neither should be prepared or kept in a vessel with a lead glaze. Lead poisoning was common in Germany in the spring of each year, when peasants ate their sauerkraut. This was not a common dish in Britain in the 1700s, but the King was certainly known to enjoy it. Other sources of lead could have added to the King's burden of this metal, such as lead decanters, glazed pottery, and even pewter tankards. Whatever the cause, it is more than likely that the King suffered from mild lead poisoning for most of his life.

The expedition that vanished

On 19 May 1845, the 59-year-old explorer Sir John Franklin set off to search for the supposed North-West Passage around the top of Canada, which was seen as an alternative route from the Atlantic to the Pacific. His two ships *Erebus* and *Terror* were well provisioned with 5 years' supply of food for the 129 officers and crew, whose quarters were equipped with central heating. Three months later, in August that year, the ships were seen in Baffin Bay, but then they disappeared. By 1848, when nothing had been heard of the expedition, other ships were sent to look for them but they returned without finding any trace, and it was not until 1850 that the graves of three crew members were discovered on Beechey Island. The bodies were those of John Torrington, John Hartnell, and William Brain who died in 1846. The ships had clearly spent some time on the island because they had discarded more than 700 empty cans.

The provisions for *Terror* are still on record and included thousands of cans of meat, soup, vegetables, and potatoes. Most of the food they took consisted of flour (30 tonnes), salted meat (14 tonnes), biscuits (7.5 tonnes), sugar (5 tonnes), spirits (2,300 gallons), chocolate (2 tonnes), and lemon juice (2 tonnes), and these were regarded as sufficient to supply this ship of 67 men for 3 years.

In 1988, Dr Owen Beattie and researchers at the University of Alberta, Canada, were allowed to exhume and analyse the perfectly preserved remains of the three men and they found such high levels of lead that it seems almost certain that the men died of lead poisoning, probably exacerbated by scurvy, despite the lemon juice that had been taken on the expedition to prevent this condition developing. The researchers were able to prove that the lead in the bodies came from the solder of the canned food that they ate, by analysing empty cans found nearby. The ratio of lead isotopes in the victims was the same as that of the lead solder, and quite different from the ratio of lead isotopes in local Inuit people. The body of Petty officer John Torrington, which was extremely well preserved, revealed levels of 600 ppm in his hair, proving that exposure to lead was high during the months preceding his death. The other bodies had slightly lower lead levels of 300 ppm but even these indicate a dangerous level of exposure.

Were these seamen really victims of the canned foods they had eaten? It is quite possible. These were the early days of this kind of food preservation and the process and technology of canning was poorly developed. The first commercial food cannery was that of Messrs Donkin & Hall of Bermondsey, London, and it began to supply the Royal Navy with canned meats, vegetables, and soups from 1812 onwards. Indeed Donkin & Hall's 'Preserved Meat' and 'Vegetable Soup' were part of the provisions of the 1814 expedition to explore Baffin's Bay in North Canada. By 1818 the Admiralty was ordering more than 20,000 cans a year, mainly of beef, mutton, veal, various soups, and vegetables.

The cans were filled through a small hole at the top, which was then sealed by having a disc soldered over the hole. They were then heated for an hour or so in boiling water but sometimes the cans were not heated long enough to kill off all the bacteria within them, and then they were found to be putrid when they were opened. The cans preserved their contents by remaining airtight but they slowly leached lead from the solder into the food they contained. In 1824 an expedition, captained by W.E. Parry, had earlier been sent to search for the North-West Passage and he took several thousand cans, two of which were found 112 years later and returned to England in 1936 for analysis. These were a four-pound can of roast veal and a two-pound can of carrots. They were opened and their contents found to be in good condition, although they had a metallic taste. They were then fed to rats without these rodents showing any adverse effects.

Nothing more was found of the Franklin expedition until 1859 when a cairn of stones was discovered on King William Island. In it was a bottle and a note to the effect that the ships had become icebound on 12 September 1846 and that they were unable to free themselves the following summer, 1847, and were still locked in the ice at the end of the following winter, 1848. Franklin himself had died on 11 June 1847, and by the spring of 1848 another 20 men had also died.

At this point the remaining crew decided to abandon the ships and walk the 150 kilometres across King William Island, pulling a boat with them in which they would then row to mainland Canada to the nearest fur-trading fort. According to the note found in the cairn they set off on 22 April 1848. When the lifeboat was eventually found it contained two skeletons and an assortment of articles that defy explanation: button polish, silk handkerchiefs, curtain rods, and a portable writing desk. Were the members of the expedition just behaving irrationally, maybe thinking these were things they could trade with the natives? Possibly. Or were they simply mad like King George III and no longer even able to think straight?

The local Inuit told stories of thin and gaunt-looking white men they had met, and they reported that they were reduced to cannibalism, and indeed some of the bones from the skeletons that were discovered bore knife marks suggesting that flesh was cut from them. Of around 400 bones that have been found, almost a quarter of them have multiple cut marks. A less gruesome explanation is that the marks were from wounds caused by Inuits who attacked them. Beattie analysed the bones for lead and measured levels of more than 200 ppm although this could only indicate a life time exposure to this metal. Again it is indicative of the men taking in a lot of lead with their diet.

Lead may not have caused the deaths of the members of the expedition but it must have seriously weakened them and there is evidence that they also suffered from scurvy. The lemon juice that they took to prevent this would retain its vitamin C for only a certain period, and after a year would be virtually useless as a means of preventing the disease. Whatever happened, the members of that ill-fated expedition certainly suffered from lead poisoning.

Lead murders

CRIMINAL poisonings with lead compounds are noteworthy because of their rarity. Indeed a person intent on poisoning someone would be unlikely to choose lead because of its uncertainty of action. Nevertheless there were murders in which it was used, such as the killing of Thomas Taylor in September 1858, when white lead was the poison, and the murder of Mary Ann Tregillis in 1882, when lead acetate was used. This salt was also the agent in the attempted murder of Honora Turner also in 1858. The lead compound that killed Pope Clement II in 1047 can only be speculated on.

The murder of Thomas Taylor

The inquest into the death of Thomas Taylor, which was held by the coroner for Gloucestershire on 27 September 1858, was reported in the November issue of the *Pharmaceutical Journal* because it was an unusual case of death by poisoning with lead carbonate. Thomas lived in Gloucestershire with his wife, Ann, and child that he had fathered by another woman. He also had a brother, Charles, who had recently been released from prison and who had gone to live with the Taylors. It was not long before quarrels broke out between Thomas and his wife, whom he accused of being too affectionate towards Charles. In fact Ann was more than affectionate; she openly stated that she preferred brother Charles and wished her husband was dead. Her wish was to be granted.

In August 1858 Thomas was seized with violent pains in his stomach which lasted for several days and for which he sought medical treatment. The doctor gave him some opium pills to kill the pain and senna water to act as a laxative. When these failed to cure him his doctor prescribed larger doses which he said could be obtained from his surgery. However, the doctor noticed that when his wife Ann went to collect more senna water she brought along a bottle that had contained the original medicine and that the dregs of the first dose were now a different colour and it tasted odd.

Thomas died on 4 September, but the doctor was sufficiently suspicious of the cause of death that he refused to issue a death certificate until he and a surgeon friend of his had carried out a post-mortem. This they did, but could find no clear cause of death. Nevertheless they took samples from the deceased's stomach and when these were analysed, four grains (about 250 mg) of lead carbonate were discovered. It would appear that Thomas had been poisoned with white lead. Lead was also found in samples taken from his liver. At the inquest, the coroner concluded that death was due to lead poisoning and the jury returned a verdict of wilful murder against Ann Taylor and Charles. She went off to prison to await trial; he was nowhere to be found. What eventually happened to the pair of lovers was not reported.

The attempted murder of Honora Turner

Another murder attempt earlier in 1858 had also been made with lead, in this case with lead acetate. This too was reported in the *Pharmaceutical Journal* and concerned a labourer, 22-year-old James Turner who had married his wife, Honora, in February of that year. The marriage was not a success and he soon left her. She retaliated by complaining to James's employers that he had abandoned her. James now wished to be free of her and so he persuaded his friend, 20-year-old Edmund Keefe to buy some sugar of lead (lead acetate), and the two men went to Honora's home. While Keefe distracted her attention, Turner slipped lead acetate into the beer that his wife and a woman friend were drinking. Both women were eventually taken ill and Honora complained to the police, charging her husband with attempting to murder her. When the police went to Honora's house they found samples of lead acetate on the floor beneath where Honora had been sitting. They arrested both Keefe and James Turner and they were sent for trial at the Old Bailey. The jury acquitted Keefe but found James Turner guilty and he was sentenced to death – attempted murder being a capital crime – but this was later commuted to life imprisonment.

Louisa Jane Taylor (1846–1883)

There are features of this case that make it unique in many ways. It was unusual in that the murderer, 37-year-old Louisa Jane Taylor, was already in police custody on a charge of stealing her victim's clothes at a time when her would-be victim was still alive and her health improving. The case is also unusual in that the murdered woman, a Mrs Mary Ann Tregillis was able to accuse her murderer at a magistrate's court hearing and make a statement before she died. Even more curious was the fact that the pharmacist who supplied several doses of the poison to the murderer was the wife of the doctor who ministered to the victim.

The early life of Louisa Jane Taylor has never been recorded but we do know she was born in 1846 and that her maiden name was Louisa Jane Scott. The only known incident of her young years that has a bearing on what was to follow was that she once attempted to commit suicide with sugar of lead. She had trained as a milliner but there is no evidence that she earned a living this way.

Our story really begins with the death of another Thomas Taylor, on 18 March 1882. He was a retired dockyard official who was living on a Government pension of £60 per year, which was about the same as that of an average working man's wage. What had induced Louisa to marry a man old enough to be her father we do not know. She had been his housekeeper and it came as a surprise to Mr Taylor's relatives that she really was his lawful wife. They refused to believe her until she produced the marriage certificate and only then was she allowed to take possession of her late husband's belongings and furniture. Sadly for her, her husband's pension ceased with his death.

Later, when she had been convicted of poisoning Mrs Tregillis, the suspicious circumstances surrounding the death of Mr Taylor were recalled. We cannot know if Louisa poisoned Mr Taylor but his relatives said she had murdered him, and his doctor suspected that lead poisoning was the cause of death. Moreover, she appeared to have a motive for killing him because she had been having an affair with an Edward Martin, a watercress vendor, although it appears that he was also married.

From March to July of 1882, Louisa rented a house in Little Heath, near Charlton, but eventually she was evicted from her home when she could not pay the rent. By this time she was heavily in debt to one and all and it may have been in desperation that she paid a visit to some old friends of her late husband, a Mr and Mrs William Tregillis. They lived in the upstairs rooms of number 3 Naylor's Cottages, in Plumstead, which they rented for 3 shillings per week, and their accommodation consisted of a front living room and a back bedroom.

Mr Tregillis was in receipt of a naval pension of £49 per year. He had worked in the customs service and was 85 years old. His first marriage, in 1856, had ended with the death of his wife in 1878 and at the time Mr Tregillis was lodged in the Barming Heath Lunatic Asylum. He blamed his wife for his acute depression and her death brought so rapid an improvement in his condition that he was soon discharged as cured. A year later he married his second wife, Mary Ann, who was then nearly 80.

Louisa visited the Tregillises on the last Thursday in July when she told them she had inherited £500. She offered to give them her old furniture, saying that she intended to buy new things for a house she was planning to buy in Charlton. Meanwhile, perhaps she could stay with the Tregillises while the purchase was completed? No doubt the Tregillises imagined that Louisa really was coming to stay for a few days, which explained the sleeping arrangements they made to accommodate her. She would sleep in the same bed as Mrs Tregillis in the bedroom, while the front room would become Mr Tregillis's bedroom at night. To begin with Louisa made herself useful to the old couple and she endeared herself to old Mrs Tregillis by always affectionately referring to her as 'mother'. Soon Martin began to call on her, telling the Tregillises' landlady, who lived downstairs that he was a relative of Louisa.

About a week after she had joined the Tregillises, on 2 August, Louisa and Mrs Tregillis were out walking when they were mugged by a youth outside Woolwich Dockyard railway station. The boy threw the old lady to the ground and in so doing her face was cut. Louisa fought him off and helped Mrs Tregillis back home and put her to bed. In fact she was never to leave it.

Nearby lived a Dr John Smith whose wife also ran a pharmacy and it was there during the week of the mugging that Louisa purchased some sugar of lead. Her reason for wanting this was never explicitly stated, although the coyness with which the subject was dealt with suggested it was of a sexual nature. While Louisa was awaiting trial she also requested some sugar of lead and, after he had examined her, the prison doctor allowed her to have some in order to stop 'internal bleeding' in other words from her vagina. This was a recognized medical use of lead acetate solution, and at her trial the only person called by her defence lawyers was the prison doctor in order to legitimize her claim for having some in her possession while living with the Tregillises.

Louisa had gone to see Dr Smith, complaining of a sore throat, for which he prescribed some medicine. It was while she was waiting for that to be made up that she struck up a conversation with the doctor's wife, saying that she was in a delicate state of health, and asked her about sugar of lead, which Mrs Smith said they had in stock and it cost two pence per half ounce (14 grams). Louisa purchased half an ounce ostensibly for her own use.

What may have precipitated her using it as a poison was the threat that she might be evicted by the Tregillises because her lack of money was about to be exposed. There was no £500 inheritance although she pretended to have deposited this with a building society, and to convince them that this really was true she said she had had a will drawn up leaving them all her worldly possessions. Mr Tregillis was given an official looking envelope bearing the letters O.H.M.S. (On Her Majesty's Service) in which the supposed will was sealed and this he locked in a drawer for safekeeping. When he looked for it at a later date, he found the lock of the drawer had been broken and the will was missing.

Louisa's web of lies was threatened when her sister-in-law called to collect 28 shillings which she had lent to Louisa, but she went away empty handed. She returned a week later on the same errand and noted that Mrs Tregillis was much worse. Again she left without her money but returned another week later, on Monday 28 August, as a result of a letter from Louisa who said that the old woman was dying. On the previous day Louisa had called the landlady from downstairs to come and look at Mrs Tregillis who appeared to be near to death. She told the landlady that the doctor had called and said that she would not last the day.

Nevertheless, she was still alive the following morning when Louisa's in-law arrived, again in the vain hope of the debt being repaid. On this occasion she found Mrs Tregillis in a terrible state: her face was ghostly white, her teeth were black, and her lips an unnatural red colour. Louisa confided that she had been told to put a white powder in the old lady's medicine each night – an instruction vehemently denied by Dr Smith – which Mrs Tregillis said made her sick. The in-law then passed a remark that may have resulted in the old lady being granted a temporary reprieve from Louisa's poisoning attempts: she said that in her opinion it did not appear that Mrs Tregillis was dying a natural death.

Dr Smith had first been called to Mrs Tregillis on Wednesday 23 August because of her poor state of health. He found her to be cold and shivering, yet perspiring, and these symptoms, together with her sallow complexion, made him diagnose malarial fever. He prescribed quinine for the fever, sodium bicarbonate from her stomach pains, and gentian bitter as a general tonic.

On his next visit, the doctor was told by Mrs Tregillis that his medicine made her sick and so he changed it. He was puzzled by the extent to which she said she was vomiting and he asked Louisa to save a sample of her vomit for analysis, but on subsequent visits he was given various excuses why this had not been done. It appears that Louisa had used up all the sugar of lead because the old lady began to recover and the doctor, who had been making daily visits to see Mrs Tregillis, felt able to call less frequently and, on 6 September, he said he would not need to come again because she was clearly getting better.

It is in the nature of those suffering from lead poisoning for relapses to occur while they are recovering, and this is what happened to Mrs Tregillis on Saturday 9 September. Dr Smith was called in and noted that her symptoms had recurred although with less shivering. He suggested that the patient be moved to the front room and he prescribed some pills for her. These seemed to help her recovery and as the days passed he could see her improving until on 16 September he again said he no longer needed to call.

Mrs Tregillis continued to recover but that recovery was not to last. On 20 September Louisa sent Martin around to Dr Smith's shop to buy another half ounce of the poison. She gave him a note asking for a two penny packet of lead acetate, which Mrs Smith happily sold to him. The note was kept by Mrs Smith and produced in evidence at the trial when it was proved to be in Louisa's handwriting. This lead acetate lasted Louisa only about a week because she went out herself to purchase another half-ounce packet a week later.

Mrs Tregillis now had a recurrence of her previous illness, although the doctor was not sent for, and during the first week of October the old lady began vomiting. On Sunday 1 October she was in a particularly bad way, so much so that Louisa had to call for help from the landlady in order to get

Mrs Tregillis back into bed, saying she had fallen out of it while having a fit. Her skin was cold, her eyes staring into space, her fingers were twitching, and her breathing was noisy and laboured. Louisa referred to it as her death rattle.

The following day Mr Tregillis drew his quarterly pension of £12 5s – and Louisa meant to have it. By now she had pawned all her own property plus various items belonging to the Tregillises and had even borrowed ten shillings from the landlady. (Indeed when she was arrested, the police found 23 pawn tickets on her, including some of the Tregillis's clothes.) After breakfast Mr Tregillis set off to collect his pension and Louisa followed him, on the pretext of buying a lobster for lunch, for which she borrowed a shilling from the landlady. Louisa met Mr Tregillis shortly after he had drawn his pension and talked him into giving her £9, saying that Mrs Tregillis had sent her to collect it so that it could be kept safe at home. What she did with the money is not certain, although she repaid the landlady the money she owed her. The rest may have gone to redeem other pawned items of the Tregillises. By the time Louisa was arrested only 9 shillings remained.

The loss of the money threatened the Tregillises with financial ruin and caused a bitter dispute between them, each convinced that the other had it. Eventually Mr Tregillis stormed out in a rage while Louisa sympathized with Mrs Tregillis, saying that she should consider committing him to the insane asylum. Quite unexpectedly a Mrs Trice, an old friend of Mrs Tregillis, called to see her and was disturbed to find her looking so ill. She was told that part of her trouble was that Mr Tregillis had lost most of his pension money. What rather surprised Mrs Trice was the callous way Louisa threw back the bedclothes to reveal Mrs Tregillis's wasted figure and especially her legs, one of which she raised with her hand and let fall saying, jokingly, 'she's going to take part in a running match.'

The following day Mrs Trice's daughter was stopped in the street by Louisa, who told her to fetch her mother to come and look after Mrs Tregillis as she, Louisa, was leaving for a new home in Chatham. Mrs Trice sent her daughter to the Tregillises' home to say that she would come round later. Louisa sent Mrs Trice's daughter out to buy some brandy into which she poured a milk-like fluid* and gave a spoonful to Mrs Tregillis, who said it

---

* If sugar of lead is added to hard water, like that of Plumstead, then it forms a milky solution.

tasted nasty and she was immediately sick. This may well have been the last time that Louisa fed lead acetate to her victim.

Things were now moving to a climax. Louisa spent Thursday packing her things ready to move out. She asked Mr Tregillis if he would like to go with her and live rent-free at the new house in Charlton. Mrs Tregillis even encouraged this, saying that she was sick of the sight of him. Poor Mr Tregillis was saddened by this turn of events and left the cottage to go for a walk, only to find on his return that his bags had been packed and a cab ordered to take him and Louisa to Charlton. By now Mrs Trice had arrived and she advised him not to go. The landlady and Mrs Trice, were beginning to suspect that something was seriously amiss and the events of the following day, Friday 6 October, confirmed their suspicions and goaded them into taking action. That morning Louisa left early and returning at 12.30 pm with Martin to pick up some of her things and even to offer Mr Tregillis a final chance to go with her. He again refused, not on the grounds of loyalty to his sick wife but because he suspected Louisa of robbing him of his pension. 'If you had not taken my turnover [pension] and a pair of boots I would have gone with you' was the reason he gave for refusing.

That same afternoon the landlady sent for Dr Smith to come and attend to Mrs Tregillis. He found her in great pain and barely able to speak. She was in a state of tremor and had lost the power of her hands and wrists. More telling, however, was his discovery of a blue line on her gums which confirmed her illness was due to lead poisoning. This was a condition that Dr Smith had met before, because he had once had a patient who was a moulder of lead bullets and who had died of lead poisoning. Knowing that Louisa had sent Martin to purchase lead acetate from his pharmacy he put two and two together and decided to call in the police surgeon.

The police were also brought into the case that same day by Mr Tregillis who told them that Louisa had stolen Mrs Tregillis's clothes. When Louisa unexpectedly arrived back at 3 Naylor's Cottages that evening, slightly the worse for drink, she was arrested on a charge of theft and taken into police custody. She was accompanied to the police station by Mrs Trice who openly accused her of trying to poison Mrs Tregillis with sugar of lead. The opinion of the police surgeon, who examined Mrs Tregillis the following Monday, was to confirm that her illness was due to lead poisoning.

The following day, Tuesday 10 October, Louisa was taken before Woolwich magistrates and charged with robbery. Because of Mrs Tregillis's weak condition, the local magistrate decided to hold his court in the front room of the cottage, where he could take a statement from her. Unfortunately she found it difficult to speak and her mind wandered. Nevertheless, she did confirm that until Louisa had come to stay she had been in good health and, more importantly, she said that she had seen Louisa putting a white powder into her medicine on one occasion and that as a result the medicine had hurt her throat and she had refused to take any more of it.

The statement that the magistrate eventually recorded was as follows:

> I am the wife of William Tregillis. Louisa Taylor has been living in this house about six months, not as a servant but as a visitor, and has slept with me all the time she had been in the house. I was always in good health till she came. I first became ill about three months ago, when I felt queer and sick and the doctor ordered me some medicine. Mrs Taylor always gave me the medicine. [. . .] It made me very sick after I had taken it. I felt sick three months ago after taking the medicine. I saw two bottles used at a time, both of this size. The doctor ordered me one bottle at a time to be used every four hours. I saw some white powder put in it by Mrs Taylor. I tasted it and said 'I can't take it, it is nauseous stuff and sour as vinegar.' [. . .] I never saw Mrs Taylor mix powder with the physic but once. I have several times before and since noticed the same nasty taste and with the same result. The vomit was of a black colour and always burnt my throat.

Many so-called fever cures in Victorian times were simply dilute solutions of nitric acid. If lead acetate were mixed with one such medicine it would react chemically to form acetic acid, and this would explain why Mrs Tregillis found the poisoned medicine tasted like vinegar, which is a solution of acetic acid.

Louisa attended the magistrate's court but had to be removed when she fainted upon hearing Mr Tregillis give evidence. The proceedings came to an end but the situation was so unusual that the magistrate forgot to sign the depositions. Although this oversight was corrected 3 weeks later, it gave the defence lawyers at the subsequent trial an opportunity to claim that Mrs Tregillis's statement could not be admitted in evidence. How near to success this move came, we shall see.

Friday 13 October really was an unlucky day for Louisa because it was then that she was charged with attempted murder and robbery at the Woolwich magistrate's court. But it was not all bad luck because it seemed that Mrs Tregillis was improving day by day, thanks in no small part to the devoted care of the landlady. She was no longer sick and even her teeth were losing their black colour. Neither patient, doctor, nor caring landlady were to know that the lead in the old lady's body had done irreparable damage and her recovery was only short lived. On Friday 20 October she began to lose her voice again and gradually became paralysed. Three days later she died.

At the post-mortem, the blue lead line on her gums was still visible. However, her brain, lungs, liver, heart, and spleen were all apparently healthy although there were dark patches on parts of her stomach and intestines. The police surgeon who carried out the autopsy was the man who had confirmed Dr Smith's diagnosis of lead poisoning. At the coroner's inquest, however, the police surgeon said that from his findings alone he did not feel justified in coming to the conclusion that death resulted from lead poisoning.

Samples of tissue from the body were sent to Guy's Hospital in London for analysis, together with a sample of the water taken from the tap in the cottage. (This turned out to be free of all but the most minute traces of lead, and although it was coming through a lead pipe it was not dissolving any of the metal on account of the natural hardness of the water.) The analyst found significant amounts of lead in each kidney and traces in the lungs, intestines, and spleen. Her liver contained 0.256 grains per pound, which is equivalent to 37 ppm, and her brain had 0.061 grains in 10 ounces, which is 13 ppm. In the stomach tissue he found 0.432 grains (27 mg) of lead and in his report he concluded that this indicated that Mrs Tregillis had been given a dose of lead acetate very recently. At the trial the analyst, Dr Stevenson, elaborated on this finding and said that it was impossible for this amount to have been there for as long as 2 weeks before death. In fact, it must have been there for 17 days because this was the time interval between Louisa's arrest and her victim's death. This would seem to suggest that Louisa had put sugar of lead into something that Mrs Tregillis consumed long after Louisa had been arrested.

The coroner's inquest was held on Friday 24 November 1882 and the jury returned a verdict of wilful murder against Louisa, whose lover was now trying to distance himself from contact with her. At the inquest Edward Martin denied even seeing Louisa after she left Charlton in July. When confronted with evidence to the contrary he said he had visited her at the Tregillises' home about five times but denied ever buying any lead acetate. Dr Smith's wife contradicted him on that score and said he had called twice at her pharmacy and each time had bought some sugar of lead. Again this seemed to jog Martin's memory and he then recollected that he had made one visit to her shop. In fact Martin the watercress seller was lucky not to be standing next to his lover in the dock.

The trial of Louisa Jane Taylor was held on 15 and 16 December 1882 at the Central Criminal Court in London before Judge Stephens. (He appears to have been sane at this time, although as we saw in the Maybrick Trial, in Chapter 8, by 1888 he was showing signs of mental instability.)

The first day of the trial was mainly taken up with the medical evidence and that of Mr Tregillis who was in the witness box for 3 hours. He told of his previous commitment to a lunatic asylum and how he had married Mrs Tregillis in 1879. He said that they were agreeable to Louisa coming to live with them as Mrs Tregillis was in poor health, although this contradicts what Mrs Tregillis said in her statement and it was not supported by the post-mortem finding. He told the court that Louisa did little to help his wife and that it was the neighbours who used to bring food for the old lady. He also accused her of stealing £1 15s from a drawer. When cross examined, he denied being in any way responsible for nursing his wife during her illness and said that he had never given her any medicine. This was later to be contradicted by Martin in his evidence, and it led to an outburst from Mr Tregillis as he vehemently denied it.

The last witness on the first day of the trial was the clerk of the Woolwich magistrate's court who admitted that some of the statements taken at the bedside hearing had not been signed until 3 weeks later. The council for the defence seized upon this oversight and asked the judge to rule them as inadmissible evidence. However, Judge Stephens called an end to the day's proceedings before coming to a decision. The local newspaper, the *Woolwich Gazette*, of Saturday 16 December carried the headline: Plumstead Poisoning Case Acquittal Expected! Things seemed to be going Louisa's way.

The second day of the trial put an end to all her hopes as Judge Stephens allowed the deathbed statement of Mrs Tregillis to be read to the court. This damning document, in which the victim recalls the way in which her murderer added powder to her medicines, sealed Louisa's fate. The only defence witness was the prison doctor. In his closing speech the defence lawyer made several key points: (1) that Mrs Tregillis may not have died of lead poisoning; (2) that Louisa was the first to send for a doctor – something she would be unlikely to do if she was poisoning the old lady; (3) that Louisa needed lead acetate for medicinal purposes of her own and made no secret of it; (4) that Louisa had no motive for killing Mrs Tregillis; and (5) that Mr Tregillis may have inadvertently given his wife a solution containing lead acetate. This last point was a valid one. If Louisa had been using lead acetate solution to douche herself with, and had stored it in an old medicine bottle, then it was just possible that Mr Tregillis had given his wife some by mistake.

Judge Stephens took 3½ hours over his summing up, and it was impeccable in its fairness. He pointed out that in Louisa's favour was the fact that she had bought the poison quite openly, and if Mrs Tregillis had taken only one dose of it then it could well have been considered an accident. But the fact that she had had several doses ruled this out. Also against Louisa's version of the story was her other behaviour towards the Tregillises which betrayed her motive and that was simply greed. She lied to them about the legacy and she robbed them. Moreover, as the person in charge of nursing Mrs Tregillis she had every opportunity to give her the poison.

The jury retired at 8.08pm and returned only 20 minutes later with a verdict of guilty and with no recommendation for mercy. Louisa's only comment was: 'I am not guilty'. She was taken away to Maidstone jail.

In the 16 days before her execution, Louisa had no visitors – not even Martin called to see her. She wrote to the Home Secretary pleading for her life but after consulting the trial judge, he decided to let the law take its course. Nevertheless, support for a reprieve did come from a most unexpected source. *The Lancet*, the authoritative voice of the medical world, said in an article after her conviction that Mrs Tregillis had suffered a heart attack a few days before she died, and that this was the primary cause of her death. The lead given to her by Louisa may have hastened her end but could not with certainty be said to have caused it. *The Lancet*'s opinion carried little weight and Louisa was hanged at 9am on Tuesday 2 January. That same day Mr Tregillis applied to the Woolwich magistrate's court for restitution of the things Louisa had pawned and was sent to negotiate with the pawnbroker.

According to the 6 January issue of the local paper, the *Woolwich Gazette*, Louisa had tried many times to commit suicide by taking sugar of lead, and that the deaths of two young women could also be laid at her door: one at Woolwich and another 'in the country' although no details were given. The newspaper even said that she may even have poisoned her husband, and quoted this as the opinion of a physician who attended him. None of these tantalizing details were further investigated.

Why did she choose such an uncertain poison? It may have been that she thought its sweet taste made it undetectable by those to whom she fed it. Her big mistake was to add it to an acid-based medicine when it would generate acetic acid with its tell-tale vinegary taste. She may then have decided to add it to brandy when its sweet taste would be masked and maybe even enhance the flavour. No doubt she found that because lead was a slow poison, producing symptoms that were easily confused with other conditions, that the victim's death would not put her under suspicion of poisoning. The wonder was that her victim, 80-year-old Mrs Tregillis, put up such a fight for life for so long.

The poisoning of Pope Clement II, 1047 AD

The human body defends itself against lead in two ways. Firstly, it resists absorbing it into the system, although some always gets through if lead enters the stomach. Secondly, the lead which passes through the stomach wall and enters the blood stream is taken to the skeleton and deposited in the bone where it can do least damage. For these reasons a person can withstand lead poisoning for many weeks until the body's defences eventually crumble under a prolonged attack.

The fact that lead becomes trapped in bone has important consequences, for it enables us to examine the lead intake of previous generations. Analysis of old skeletons has shown how the exposure to lead increased over the centuries and then fell sharply in the last century. The present level of lead in humans is lower now than it has been for nearly a thousand years, and despite the enormous increase in the use of this metal. The reason is that lead has virtually disappeared from the home environment.

Examination of the lead in old bones has brought to light some interesting discoveries. None was more surprising than the analysis of the bones of Pope Clement II who died mysteriously in 1047 AD. His remains lie in a stone coffin in Bamberg, Germany, and in 1959 samples of his bones were taken for analysis by W. Specht and K. Fischer. Their findings were published in the German journal of forensic science, *Archiv für Kriminologie* [vol. 124, p. 61, 1959] and they showed a lead level far larger that normal, thereby confirming a long-held belief that he had died of poison. But who would want to poison a Pope? The state of the Roman Catholic Church in the eleventh century reveals at least one such person.

The first century of the second millennium, the 1000s, was noted for the corruption of the church in Rome. Pope Benedict IX was even elected to the Papacy while he was still in his teens in 1032 AD, but his licentious behaviour became so notorious that the citizens of Rome ejected him from office in January 1045. In his place they elected Sylvester III, but the revolution was short-lived. Four months later he too was removed in a counter-coup, which reinstated Benedict. However, Benedict preferred money to spiritual power and he sold the Papacy to his godfather, who became Pope Gregory VI.

The following year, 1046, both of the previous occupants of the papal throne returned to Rome and demanded to be reinstated. Rome now had three claimants to the Papacy, and in desperation its citizens appealed to the German King Henry III to resolve the situation. He came to Rome and brought his own claimant, who was crowned Pope Clement II. He in return crowned his master Holy Roman Emperor on Christmas Day 1046.

Clement II was no mere puppet. He proceeded to carry out a programme of reforms, but these were not popular. He began by calling a Council of Rome which banned the sale of official positions which had been a lucrative business for some of the city's leading families. His reforms were brought to a sudden halt by his untimely death in the October of the following year, 1047, after he had held office for only nine months. Popular rumour was that Benedict's agents had poisoned him. In any event Benedict reappeared in Rome the next month and was reinstated as Pope. His triumph was short lived. In July 1048 Emperor Henry III expelled him again and a new Pope (Damascus II) took over.

Analysis of Clement II's bones showed such a high level of lead that it is certain that he died of lead poisoning, but whether this was accidental or deliberate will always remain conjecture. There was no lack of motive on Benedict's part for wishing Clement II dead, but would such a man have chosen a lead compound with which to poison him? Possibly. As we saw above, lead can easily be added to a person's drink as lead acetate and not be noticed, but lead acetate was unknown at the time of Clement II. Alternatively wine could have been doctored with litharge (lead oxide, PbO) which would quickly dissolve by reacting with other components of the wine, especially if it was slightly acidic as it often was. The sweetening effect of this process was known.

Another explanation is that Clement II may have poisoned himself by drinking too much wine that had been sweetened in the normal manner. The

German vintners were particularly fond of this method of improving the quality of their wines and Clement was particularly fond of German wines which he had specially brought to Rome. If Clement II drank his native wines then he may well have succumbed to accidental lead poisoning, as did many others down the centuries who drank German wines. The practice was later outlawed in Germany.

Specht and Fischer worked for the Bavarian Criminology Department and they were given permission to open the stone sarcophagus in which Clement II's body rested in the Cathedral of Bamberg, to which it had been taken in 1052. They hoped to end the speculation as to whether the Pope had really been poisoned. Even at the time of Clement II's death it was rumoured that he had been poisoned, and such was the fear of meeting a similar fate that the Holy Roman Emperor had problems persuading other bishops to go to Rome.

Specht and Fischer collected various samples from the sarcophagus for testing including pieces of dried tissue, a rib bone, hair, and samples of clothing. The finding of large amounts of lead in his rib bone was proof that he had died of lead poisoning. The weight of rib that was analysed was 1.8652 mg and it contained 936 picograms of lead of which 82.8% was in the outer bone, 6% in the middle bone, and 11% in the inner bone. The lead level in the bone was 50 ppm, and far in excess of what would normally be present. They were able to show that this lead had not come from the environment within the sarcophagus because there was no lead to be detected in the remains of Clement's clothing.

Specht and Fischer deduced that Clement had been fatally poisoned with lead and that this had been taken repeatedly and over a period of time. They concluded that his remains revealed a pattern of lead poisoning similar to those who had been exposed to lead as a result of their occupation and who had died of this cause. While Clement II was clearly a victim of lead poisoning, it is not now possible to prove that he was deliberately killed by it. We know that an intake of 5–10 mg of lead per day will kill within 3–4 weeks and an intake of 1–3 mg of lead per day will kill within three months. Such quantities of lead could easily be administered in wine, and there would have been those in the Vatican who were in a position to doctor his drink. It would appear to have been a case of murder by person or persons unknown, although we can be fairly certain who the instigator was.

Thallium

Driving you hairless

For more technical information about the element thallium, consult the Glossary.

WILLIAM Crookes named thallium after the bright green colour it produced when its salts were put into a Bunsen burner flame. He compared the colour to that of a fresh green shoot, so he based its name on the Greek word for this, which is *thallos*. Its deadly nature was not at first appreciated and it became part of the treatment for ringworm of the scalp, given in relatively large doses to children because it caused their hair to fall out, the better to treat the disease. Meanwhile others were using it to kill vermin, and always thallium brought tragedy in its wake.

Agatha Christie built one of her murder mysteries around thallium poisoning. In 1952 she wrote *The Pale Horse*, in which the murderer used it to dispose of people's unwanted relatives and disguised his activities as black magic curses. The plot involves a murdered priest and a pub owned by three modern-day witches.* Christie described the symptoms of thallium poisoning very well: lethargy, tingling, numbness of the hands and feet, blackouts, slurred speech, insomnia, and general debility, and she is sometimes blamed for bringing this poison to the attention of would-be poisoners. However, her book was responsible for saving the life of one young girl as we shall see. In any case Christie was not the first mystery writer to employ this deadly agent.

In *Final Curtain*, written in 1947, the novelist Ngaio Marsh had her villain

---

* It was turned into a movie in 2003 starring Colin Buchanan and Jayne Ashbourne and directed by Charles Beeson.

using it. The murder to be investigated was the death of Sir Henry Ancred who had been poisoned with thallium acetate which had been prescribed in the treatment of his granddaughter's ringworm. Marsh clearly had no knowledge of how thallium worked in that she imagined that those poisoned with it would drop dead in minutes. Would-be murderers seeking to emulate her villain would have been very puzzled when their intended victims appeared to suffer no ill effects, although this disappointment might only have lasted a few days, and then they would have been fascinated at the many symptoms it produced.

Thallium in the human body

We all have a little thallium in our body, probably no more than half a milligram, and the level in blood is only 0.5 ppb. The average person takes in about 2 micrograms of thallium a day as part of the food we eat and this accumulates in the body over time and most ends up in the skeleton. Indeed thallium finds its way into all tissues except fat, and it can even pass the placental barrier. There is no biological role for thallium although some marine organisms appear deliberately to concentrate it, but for what purpose is unclear. Thallium is like lead in that it is a cumulative poison and, like lead, it attacks the nervous system. Thankfully we never accumulate enough to affect our health. Others have not been so lucky and have had to cope with excess thallium, given accidentally, deliberately, or even medicinally. Then it slowly begins to affect those parts of the metabolism that depend on potassium, and these include the brain, nerves, and muscles. Water-soluble thallium salts are readily absorbed by mucous membranes such as the mouth, stomach, and intestines, and they can even penetrate through the skin.

Why does the body accept thallium so easily? The reason is that the positively charged thallium ion (Tl^+) is almost the same size as the potassium ion (K^+) which is essential to living cells. Tl^+ is sufficiently similar to that of K^+ for it to get into the cell but once inside the cell the slight differences that there are become apparent and the cell's function is impaired. Thallium mimics potassium so effectively that it can displace it at all sites around the body but most damagingly along the central nervous system which soon begins to malfunction. Thallium eventually affects hair follicles and these become unable to produce more hair. The existing hair drops off, and this occurs all over the body.

Studies using the radioactive isotope thallium-204* can trace the movement of thallium in the body, showing that it accumulates in the bones, kidneys, stomach wall, the intestines, the pancreas, and the salivary glands. The hair, eyes, and tongue also contain considerable amounts, while the muscles and liver have lower concentrations. Excretion is mainly through the faeces and urine with the former predominating. Once it has penetrated the body, thallium tends to form thallium chloride, which is not very soluble and consequently it takes the body a long time to eliminate thallium. The studies with thallium-204 showed that it takes at least a month to remove half of a given dose, but some is released so slowly that it can still be detected in the urine after three months.

The biochemistry of thallium has not yet been completely researched, so its mode of operation within the body is still not fully understood. In addition to mimicking potassium, it interferes with the operation of the B vitamins and with calcium and iron. Comparison of the effects of thallium with those of thiamine (vitamin B1) deficiency show such remarkable similarity that it would seem that the metal is somehow interfering with the body's thiamine metabolism. Another vitamin that thallium disrupts is riboflavin (vitamin B2) which is also involved in energy production within the body.†

Thallium upsets the sugar metabolism of the body and produces symptoms of diabetes. In addition to all of this, men are affected sexually and become impotent. Its most dangerous impact is on the central nervous system, particularly damaging organs that have high energy requirements, such as the skin, testicles, and heart.

A fatal dose of thallium for an adult is around 800 milligrams, which is less than a quarter of a teaspoonful, and yet doses of 500 milligrams of thallium salts were prescribed medically as a pre-treatment for ringworm.‡ Only when all the hair had been removed was it thought possible to eradicate the fungus. Hair loss would begin after 10 days or so, but such hair loss would today be taken as an indication that a person was suffering from

---

* This has a half-life 3 years 40 weeks and emits beta-radiation which does not damage the body.

† At one time riboflavin used to be separated from milk by adding thallium acetate, which caused it to precipitate as an insoluble material.

‡ Ringworm is highly contagious and is often transmitted through contact with farm animals like cows.

near-lethal thallium poisoning. A lethal dose of a thallium salt produces the following effects:

Day 1 – no symptoms at all or mild symptoms typical of a cold or flu.

Day 2 – gastroenteritis, pins and needles in the feet, and possibly diarrhoea.

Day 3 – band-like pain around the body, joint pains, feet become very sensitive to touch, and sleep is almost impossible.

As the days progress, the nervous system becomes more and more affected, and these symptoms become intensified. Later there is paralysis of the muscles that control talking, swallowing, and the movement of the tongue and lips. The eyes become inflamed and blindness may result. Paralysis of the face and mouth produces a mask-like feature and the patient is unable to speak. The skin becomes grey and scaly and a rash may appear on the hands. Profuse sweating occurs and this may lead to an offensive odour from the palms of the hands and soles of the feet. Thallium can lead to excessive urination, although this removes only a little of the poison. Victims suffer degeneration of the heart, liver, and kidneys and may die from respiratory disorders such as paralysis of the lungs, pneumonia, or heart failure. Blood pressure is high due to thallium stimulating the muscles of the arteries. Side by side with the physical manifestations are the mental disturbances which may range from acute depression and a wish to die, through hallucination to epileptic fits.

It is easy to see why thallium poisoning was often mistaken for so many other ailments and especially in societies where thallium poisoning was extremely rare. As we shall see in Chapter 16, only one of the 43 doctors who examined the victims of poisoner Graham Young correctly diagnosed thallium poisoning. It is almost impossible to diagnose it from its symptoms alone, not even when a post-mortem is carried out. Typical autopsy findings are widespread degeneration of the peripheral nerve cells, but heart, intestines, liver, spleen, and pancreas may appear normal.

A case of misdiagnosed poisoning with thallium sulfate was reported in 1977, but the victim was saved thanks to Agatha Christie's novel *The Pale Horse*. A 19-month-old girl living in Qatar had been taken very ill and the doctors there had been unable to diagnose the cause of her illness, so her parents took her to London for specialist treatment. It was clear she was suffering from something serious – by the time she arrived at the hospital she was semiconscious – although it was unclear what was wrong with her. At the Royal Postgraduate Medical School at Hammersmith Hospital she was referred to Dr T. G. Matthews and Professor Victor Dubowitz of the Department of Paediatrics and they began to examine her in the conventional manner taking blood samples, carrying out a lumbar puncture, and giving her a full body X-ray. They found nothing that would account for the child's condition although an EEG (electroencephalogram) of the brain was clearly abnormal. During the next few days she got progressively worse, her blood pressure was high, her heartbeat rate was 200 beats per minute, and she was breathing irregularly. There seemed no hope for her until a nurse Marsha Maitland heard the doctors talking about the child and she suggested that she might be suffering from thallium poisoning. Maitland drew their attention to the book she had been reading, which was *The Pale Horse*, and they were soon convinced that they were dealing with a case of thallium poisoning because the child's hair was starting to fall out.

The doctors immediately contacted Scotland Yard and were able to rush a sample of urine to a forensic laboratory where thallium was shown to be present at 3.7 ppb, ten times the normal level. The child was immediately put on a course of the antidote potassium ferric ferrocyanide and over the next 2 weeks she stabilized as the level of thallium in her urine fell. After 3 weeks she began to show clear signs of improvement and a week later she was discharged into the care of her parents who took her home to Qatar. A follow-up assessment four months later showed her almost back to normal. And the source of the thallium that nearly killed her? It was a pesticide that they had been using to kill the cockroaches and rats which infested the drains and septic tank of their home. The child was thought to have found it under the kitchen sink and consumed some of it.

The only certain way to prove that someone is suffering from thallium poisoning is to analyse their blood, faeces, or urine, and to prove that death has been caused by thallium it is necessary to perform chemical analysis on body tissues and bone. Thallium may be so widely dispersed in the body that its concentration in tissue is likely to be low. The level of thallium in a person who has been killed by it will vary from organ to organ but typically levels of 8–10 ppm will be found in the liver, muscle, and bone, with slightly lower levels in the heart, kidney, and lungs, and only around 2 ppm in the brain. It affects this last organ quite dramatically in those who have been poisoned with a large dose, but can it also affect those who have been exposed to non-lethal doses? The eccentric behaviour of one of its discoverers would seem to suggest that it might.

The discovery of thallium

William Crookes (1832–1919) was a chemist at the Royal College of Science in London when he was asked in 1861 to investigate a batch of sulfuric acid that was contaminated with an unknown impurity. The first thing he did was to carry out a flame test on it, and this merely involved dipping a piece of platinum wire into the acid and placing it in the colourless flame of a Bunsen burner. There was a bright green flash which lasted only a second or so but it was enough to tell Crookes that the acid was indeed contaminated. What it also told him was that there was a new element present because this green flash was unlike any he had ever seen.

Of course a simple flame test was not *proof* that he was dealing with a new element; other elements, namely copper and barium, give a green flame in such a test. However, he recognized that this was a different green to the ones they produced. Crookes confirmed that it really was a new element by measuring the green light using a spectrophotometer and there he could see that the green was in a different part of the spectrum. (We now know that the actual wavelength of this line is at 535 nm.) Crookes rushed to announce his discovery in the 30 March issue of *Chemical News*, a weekly publication that he edited, and he gave the new element the name by which it was to be known: thallium. He began to investigate its chemistry, but over the next year he only managed to make small amounts of a few simple compounds.

Unbeknown to Crookes, a French physicist, Claude-Auguste Lamy (1820–1878) of Lille, had also observed the new green line in a flame test he carried out on a strange deposit that had been scraped from the walls of a lead chamber in which sulfuric acid was manufactured. He too realized that it must be due to a new element and he carried out a more thoroughgoing investigation into what it might be. He even managed to extract a sample of the metal itself and cast a small ingot of it. He informed the French Academy of Sciences of his research findings and they credited him with the discovery. He then sent the ingot to the 1862 London International Exhibition, where the organizers displayed it prominently and acclaimed it as a new metal. They even awarded Lamy a special medal for chemical innovation. Crookes was furious when he learned what had happened and over that summer *Chemical News* led a campaign for the award to be withdrawn from Lamy and given to Crookes instead. Accusations and

William Crookes was a man of many parts. He had been born in Regent Street, London, in 1832, the son of a tailor. His father eventually left William enough money to allow him to follow his interests in chemistry and photography, and even to build his own private laboratory. Crookes made forays into many other fields of science and was responsible for early discharge tubes, which others later used to discover X-rays, and cathode rays, and which were mass produced in their billions in the 20th century for television sets. He was knighted in 1897, awarded the Order of Merit in 1910 (given personally by the reigning sovereign), and became President of the Royal Society, the highest honour a British scientist can achieve. He died in 1919. If thallium affected him in the 1860s it did him little permanent harm.

counter-accusations were made until eventually the exhibition committee felt obliged to award Crookes a medal also, and indeed the discovery of thallium is now generally attributed to him.

Crookes also became famous for other research that he carried out, including developments in photography and physics. He became infamous for devoting his time to investigating spiritualism. Could thallium have affected his mental processes? He began to attend séances and went so far as to validate the appearance of spirits conjured up by a particular young and attractive medium. He even took photographs of the spirit that materialized in her séances. There were rumours that his interest may have been more in the medium than in the manifestations, and it all happened while Crooke's wife was pregnant with her tenth child.

In the years following its discovery, thallium was found to be quite widespread in nature; it was detected in spring waters, tobacco, sugar beet, and wine. Indeed all plants are likely to contain this element because it closely resembles the potassium that they require. A thallium mineral was found in Sweden in 1866, and named crookesite in honour of Crookes.* In all cases the characteristic flash of bright green when a sample was put into a flame enabled it to be easily detected. Crookes devoted the 10 years following the discovery of thallium to investigating it and especially to measuring its atomic weight, which is 204.

---

* This is a a copper thallium selenium mineral with the chemical compositions Cu_7TlSe_4.

Thallium in nature and its uses

Traces of thallium find their way into most plants and thence into the food chain. Vegetables and meats have 0.02–0.12 ppm of them, amounts too small to effect human health. Most roots have no difficulty in absorbing thallium from the soil in which they grow and the more thallium there is, the more they absorb. Some plants take in quite a lot: pine trees can have 100 ppm, while some flowers have reached 17,000 ppm (1.7%). In 1980, vegetation growing around a cement works in Germany was found to contain high levels of thallium, this being emitted from the cement kilns in which the company was developing a new type of cement using rock which had an unsuspected high level of thallium. Cabbages in fields around the works were found to have the highest levels of thallium, with up to 45 ppm (fresh weight), and so had grapes with 25 ppm. Even local hens were laying eggs with more than 1 ppm.

Thallium is not a rare element: it is ten times more abundant than silver. In soils it can range from 0.02 ppm to more than 2 ppm but is mostly around 0.2 ppm. There is very little thallium in seawater, the level there being only 10 ppt, and in the atmosphere there is almost none at all. There has been no significant contamination of the planetary environment by thallium from industry, unlike that caused by the other heavy metals mercury and lead. It has been estimated that around 600 tonnes a year are emitted from smelters and metal processing industries, and a similar amount from coal-fired power stations.

The thallium required by industry is obtained mainly as a by-product of zinc and lead smelting. World production of thallium compounds is around 30 tonnes per year, and of thallium metal itself less than a tonne. Indeed no useful alloy of thallium with any other metal has ever been fabricated. Some thallium is used for making thallium oxide which is needed to make special glass for high-refractive lenses, and some is used for chemical research. Thallium sulfate is still sold in developing countries where it is permitted as a pesticide, although this use is banned in Western countries. It is particularly effective when dissolved in sugary syrup and laid as bait for rats, cockroaches, and ants. Some thallium is turned into thallium sulfide, selenide, and arsenide for photoelectric cells, and some into thallium bromide-iodide crystals for infrared detectors. Research chemists use thallium (III) nitrate as a particularly selective oxidizing agent.

The medical uses and commercial abuses of thallium

Thallium salts were once part of the medical pharmacopoeia, and used to remove hair. This unusual effect of thallium was discovered by accident in the 1890s when thallium was tested on tuberculosis patients as a cure for night sweats. It didn't stop them having them, but their hair fell out. Dr R.J. Sabourand, the chief dermatologist at the St Louis Hospital in Paris, reported this side effect in 1898 and for a while he used it specifically to remove body hair from those with ringworm but he gave up using it because it was too toxic. Its use was revived in the early 1920s at a recommended dose of 8 mg per kg bodyweight and it became the standard treatment for hair removal for 30 years, even though it was reported that around 40% of those given it experienced other side effects, generally very mild ones, although these were reported to disappear after 3 weeks. Another analysis of 500 patients, carried out by Drs Lourier and Zwitkis, was more reassuring in that no serious symptoms of thallium poisoning were observed although a quarter of those treated with it experienced pain in the legs and upset stomachs. (Once all the thallium has been excreted from the body, the hair will begin to grow again and return to its normal state.)

There seemed little cause for concern and thallium acetate was even regarded as sufficiently safe to be sold in over-the-counter products for removing unwanted hair. It was also used to remove unwanted relatives. It could be purchased as Celio cream or Koremlou cream and these became particularly popular in the 1930s. They contained 7% thallium acetate and a typical 10 g tube would contain 700 mg of the active agent.

The reaction to thallium varies markedly from person to person. A medical dose of 1,200 mg of thallium acetate has proved fatal to some in the past, yet others have survived self administered doses of more than twice this amount. One 10-year-old boy died after eating only 200 mg of it. One individual who tried to kill herself by eating three tubes of Celio cream survived. Most of those who tried to kill themselves generally ate a single tube of this preparation, which should not have been enough to kill them, but for some it proved to be a fatal dose. Whether a would-be suicide or accidental victim of thallium poisoning survived depended primarily on thallium poisoning being recognized as such, and quickly. If it went undetected then even the most intensive care could not succeed in saving life because there was no antidote for the poison until the early 1970s.

Accidental overdoses of thallium acetate led to several fatalities in the 1920s and 1930s, and as a result thallium treatments were phased out and ceased being used after the 1950s. In one incident a group of boys in Budapest who had ringworm were given 5,000 mg doses instead of 500 mg and they all died. The same thing happened at an orphanage in Granada, Spain, when 14 out of 16 children died who were given overdoses, in that case the cause was the pharmacist's weighing scales which were faulty. Of those who died, none showed loss of hair and of the two who survived, the hair loss did not begin until as late as a month after the thallium acetate had been taken.

Thallium sulfate was introduced as a pest control agent in the 1920s, and this too inevitably led to accidental deaths, suicides, and murders. Many members of an extended Mexican family were affected by eating tortillas made from a bag of Thalgrain, which had been stolen. Thalgrain consisted of barley coated with 1% of thallium sulfate and was used to kill squirrels. The women who had prepared the tortillas were suspicious of the grain because it was an unusual colour and appeared to be coated with something, but they decided to cook with it anyway. As a result, of the 31 people who were present at the event, 20 were taken ill and six died. Five died within 2 weeks but one member of the family lingered for a month before expiring.

A particularly large outbreak of thallium poisoning occurred in Guyana, South America, in the early 1980s when hundreds of people appear to have been affected, and 44 died. It all began when the Guyana Sugar Corporation imported 500 kg of thallium sulfate from Germany and used it to kill rats that were infesting their fields of sugar cane. For 2 years nothing untoward happened but then in 1983 doctors at St Joseph's Hospital in the Guyanese capital Georgetown began to treat people suffering from thallium poisoning. What started as a slow trickle of cases grew month by month until more than a hundred people needed treatment, and when a prominent Georgetown family reported sick then clearly something had to be done. By now there was widespread alarm as to the extent of the poisoning.

Tests on the milk which the family regularly drank revealed it was the source of the poison and checks at the farm from which they got their supply revealed sick cows suffering from thallium poisoning. They in their turn had been poisoned by eating molasses laced with thallium which had been used to deter farmers from letting their cattle wander into a nearby sugar plantation. Although the cows that ate the molasses were sickening, they continued to produce milk and this was sold to the general public. Some reports claimed that thousands of Guyanese people were affected by thallium but the US Centers for Disease Control, who were asked to investigate, found that many of the blood tests on those who thought they were suffering from thallium poisoning gave misleading results indicating levels of the metal that were totally incorrect.

Sometimes thallium has poisoned people, even to the extent of making their hair fall out, although no source of contamination could be identified. More than 300 people living in the Ukrainian town of Chernovtsy suffered this fate in 1989, and investigators discovered that the soil in the town was heavily polluted with the metal. The residents believed it had fallen in heavy rains, but a more likely explanation is that people in the town were using thallium compounds as home-made petrol additives to boost low grade fuels.

There are still some uses of thallium in medical treatment even today but patients are not given anything approaching a toxic dose. The radioactive isotope thallium-201, which has a half-life of 73 hours, is used in the diagnosis of heart disease. This displaces some of the potassium in the heart muscle, but only if there is an adequate supply of blood to carry it there, and then the penetrating γ-rays that it emits can be monitored from outside the body. The patient is given an injection of the isotope and then scanned by a scintillation counter before and after physical exercise. The uptake of thallium-201 by the heart, and its distribution within the heart, will reveal the extent of the damage to that vital organ.

Thallium as a homicidal agent

As a murder weapon, thallium sulfate has its attractions. It is soluble in water, gives a colourless solution that is virtually tasteless, and what little taste there is can easily be masked by other things, such as tea, coffee, or cola – and a fatal amount can be given in a single dose. Its symptoms are delayed and as we have seen they can be mistaken with those of other diseases. It seemed such an ideal murder weapon that corrupt regimes used it to dispose of unwanted individuals. Indeed there was even a plot to poison Nelson Mandela with thallium when his term of imprisonment at Poorsmoor came to an end in 1990. This nefarious plan was revealed during the trial of

51-year-old Dr Wouter Basson in South Africa in April 2002. Witnesses said that Basson was responsible for making poisons as part of Project Coast, the code name for a scheme devised by the former apartheid government of that country, which intended using them against black activists and African National Congress leaders. According to witnesses, the plan was to put the thallium into Nelson Mandela's medicine a day before he was released from jail. Thankfully by 1990 it was apparent that apartheid in South Africa was doomed and Mandela was released unharmed. Basson had been accused of murder, fraud, and drug dealing, but he was acquitted.

Thallium may appear to be an ideal poison but it has two major disadvantages. The first one is that a victim will recover from a less-than-fatal dose and suffer hair loss, which would give the game away. The second is that forensic analysis can reveal its presence after death and even after cremation because some thallium migrates to the bones and is there retained. Nevertheless, unless thallium poisoning is suspected, a murderer might well escape detection, and this is what happened in Austria in the 1930s and Australia in the 1950s.

Martha Löwenstein was born in 1904 and adopted by a poor family in Vienna. When she was 15 she took a job in a fashionable dress shop and there she was approached by an elderly gentleman Moritz Fritsch who was the owner of a department store who was struck by her good looks and figure. He paid for her to attend a finishing school but she also became his mistress. He took her on trips to England and France and eventually made a will in her favour, which left his house and considerable fortune to her when he died in 1924. Moritz's ex-wife and relatives were outraged and accused Martha of poisoning him. The authorities were not convinced there was a case to answer and refused permission for his body to be exhumed. Considering what subsequently happened it now seems likely that he was poisoned.

Martha had in fact been two-timing Moritz and having an affair with an Emil Marek whom she married a few months after Moritz had died. The wealth she had inherited was soon gone so she and Emil devised a plan to defraud an insurance company. Emil was insured against having an accident for £10,000 and no sooner had he paid the first instalment than, while cutting down a tree with a hatchet, he severed his leg so badly it had to be amputated below the knee. Unfortunately the surgeon who examined him realized that the three deep cuts Emil had sustained could not have been accidental wounds but self inflicted – in fact they were made by Martha –

and the couple had to settle for only £3,000. This too was soon spent. Life then became hard, so much so that Martha was reduced to selling vegetables from a street barrow in order to support her invalid husband and their two children, a baby daughter and a son.

She poisoned Emil in July 1932 and their baby daughter Ingeborg, a month later. Free of these impediments, Martha became the companion of an aged relative, Susanne Löwenstein, who was impressed by her kindness, made a will in her favour, and quickly thereafter died. This inheritance soon ran out so Martha took in lodgers, one of whom, Frau Kittenberger, soon died although her death brought Martha only £300. Then she tried another insurance fraud claiming the house had been robbed and valuable paintings stolen, but the police tracked down the firm she had hired to remove them. It was Frau Kittenberger's son who finally brought Martha to justice by insisting his mother had been poisoned and demanding her body be exhumed and analysed. Sure enough, it contained thallium. Then the bodies of Emil, Ingeborg, and Susanne were exhumed and also found to contain thallium. By this time Martha's son was seriously ill with thallium poisoning but he was taken to hospital and recovered. Martha pleaded that she had never had thallium in her possession but the authorities tracked down a pharmacist who had sold her the poison. At her trial Martha was found guilty, sentenced to death, and was beheaded on 6 December 1938, the death penalty having been restored when Hitler took control of Austria in March of that year.

Another famous case of thallium murder was that of Mrs Fletcher who was tried in 1953 in New South Wales, Australia, accused of murdering her husband by feeding him Thalrat rat poison.* He died after 11 days of agony including hair loss and excruciating pain in his arms and legs. His condition was undiagnosed during his life-time but post-mortem analysis of his remains revealed 100 mg of thallium in his body. Mrs Fletcher's first husband, Mr Bulter, had died in 1947 of similarly mysterious symptoms and the suspicion of poisoning was such that some of Mr Bulter's organs had been tested for arsenic and lead after his death, but none was found. Mr Bulter's remains were now exhumed and re-analysed for thallium; it was detected in substantial amounts. Mrs Fletcher was found guilty of murder and imprisoned.

Two series of thallium deaths occurred in The Netherlands. One murderer

---

* This consisted of a paste containing 2% thallium sulfate.

used the poison that had been put down in a food factory to kill rats and with it he disposed of the works manager and three foremen. He was never discovered because the deaths were not properly investigated for political reasons. The year was 1944 and the country was still occupied by the German army. No one wanted to tell the police because their response might well have been to inform the Gestapo and they came down heavily on those suspected of sabotaging the war effort and for which the consequences were dire. The second case in Holland was that of a woman who used Celio paste to dispose of members of her family. She had killed seven of them before she was arrested. Six other victims recovered from the doses she had given them and it was one of these who was correctly diagnosed as suffering from thallium poisoning which led to the woman's arrest. Her other victims had been diagnosed as having encephalitis, brain tumour, alcoholic neuritis, typhus, pneumonia, and epilepsy.

In 1964 Robert Hausman and William Wilson wrote an article in the *Journal of Forensic Sciences* [vol. 9, p. 72] about the alarming increase in cases of thallium poisoning in San Antonio, Texas. They attributed this to the sale of thallium-based rodent killers. They had examined the hospital records of the three largest hospitals in that city and found 52 cases of thallium poisoning in the previous 8 years. Of these 29 were accidental poisonings, mainly of children under the age of 4, 17 were attempted suicides of which two were successful, and six were attempted murders of which five resulted in death. Neither the perpetrators nor the victims were identified in the article, but one of the cases does show how those poisoned with thallium are not always properly diagnosed.

A Mrs B had bought a 39-cent bottle of rat poison in October 1961 and used it to murder her sister's husband, a Mr P, a 66-year-old insurance agent. The bottle contained a 1.3% solution of thallium sulfate and this she added to bottles of water that were stored in the fridge at her sister's home with the result that the couple became ill, Mr P particularly so. Because he was a Christian Scientist, he did not seek medical aid, and was only rushed to hospital when a friend of the family informed the medical authorities of his plight. Mr P died of thallium poisoning, but was mistakenly diagnosed as having burst blood vessel in his heart. He was buried on 22 November with Mrs P attending his funeral on crutches as she was also suffering from thallium poisoning.

Mrs B accompanied her sister to her home afterwards and used the opportunity to add more of the poison to her drinking water with the result that the following day Mrs P was discovered by the family friend to be so ill that she too required hospitalization. (Another couple who had gone back to the house after the funeral were also discovered to be suffering from thallium poisoning.) Mrs P died on 30 November, but by then the cause of her condition had been correctly diagnosed and the analysis of her liver revealed abnormally high levels of thallium. Mr P's body was disinterred and analysed and his organs also showed high levels of thallium. Food from their refrigerator was removed to the laboratory and the poisoned water discovered. Mrs B was arrested but found to be insane and was committed to a mental institution. She said she had murdered her sister and brother-in-law because she believed they interfered too much with her private life. In fact it was later discovered that she was suffering from a brain disease.

Too clever by half

A more recent case was that of George James Trepal. He has an IQ in excess of 150 and we know this because he was a member of the elite group Mensa who only admit people to their ranks on the basis of a high IQ. George Trepal is currently awaiting execution in a Florida prison after being found guilty in June 1991 of putting thallium nitrate in the Coca-Cola of his neighbours, thereby poisoning a whole family and killing one of them. What drove Trepal to such action was the endless noise they generated which took the form of loud music and barking dogs. His first attempt to get them to move house was to pin a death threat to their door, which they thoughtlessly ignored. His second attempt was to sneak into their kitchen and add thallium nitrate to bottles of Coca-Cola.

The year was 1988, the place was the small community of Alturas, and the family that was causing 39-year-old Trepal so much grief was the somewhat dysfunctional Carr family. This consisted of husband Pye, who spent most of his time with his girlfriend Laura Ervin, his wife Peggy, their children, and a number of dogs. Peggy was the person who drank most of the doctored Coca-Cola and within a few days she was displaying the classic first symptoms of thallium poisoning: her fingers tingled and the soles of her feet were extremely painful. She slowly became worse and eventually she went into Winter Haven Hospital where her hair fell out. She sank into a coma, and within a few weeks she was dead. Meanwhile Mr Carr and son Travis were also suffering from the effects of drinking the Coca-Cola although not so severely. Duane Dubberley, a visitor to the Carrs' home, also went down with thallium poisoning. Tests on all three revealed thallium in their urine and blood.

The police were baffled and enquiries around Alturas proved fruitless, the only person offering any clue as to why the Carrs should have been poisoned was their next-door neighbour, Mr Trepal, who said that clearly somebody wanted the family to move out, although he didn't say who that somebody might be. However, the police had already deduced who that person was; the problem they faced was in proving it, but they found a rather ingenious way of doing this by playing on his vanity. They sent one of their officers to attend the annual Murder Weekend that Mensa organized for its members. This popular event involved a fantasy murder which they had to solve using various clues. Trepal had even written a booklet for the event in which he offered the curious advice that when someone received a death threat they should throw out all their stored food and take care over what they ate.

While he was at the Murder Weekend, Trepal became friendly with a new participant, Susan Goreck, with whom he discussed how it would be possible to poison someone without their knowledge and what agents might best be used. Bright though Trepal was, he wasn't bright enough to deduce that his new companion was an detective working for the Polk County Sheriff's Office. As a result of what she learnt, the police raided Trepal's home and discovered in his garage a small bottle containing around half a gram of thallium nitrate. (It later turned out that Trepal had worked as a chemist in a drug laboratory and therefore had easy access to thallium nitrate, which is used as a chemical reagent.) Trepal was arrested, indicted by the grand jury of Polk County, Florida, on 5 April 1990 on a count of first-degree murder and several counts of attempted murder. His 4-week trial began on 7 January 1991. There were 80 witnesses for the prosecution but no defence witnesses. The jury found Trepal guilty on all counts, voted 9 to 3 in favour of the death penalty, and this sentence was passed on him on 6 March.

Thereafter began the long-drawn-out appeals that meant George Trepal was still awaiting execution 13 years later. His lawyers have tried to show that Mr Carr had a motive for wanting rid of his wife, that the police may have planted the bottle of thallium in the garage, that the analysis of its contents was unreliable, and that there were traces of thallium under the Carrs' kitchen sink (which may have been traces of roach or rat poison used at an earlier date). A curious line of evidence that did not emerge at the trial was that Peggy Carr was earlier admitted to Bartow Hospital, and on that occasion a higher-than-expected level of *arsenic* had been discovered in her urine. However, after a few days she recovered and left hospital. There was even a suggestion that arsenic was also detected along with thallium the second time she was admitted to hospital. No doubt this case will continue to fascinate for years to come.

Saddam Hussein's secret weapon

Thallium sulfate became the preferred weapon for disposing of people who opposed Saddam Hussein and it seems likely that dozens met their death this way in the years in which he ruled Iraq. At one time he had been the head of internal security, and with the help of his half-brother Barzan Tikriti they turned the Mukhabarat intelligence service into a formidable weapon of terror. Somewhat ironically Barzan eventually became head of Iraq's permanent mission to the United Nations and by 1992 was heading its delegation to the Human Rights Commission. Meanwhile Mukhabarat agents were disposing of opponents of the dictator both inside and outside Iraq and this they did with thallium sulfate.

Barzan had set up a Medical Poisons Unit in Baghdad University's Medical College in 1978 under the direction of two prominent doctors, Ala Khalidi and Muayad Umari. The unit struck its first victims a year later when the religious scholar Mohsen Shubbar was poisoned with thallium sulfate, but survived. However, in 1980 Salwa Bahrani, a prominent Shia, had his yoghurts laced with the same poison and he died a long and painful death in May. Another man who was murdered was Majidi Jehad, who said he was given a drink of orange juice at the police station in Baghdad where he had gone to collect his passport, which he needed in order to visit Britain. Soon after his arrival in London he was taken ill and was admitted to hospital where he died.

In the early 1980s the Iraqis most likely to be poisoned were dissident scientists and clerics living within its borders, but later that decade it was used against opponents of the Saddam regime who lived abroad, and 44-year-old Abdullah Ali, an Iraqi dissident living in the UK, was killed this way in 1988. He had moved to London in 1980 with his wife and two children and had set up in business as a publisher, although his company eventually went into liquidation with large debts but substantial assets.

The inquest into Ali's death was held by the Westminster coroner, Paul Knapman, and he put down the cause of death as bronchopneumonia due to thallium poisoning and said that Ali had been murdered by persons unknown. Ali became ill when he went to the Cleopatra restaurant in Notting Hill Gate in London early in the New Year 1988, with three men who had come from Baghdad on business matters. There were hints that these were men to whom his company owed money, which gave them a motive for a revenge killing. In his deathbed statement he said that he suspected they had laced his vodka when he left the table to go to the lavatory. The following morning he awoke feeling ill and went to see a doctor; he died 15 days later on 16 January 1988 at St Stephen's Hospital, Fulham.

An alternative theory aired at the time was that Ali was the victim of the Iraqi 'Mata Hari', an attractive woman agent of the Saddam regime who was credited with several murders. Newspaper reports even gave her a name, Narmeen Hawaiz, and Amnesty International said she had been forcibly recruited into the Iraq secret service in return for which her husband would be freed. In London she was alleged to have killed 37-year-old Adnam Al Mifti, 38-year-old Sami Shorash, and 40-year-old Mustafa Mahmoud, all members of the Patriotic Union of Kurdistan, a militant anti-Saddam group.

In 1992 two high-ranking Iraqi army officers, Abdallah Abdelatif and Abdel al-Masdiwi, fell from favour and just as quickly fell ill. They escaped to Damascus, were granted emergency visas by the British Foreign Office, and flown to London where thallium poisoning was diagnosed, and treated successfully. The same treatment was given to resistance leader 31-year-old Safa al-Battat who fell ill while visiting the headquarters of the Kurdish resistance fighters and he suspected that a cola drink had been poisoned. He travelled to Britain via Syria and was successfully treated for thallium poisoning at a hospital in Cardiff. He believed that a Saddam agent had infiltrated the Kurdish guerrilla camp.

Antidotes

Thallium mimics the nutritionally essential element potassium, and as a result it passes through the gut wall and into the bloodstream. The body is not fooled for long by thallium and excretes it into the intestines. This is not particularly effective, since a little further along it is once again mistaken for potassium and reabsorbed. The cure for thallium needs to break this cycle of excretion and re-absorption, and the best cure is Prussian blue, the dye of blue ink. This is a complex salt of potassium, iron, and cyanide and was suggested as an antidote in 1969 by a German pharmacologist, Horst Heydlauf of Karlsruhe, at a time when thallium poisoning was believed to be incurable.

Before the introduction of Prussian blue, various other treatments were tried such as Dimercaprol which in some cases met with apparent success, but it is not now recommended. Dithizone or Dithiocarb (see Glossary) were more successful when administered at the daily rate of about 25 mg per kg bodyweight. This latter compound had been used in mobilizing nickel and copper in the body and thereby facilitating their excretion and in 1959 it was shown to protect mice against poisoning by thallium sulfate. It was first used in 1962 in a case of a human thallium poisoning and it saved a life. The ability of Dithiocarb to increase thallium excretion via the urine was demonstrated in 1967 in the case of an 18-year-old college student who consumed 375 mg of thallium sulfate when she tried to abort herself (in fact it turned out she was not pregnant). Both Dithiocarb and Dimercaprol were tried as chelating agents. Tests showed that, with no chelating agent, excretion of thallium through her urine was only 1.7 mg per day, whereas with Dimercaprol it was no better at 1.6 mg per day, but with dithiocarb it went up to 6.1 mg per day. The young woman eventually recovered and left hospital after 7 weeks but she suffered most of the symptoms of thallium poisoning ranging from the short-term vomiting and diarrhoea, through the long-term hair loss, blindness, hallucination, and severe headaches and coma, not to mention the terrible pain. Her case was reported in the *American Journal of Medical Science* [vol. 253, p. 209] in 1967 by a Dr F.W. Sunderman.

Dithiocarb is far from being an ideal antidote. It forms a water-soluble complex with thallium which is why it helps in removing thallium from the body. Unfortunately by solubilizing the thallium it becomes more mobile in the body and more of the metal moves to the central nervous system as well as being excreted. In other words it poisons yet again during chelation therapy.

Whereas chelating agents such as Dithiocarb speed up elimination of thallium via the urine, they do not affect elimination via the faeces. Thallium tends to collect in the intestine walls and from there it passes into the faeces.

Unfortunately the process is reversible so that a lot of the thallium that could be excreted this way passes back into the bloodstream. This re-absorption is prevented by giving the person Prussian blue (see Glossary), whose chemical name is potassium ferric ferrocyanide. This is administered by duodenal tube at the daily rate of about 300 mg per kg bodyweight and it exchanges its potassium ions for thallium ions to which it clings more strongly and carries it out of the body.

One of the first persons to be treated with Prussian blue was a woman aged 26 who had been admitted to a South African hospital in 1972 with thallium poisoning, having taken an estimated 700 mg dose of thallium. She was treated with four daily doses of 3.75 grams of the dye for 13 days and successfully recovered. Cases of thallium poisoning in which people have taken a more-than-lethal dose of thallium sulfate have been dealt with very successfully by the administration of Prussian blue. This can be given by mouth but is best administered via a tube directly into the duodenum because thallium poisoning hampers the progress of material from the stomach downwards by closing the pylorus, the sphincter at the base of the stomach. The bowels also need to be lubricated to ensure continual passage of the Prussian blue.

If it is suspected that someone has been poisoned with thallium then analysis of blood or urine, will show levels higher than would normally be present, and the victim can be treated with the antidote and their life saved. Modern methods of analysis can detect the tiny amounts of thallium present in body fluids and tissue at concentrations of the order of less than 1 ppm. Analysing for thallium in a corpse can be carried out by taking a sample of tissue or bone and converting it to dry ash by heating in a microwave oven. The thallium present can then be measured.

The organs of people who have died of thallium poisoning contain concentrations of the order of several parts per million and can be as high as 120 ppm in the intestines. Because the body retains thallium for a long time, and because it is not naturally present, there are no problems in proving it to have been administered once its presence has been detected. Perhaps the most spectacular detection, and one which shows the sensitivity of modern chemical analysis was the analysis of the cremated remains of Bob Egle. His murder will be discussed in detail in the next chapter.

Graham Young

THE one multiple murderer whose name will for ever be linked to thallium is that of Graham Young. As we saw in the previous chapter, victims of thallium poisoning were generally thought to be suffering from some other condition and treated accordingly, so there was little in the way of evidence that we can use to follow the effect this metal had on them. In Young's case there were several victims whose illnesses were carefully recorded and we can reconstruct the way that Young administered the poison, although it is difficult to deduce why he chose one person to die and not another.

Young used two metal poisons: antimony and thallium, the former to punish, the latter to kill. With thallium acetate he murdered his stepmother, Molly Young, when he was a boy of 14, and later he murdered workmates Bob Egle and Fred Briggs. He fed antimony sodium tartrate or antimony potassium tartrate to all and sundry and thallium acetate in sub-lethal doses to some people. Altogether 13 people, and maybe more, felt the repressed wrath of Graham Young.

Formative years

Graham Young was born in the less-than-fashionable London suburb of Neasden on 7 September 1947 and his mother, Margaret, died of tuberculosis 15 weeks later on 23 December. His father, Fred Young, was obviously not capable of managing a single parent family and Graham was passed to Fred's sister and her husband who lived nearby at 768 North Circular Road. Graham's 8-year-old sister Winifred went to live with her grandmother. Despite the care of his aunt, baby Graham was already displaying a common outward sign of the emotionally disturbed child: excessive rocking to-and-fro in his cot.

Whether his aunt could ever have supplied all the love of a mother is unlikely, especially as Graham taxed her patience by being a poor sleeper. Whatever chance of emotional stability he had was upset by his having to go to hospital for an operation on his ears. When his father found a new wife both Winifred and 3-year-old Graham went to live back home. By now Graham was a very withdrawn little boy and his childhood years were made even more miserable by his stepmother, whom he openly resented, and who returned his animosity. On one occasion she smashed up his collection of model aeroplanes in retaliation for some misdemeanour. She clearly felt him a burden, sometimes locking him out of the house, ostensibly to prevent his raiding the larder, or leaving him waiting outside the pub where she played the accordion. Graham was called 'Pudding' by members of the family because he was fat and awkward. Not surprisingly he became more withdrawn and secretive.

At school Graham showed promise and passed the examinations that allowed him to go to the local high school. His father brought him a chemistry set as a reward. His other pastime was reading about subjects that particularly interested him: crime and especially murder by poison; medical science; the occult and black magic; and the Nazis. Graham's interest in chemistry drove him to collect discarded bottles from a local pharmacist's refuse bin. He also stole poisons from his school laboratory and is reputed to have used them to poison the family cat, of which his stepmother was particularly fond.

The child poisoner

In 1960, the 12-year-old Graham was ready to take practical steps to relieve the knocks he was taking from the world, and particularly from his step-mother. To begin with he merely resorted to black magic curses and he made an effigy of her which he stabbed with needles. Needless to say, in the hands of a novice these efforts were unsuccessful, so he moved to the more reliable magic of chemistry. In April 1961, he bought 25 grams of antimony sodium tartrate from a local pharmacist and he signed the poisons register as M.E. Evans. Unfortunately for Graham, Molly discovered the poison in his bedroom and reported it to her doctor. She also complained to the pharmacist for selling it to him. A big family row ensued and Graham was told he must never buy poison again.

Thwarted by Molly, Graham transferred his custom to another pharmacist and hid his collection of poisons in an old shed on some allotments near the Welsh Harp reservoir near his home. He managed to buy various toxic substances, including atropine, digitalis, aconite, and thallium acetate. Poisons were not his only interest in chemistry; he was also interested in pyrotechnics, and bought various chemicals to make fireworks. However, his attempts in this direction led to disaster in that he set fire to the shed and the police were called in to investigate. They were puzzled by the cache of chemicals they discovered there, and although some of them were listed poisons this was not followed up.

Throughout 1961 Graham administered poison to his family and school friend Chris Williams. His usual dose was 100–200 mg of sodium antimony tartrate, just enough to produce the symptoms generally associated with gastroenteritis, food poisoning, or the ubiquitous stomach 'bug' that is often blamed for upset stomachs. Graham poisoned Chris by persuading him to skip his school dinner and share his sandwiches. Thus, on successive Mondays in May 1961, Chris was treated to a meal laced with antimony and this produced violent and copious vomiting. At a later date, the two boys went for a day's outing to London Zoo and on that occasion Graham offered Chris some lemonade which he said contained a special powder to help him recover from this recent bouts of sickness; the result was even more vomiting. Chris's family doctor could find no cause to account for his sickness, which was accompanied by chest and stomach pains, terrible headaches, and cramp in his limbs. He referred the boy to the local hospital where migraine was diagnosed and the cause thought to be psychosomatic. Luckily Chris was able to avoid Graham on most days. Graham's family were less fortunate – they had to live with him every day.

In November 1961, Graham put about 50 mg of atropine in his sister Winifred's breakfast cup of tea. She drank only part of it because of the taste. On the way to work that morning she became giddy and on arrival at the music publishers in Denmark Street, London, where she worked, she was so ill they sent her immediately to the nearby Middlesex Hospital where atropine poisoning was diagnosed. (This poison, commonly called belladonna, comes from the deadly nightshade and was once used as a cosmetic to dilate the pupils of the eye to give a fashionable doe-eyed look, and was especially favoured by actresses and models to counteract the effect of bright lights.) That evening there was the inevitable big row at home with Graham lying about his part in the affair and blaming chemicals in Winifred's shampoo for the poisoning she had experienced.

Graham's poisonings of his friend Chris and his sister Winifred were merely sideshows to the main event: the poisoning of his stepmother. Graham began slipping antimony sodium tartrate into Molly's food early in 1961 and continued with this line of action up to February 1962. Whether Graham had a plan to administer small doses, to give the impression of a recurrent natural illness and then finish her off with one big dose, is not certain, but this is in effect what happened. Plagued by constant stomach upsets, Molly was even admitted to hospital with a suspected ulcer. While there, she improved rapidly and was discharged, only to fall victim to the same complaint when she returned home. She gradually began to waste away, nor was her health improved when she suffered a damaging blow to the head when the bus she was travelling in was involved in a road accident in the summer of 1961. (Her eventual death, nine months later, was attributed to this blow to the head, which caused a bone prolapse at the top of her spine.)

Graham eventually gave Molly a fatal dose of poison on Good Friday, 20 April 1962. This time he chose thallium acetate as the agent and fed her 1300 mg with her evening meal. The next day, Molly awoke with a stiff neck, and pins and needles in her hands and feet. However, she went shopping in the morning while Fred went out to the local pub where he had his lunch. He retuned home to find his wife writhing in agony in the back garden and Graham watching her from the kitchen window. Fred took her to the doctor who immediately sent her round to the hospital where she unexpectedly died later that afternoon.

Thallium poisoning was not revealed at the post-mortem and death was put down to the prolapsed bone. Molly was cremated at Golders Green Crematorium on Thursday 26 April, and her ashes were scattered. There can be no doubt that thallium acetate was the cause of her symptoms on that fateful Saturday because in 1971 Graham confessed to poisoning her food with it on the day before she died.

One down, one to go. Although Graham had begun to poison his father with antimony sodium tartrate before Molly died, he now began to concentrate on him more systematically. He had first given his father poison when Molly was in hospital the previous summer, and like Molly, his father began to lose weight. He went to the doctor, who referred him to the hospital, but there they could find nothing wrong with him which would justify his being admitted as an in-patient and so he was sent home. The doctor put him on a light diet but even the Bengers Invalid Food that he ate was doped with antimony sodium tartrate. Consequently further attacks of vomiting followed and Fred Young got noticeably worse. Again he went to see his doctor and again he was sent to the hospital where this time he was correctly diagnosed as suffering from an irritant poison, either by arsenic or antimony, and tests quickly proved it was the latter that was causing his illness.

Graham at this time was staying with his aunt and she accused him of poisoning his father, which of course he denied, and although she searched his room for poisons all she found was a model of a human figure transfixed with needles. Knowing of Graham's attempts to poison his sister, she told the doctor that she suspected him of being responsible for his father's condition. The trap was about to close on Graham, although what finally snapped it shut was his behaviour at school.

Although he was intelligent, Graham's performance in most subjects at school was mediocre, with the exception of science. At the John Kelly Secondary School that he attended his classmates called him the 'Mad Professor' and he would tell them of his ambition to become a famous poisoner. He even attempted to chloroform another boy in the school laboratory. Graham started to take poisons to the school and his obsession with the subject aroused the interest of his teachers. The science master informed the head teacher of his suspicion that Graham was responsible for the illness of his friend Chris Williams. The head telephoned Graham's family doctor, Dr Wills, who told him of the mysterious illnesses which had been afflicting the Young family. Acting on this information, and Graham's obvious oddness, the doctor and the headmaster called in psychiatric help.

On Tuesday 22 May 1962, a psychiatrist went to the school to see Graham, posing as someone from the Child Guidance Unit. With a little prompting from the psychiatrist, Graham was led into a discussion of his progress in chemistry. It was intimated that he might even be clever enough to qualify for university. Remarks along these lines led Graham into betraying his chemical knowledge, almost all of which involved poisons. By the end of the interview, the psychiatrist was in no doubt that Graham Young was a psychopathic poisoner. The following day the police were informed of the situation and arrested Graham as he arrived at school. A bottle of antimony sodium tartrate was found inside his shirt. When questioned, Graham confessed to poisoning his father and revealed the hiding places of his stock of poisons which he kept partly in a hut near the Welsh Harp reservoir and partly in a hedge nearby. Other poisons and chemicals were found hidden in his bedroom. Graham was transferred to Ashford Remand Centre prior to his appearance at the Central Criminal Court. While he was there he attempted suicide by hanging himself with his tie, but his attempt failed, which might be seen as somewhat unfortunate in view of what later transpired.

On 5 July 1962, Graham's trial began and he was charged with the attempted murder of his father, sister, and school friend Chris Williams. He pleaded guilty. Dr Christopher Fysh, senior medical officer at Ashford, and scientist Dr Donald Blair, advised that the only safe place for Graham was Broadmoor, the famous institution for the criminally insane. Graham's father had informed the authorities that if Graham was released he would have no home to go to; the house on the North Circular Road had been sold and Mr Young had gone to live with his own sister. The judge committed Graham to 15 years' detention in Broadmoor. The trial was over almost as soon as it had begun. Nothing was said about Molly's death, and nothing was said about thallium acetate.

Broadmoor

Graham, still only 15-years-old, was the third youngest inmate ever to be committed to Broadmoor. To begin with he was uncooperative, but when he was about 18 he realized that in order to gain an early release he would have to be cured of his psychopathic obsession with poisons – or at least appear to be cured. By the autumn of 1970, and after 5 years of presenting himself as a reformed character and behaving normally, he was sufficiently convincing for the doctors there to recommend his release to the then Home Secretary, Reginald Maudling, who granted it.

But then why should he refuse the application? Here was a young man who had been convicted of poisoning his father, sister, and friend when he was barely at the age of criminal responsibility. Officially he had killed no one and all that he had done was to injure them slightly. Other disturbed adolescents have inflicted much worse injuries on people and not been given such a harsh sentence. His mental condition had apparently improved dramatically and unless he was allowed back into the world soon there was a danger that he would become institutionalized and incapable of every returning to earn his living and live a normal life.

The two men responsible for recommending Graham's rehabilitation were Dr Patrick McGrath, medical superintendent, and Dr Edgar Udwin, who was his personal psychiatrist. Graham was sufficiently well read in medical matters to hoodwink them into believing he was cured, although he did occasionally behave in ways that suggested otherwise. Certainly the nursing staff and other patients at Broadmoor knew only too well of his continued obsession with poisons.

Four poisoning incidents occurred at Broadmoor during Graham's stay there. The first was the death by cyanide of 23-year-old ex-soldier John Berridge who had shot and killed his parents. The verdict on Berridge was that he had committed suicide. The source of the poison that killed him was never discovered, although the rumour was that Graham had extracted it from the leaves of the laurel bushes that grew in the grounds. These are a possible source of cyanide and Graham was later to confess to poisoning Berridge. In any event the laurels were cut down.

Graham did nothing to allay the suspicions of the other inmates about his role as a poisoner. In fact he revelled in the notoriety, labelling his personal supplies of tea, coffee, and sugar with the names of well-known poisons. When Graham was given a green card which allowed him to wander the grounds of Broadmoor unsupervised the nursing stall were so horrified that they complained to the tabloid newspaper the *Daily Sketch* which in August 1963 took up the affair. However, Graham's treatment was no different from that on any other inmate and later that year the newspaper was forced to publish a personal apology to Dr McGrath for doubting his professional competence.

Perhaps to allay suspicions and to prove to the staff that their fears were unfounded, Dr McGrath appointed Graham as staff tea-boy. He lost this job, however, when he put Harpic lavatory cleaner in the tea one day. When he tried the same trick on the other inmates by putting Mangers sugar soap in their tea urn he was more severely reprimanded by the authorities who put him in the 'cooler' for a spell, and by the inmates themselves, who beat him up. These, and other incidents, should have been seen as warning lights, but they were ignored.

While he was in Broadmoor, Graham became politically active. Always an ardent admirer of Hitler and his Nazi movement, he decorated his room with Nazi mementos, wore a brass swastika medallion he had made himself, and fantasized about himself as a concentration camp commandant and about mass exterminations using poison gas. Graham joined the Broadmoor branch of the National Front, the political party of the extreme right in 1960s. One day he had a visit from an old lady who was a member of the party, who gave him a brick she had brought back to England from the remains of Hitler's mountain retreat at Berchtesgaden.

Good behaviour and an apparently normal attitude to life earned its reward. On 16 June 1970 Graham was sufficiently sure of parole to write to his sister Winifred: 'another few months and your friendly neighbourhood Frankenstein will be at liberty again!!!' The Home Secretary eventually agreed to a release and, in November, Graham was allowed to spend a week with his sister and her husband Dennis Shannon in Hemel Hempstead. Everything went well. Naturally, Graham was rather reserved, but when released for another week at Christmas he was more sociable and there was no indication that this progress would not continue when he was discharged full time on Thursday 8 February 1971. In fact the 23-year-old Young had changed not at all: he was merely an older poisoner.

Not long before his release he boasted to a nurse that he would revenge himself on society by poisoning one person for each year he had been incarcerated in Broadmoor. As he had told his school mates, so he had told fellow inmates: his ambition in life was to go down in history as a mass poisoner. What he didn't confide in anyone was how he intended to do this. His knowledge about thallium made him certain that this was the perfect poison and its success in killing his stepmother made his ambition more than a mere boast. Thallium acetate is soluble, tasteless, colourless, and odourless. Its delayed action, confusing symptoms, and the almost total ignorance of it in the medical world in Britain made it the perfect agent. And just to confuse matters, Graham would dispense occasional doses of antimony sodium tartrate to those around him.

The adult poisoner

When Young left Broadmoor in early February 1971 he went first to spend the weekend at his sister's home. Winifred was the only person in the outside world to whom he could turn. His father had retired and moved to

Sheerness to live with his sister and her husband Jack, and they had had no communication with Graham for many years. Nor did they want anything to do with him now.

The following week Young turned up at the Government Training Centre in Slough, where he began a three months' course in storekeeping. This was well within his capabilities: at Broadmoor he had spent some of his time studying and had managed to pass some of the standard school examinations of the day. While at Slough he lived at a hostel in Chippenham about 6 miles away, and there he became friendly with 34-year-old Trevor Sparkes. Why he poisoned Trevor is something of a mystery, and at the second trial he was actually acquitted of this crime, although he later admitted to giving him antimony sodium tartrate.

Sparkes's trouble began on the evening of Wednesday 10 February when he had a conversation with Young, who gave him a glass of water. During the night he was violently sick and had diarrhoea, which persisted for 4 days, accompanied by pains in his testicles. When he next played football the following Saturday he was still unwell and after a few minutes of play he had to leave the field. About 6 weeks after this incident Sparkes spent an evening with Young and drank some wine, after which he was again very ill. He went on Thursday 8 April to see his doctor who diagnosed a urinary infection. Luckily Sparkes left Slough on Friday 30 April and never saw Young again until his trial 15 months later. However, his troubles persisted throughout the summer of that year, when he was twice examined in hospital where he was treated for strain and muscular troubles, and it was not until the autumn of that year that he began to recover.

If Young had given Sparkes antimony potassium tartrate in a glass of wine where had he got it from? His first attempt to purchase this poison was made on Saturday 17 April, more than 9 weeks after Sparkes's first bout of sickness and over a week after his second bout. The pharmacist Albert Kearne who worked at the London chemists of John Bell & Croyden of Wigmore Street refused to let him have any without written authority. The following Saturday Young returned with a handwritten note on Bedford College notepaper requesting 25 g of antimony potassium tartrate. When asked what it was needed for he replied that he was carrying out qualitative and quantitative analyses, and indeed there are legitimate uses to which this compound can be put. He was sold it as he appeared to be a *bona fide* research student of that college, which was part of London University and only a short distance from Wigmore Street. Young returned on Wednesday 5 May to purchase more antimony potassium tartrate and in addition some thallium acetate. By then Young had secured a job as an assistant storekeeper and he left Slough on the following Saturday.

Although it has been claimed that several residents of the hostel where he had been staying went down with a stomach upsets during Young's stay there, it may have been a genuine infection rather than poison. Alternatively it may had been that Young had obtained antimony sodium tartrate from a local pharmacist who was never identified. Nevertheless, Young certainly had poisons in his possession during his final 2 weeks at Slough and could conceivably have used them on the residents of the hostel.

During his training at Slough the authorities kept an eye on his progress and he was seen three times by a psychiatrist who reported that he was readjusting to life in the outside world and his training was progressing satisfactorily. Young had also reported weekly to the local probation officer. On 14 April, he applied for a job as assistant storeman at a firm in Bovingdon, near Hemel Hempstead. He claimed in his letter of application to have studied chemistry and toxicology. The firm was owned by John Hadland and manufactured optical lenses and specialist photographic equipment. On Friday 23 April, Young was interviewed at the firm by the managing director, Mr Godfrey Foster, who naturally enquired about his background and the fact that this was his first job. Young explained that he had had a nervous breakdown following the death of his mother in a car accident. On the following Monday, Young's progress report from Slough, together with his medical report from Dr Edgar Udwin was sent to Hadland's. This report, dated 15 January 1971, stated that Young has suffered 'a deep going personality disorder . . . he has, however, made an extremely full recovery and is now entirely fit for discharge . . . he is above average intelligence . . . he would fit in well and not draw attention to himself in any community.' The report made no mention of his criminal record or his stay in Broadmoor. Any doubts about Young were allayed and Mr Foster offered him the job, at £24 per week, starting Monday 10 May at 8:30 am. Young accepted. So it was on the fateful Monday morning that a neatly dressed, quietly spoken, young man was introduced to a group of fellow workers, four of whom he would give thallium acetate to, of whom two would die, and four others he would poison with antimony.

Newhouse Laboratories was a thriving firm employing 70 people. It had been built up by John Hadland and he had expanded the company by taking over Newhouse Farm on the edge of a disused wartime airfield at Bovingdon, a village near Hemel Hempstead. The firm had the distinction of producing the Imacon camera, which was capable of taking photographs with an exposure time of a sixty-millionth of a second. It was a strange coincidence that Hadland's was one of the few companies in Britain to use thallium legitimately because this metal imparts a high refractive index to the glass that was used as lenses. Glass containing thallium is harmless because the metal is firmly bonded within the body of the glass. There were no thallium compounds as such in the stores at Hadland's. When Young wanted thallium acetate he had to go to London to purchase it.

Hemel Hempstead was where Winifred lived, and to begin with Young lodged with her. He still had to make weekly visits to the local probation officer, who advised him to become more independent and find a place of his own. This suggestion was welcomed by Young and he soon found the perfect accommodation at 29 Maynard Road, Hemel Hempstead. Here he rented a small bed-sitting room at £4 per week (no cooking allowed) in a house owned by a Mrs Mohammed Saddiq from Pakistan. Young had little contact with her, or with a fellow lodger, but that was what he wanted. He now had his own room which he could decorate with Nazi insignia and in which he could store his collection of poisons. During the next six months he had no visitors to his room, but that was just as he wanted it to be.

To outward appearances, Young led a normal if somewhat lonely bachelor's life. (He appears not to have been gay but was unable to form relationships with women.) He worked five days a week at Hadland's, visited his sister and her family regularly, where he was sure of a good meal, otherwise he ate at a local fast food restaurant. At weekends he sometimes went to St Albans to visit his cousin Sandra, and just occasionally he paid a visit to his father in Sheerness. When visiting St Albans he also went to a local pharmacist where he bought more poison.

Young had one or two odd traits, such as cleaning his teeth every time he ate anything and he carried a toothbrush around with him for this purpose. He was also fanatical about killing insects. He smoked and drank like most working people but his topics of conversation were a little odd, dwelling as they did on death, war, the Nazis, and the occult. Such was his knowledge of medicine and chemistry that his workmates thought he had failed his exams at a university medical school and that was why he had had to leave. By the end of that May, he had settled in at work and had built up a stock of poisons in his room in Maynard Road and was now ready to strike. His first victim was the head storeman, 59-year-old Bob Egle.

The murder of Bob Egle

Young's *modus operandi* was to add the poison to that person's morning coffee or afternoon cup of tea. He was able to do this without arousing suspicion because it was one of his duties as the junior storeman to fetch the drinks from the tea trolley in the corridor. Every person had a differently patterned mug which made it possible for the poison to be fed with certainty to the intended victim. Moreover, while he was collecting the drinks he was out of sight of anyone for part of the journey between the tea trolley and the stores. Young carried poisons about with him and could easily slip a dose to anyone who upset him.

The stores at Hadland's consisted of the service area, and a back room, which is where Young worked supervised by Bob Egle. Young's co-workers in the stores were Fred Biggs, who was a local councillor, Ron Hewitt, and Diana Smart. Bob Egle got his first dose of antimony sodium tartrate on Thursday 3 June. This made his ill and he went home where he stayed in bed for three days with sickness and diarrhoea. He was back at work on the following Monday but had a recurrence of his symptoms again that week, and the following week as well. Consequently he and his wife decided to take a holiday and went to Great Yarmouth on the Norfolk coast for a week from Saturday 19 June to 26 June. He appeared back at work the following Monday with his health much improved.

On the Friday of the week that Egle was on holiday, Young went to the Wigmore Street pharmacists and bought 25 g of thallium acetate. On the Monday of Egle's return he put a fatal dose of this into his afternoon tea. By the following day the thallium was beginning to take effect and in the afternoon Egle was complaining that his finger ends had gone numb. In the evening he deteriorated rapidly with pains developing in his back, and his feet had lost all sensation. He had a painful night and could not sleep, and at 6.30am his wife called the doctor because he seemed so ill. He diagnosed peripheral neuritis and prescribed tablets but Bob was unable to keep these down. The doctor was called again as Bob's condition deteriorated and he was transferred to the West Hertfordshire Hospital at Hemel Hempstead. By

Thursday 1 July, his condition was so bad that he was removed to an intensive care unit at St Albans City Hospital. Despite all the efforts of the doctors and nurses there, which twice included restarting his heart, he gradually became paralysed and died on Wednesday 7 July. A post-mortem revealed the cause of death as being due to pneumonia consistent with acute polyneuritis, of a form known as Guillain-Barré syndrome which is caused by an autoimmune attack of the myelin sheath that surrounds the nerves. Because of the unusual nature of the case, one of Egle's kidneys was removed and preserved. When later this was analysed for thallium a concentration of 2.5 ppm was detected.

Throughout Bob Egle's illness, Young made several enquiries about his progress. The Managing Director, Godfrey Foster, even chose Young to accompany him to Egle's funeral as a representative of Bob's workmates in the stores. Their journey together to the funeral and then to the crematorium and back to Hadland's gave Young an opportunity to impress Foster with his medical knowledge, to the extent that he was temporarily put in charge of the stores as head storeman for a probationary period. This was partly done out of necessity since the other man in the stores, Ron Hewitt had left the previous Friday, two days after Egle's death.

During Egle's absence Hewitt had been at the receiving end of Young's attention. He had been suffering intermittently for the past month, starting on 8 June when Young first put antimony potassium tartrate in his tea. This produced stomach pains, diarrhoea, and a burning sensation at the back of his throat and Ron went home ill. The following day, Wednesday, he saw a doctor who diagnosed food poisoning. For the remainder of that week, Ron suffered from vomiting, stomach pains, and diarrhoea but was sufficiently recovered to return to work the following Monday. During the next 3 weeks he had 12 similar episodes which only came to an end when he left Hadland's for another job. It seems strange that Young should have targeted Hewitt this way as he was leaving the company anyway and in the last month was simply working out his notice. Indeed Young had been hired specifically because Hewitt was leaving.

After only two months in the job Young had risen purely by his own, somewhat unorthodox, efforts to a position of some responsibility in the company. Perhaps he seriously intended to make a go of his new role and directed his efforts to being a good storekeeper, although he generally got things into a muddle. Nevertheless, for the next three months he restricted himself to the occasional small dose of antimony sodium tartrate in Diana Smart's tea – just enough to make her sick and send her home early, especially when she annoyed him.

During this period, Young became a bit more sociable and even tried to help Fred Biggs who worked part-time. When Fred complained of insects in his garden, Young offered him first nicotine, and then thallium acetate. Fred accepted a packet of the latter, containing 15 g of the poison, and took it home but never used it. The next lot of thallium acetate he got was in his tea.

Autumn of poison

After the relative tranquillity of the summer months, the Bovingdon 'bug' began to strike again in earnest in the autumn of 1971. The sequences of illness in the stores had not gone unnoticed at the factory and people talked of it in terms of a microbe that was afflicting the workers, naming it the Bovingdon bug, thereby linking it to a local outbreak of gastroenteritis. Young was in a position to disguise his poisoning because the Bovingdon bug produced symptoms similar to those of mild antimony poisoning.

Diana Smart in particular seemed most susceptible to the bug, although in her case it had an unpleasant side effect of bad body odour, so much so that her husband, who also worked at Hadland's, found it unbearable to sleep in the same bed with her. (The doctor treated her smelly feet as a case of athlete's foot.) Her constant battering by antimony sodium tartrate made her depressed and listless. After Young's conviction, The Criminal Compensation Board eventually awarded her £367.

About the beginning of October, Young began to put thallium acetate in the drinks of his fellow workers and his first victim was David Tilson who was a clerk in the import-export department. He received a dose in his mid-morning cup of tea on Friday 8 October. He found that his tea was very sweet, contrary to his taste, and so he only drank a little of it. Young had tried to camouflage any detectable taste that thallium might have imparted to the drink by adding sugar. The result was that Tilson had taken only a small amount of the poison. Even so this produced some symptoms and had he drank all the tea he would no doubt have ended up as Egle had: dead.

On the following day, Saturday, he felt pins and needles in his feet and by Sunday his legs had gone numb. He saw his doctor on Monday and was back at work by Wednesday even though his legs were still a little stiff. He worked for the remainder of that week and then got a second dose of thallium acetate on the Friday. Young had foreseen the need for Tilson to have a second dose because his original plan involved a visit to David in hospital following his first dose, which was where he would have been had he drunk all the sweetened tea. Young planned to take a small bottle of brandy for David, laced with more thallium acetate. Instead he gave Tilson his second dose in his tea at work.

That weekend, Tilson's legs got progressively worse and chest pains developed which made breathing difficult. He saw his doctor again. By the Monday he was unable to sleep and could not bear the weight of bedclothes on his body. His condition deteriorated rapidly and he was admitted to St Albans City Hospital on Wednesday 20 October. There he started to improve and was sufficiently recovered to be discharged on the Thursday of the following week. The next day, 29 October, his hair began to fall out and within two days he was almost bald. On Monday 1 November he was readmitted to hospital where a Dr Cowan questioned him about his habits and diet on the assumption that he had been poisoned, but this line of enquiry revealed nothing. He was again discharged from hospital five days later and convalesced from several more weeks before returning to work. His only long-term disability was to be impotent and the Criminal Compensation Board eventually awarded him £460.

The day before Tilson left hospital, Fred Biggs was admitted and the day David Tilson left hospital on 5 November, another Hadland employee, Jethro Batt, was taken there, suffering from the same complaint. We will consider Biggs's murder in a minute and deal with Batt's poisoning first. Batt was 39 years old and lived in Harlow which was quite a long way from Bovingdon. He had an arrangement whereby he worked late so as to avoid the rush hour and he would make himself a cup of coffee during this time. He became friendly with Young and would give him a lift back to Hemel Hempstead if he too was working late.

On Friday 15 October, and the day that Young put the second lot of thallium acetate in Tilson's tea, Jethro found Young working late in the stores and Young made him some coffee. However, he found this too strong and drank only a mouthful. But, like Tilson, he had taken enough poison to produce the typical symptoms of thallium poisoning. The drink made him feel queasy and shortly after he arrived home he decided to go to bed. On the next day, which was a Saturday, his legs felt strange and by the Sunday they felt numb and stomach pains had developed. On the Monday morning he went to see his doctor who diagnosed influenza.

The symptoms that Tilson and Batt displayed were very similar to those who responded adversely to thallium in the days when it was given as pre-treatment for ringworm. In those days the typical dose for an adult was 500 mg. If both Tilson and Batt had drunk a quarter of a mug of the doctored tea or coffee, before discarding the rest as undrinkable, then it would appear that Young had added 2 g to each drink, which would certainly have been a fatal dose. Clearly his intention was to murder them. Young later admitted that he had administered 4 g of the poison in two doses to Batt. Such a quantity would definitely kill but thankfully he had only taken a fraction of this dose, enough to explain his symptoms but not enough to kill him.

By Thursday 21 October, Batt was unable to get out of bed, his feet hurt, and he had pains in his stomach and chest. As the days went by the thallium began to affect his brain, he became delirious and had hallucinations. His depressed state of mind became so acute that he contemplated suicide. By the end of the second week of his illness his hair had begun to fall out and on the Friday of the third week he was admitted to hospital. Like Tilson, Batt eventually recovered but he too was rendered impotent. Being a married man, this was obviously a more serious handicap than David's loss of virility and consequently Batt was awarded £950 by the Criminal Compensation Board.

Murder of Fred Biggs

With David Tilson and Jethro Batt out of the way, and likely to remain so for quite some time, Young now reverted to using antimony sodium tartrate. In the last 2 weeks of October he gave doses to Diana Smart and 56-year-old Fred Biggs, both of whom went down with the Bovingdon bug again. Unfortunately Biggs was finding Young's inability to cope in the stores an irritation and this let to confrontations between the two men. There was only one answer as far as Young was concerned and that was thallium acetate. The opportunity arose on Saturday 30 October.

Fred, aided by his wife, came in to help Young with the annual stocktaking and Graham had been given the key to the tea cupboard so that he could make drinks for the three of them. He gave Fred three doses of thallium acetate that day, which he referred to as his 'special compound' in his 'diary' which was a loose-leaf notebook in which Young kept an account of his poisoning attempts. This diary was the most damaging piece of evidence against him at his trial and one can only assume he wrote it because by then he believed he would never be caught.

The Saturday evening of the stocktaking, Fred took his wife for a night out in London, but by Sunday he was decidedly ill. On Monday he stayed in bed with pains in his chest. On Tuesday the doctor was called and found Fred's feet to be very painful, and on Wednesday he was admitted to Hemel Hempstead General Hospital. Hadland's was now abuzz with speculation. David Tilson, Jethro Batt, and now Fred Biggs all had the more virulent form of the Bovingdon bug. Young talked of nothing else, but then he was known to be interested in medical subjects, and sometimes he even talked about poisons and on one occasion he joked about putting poison in the tea.

As Fred Biggs deteriorated he was transferred first to the Whittington Hospital in North London, and then to the National Hospital for Nervous Diseases, Queen's Square, London. Everything that could be done for him was done, but of course the staff were fighting in the dark. The degeneration in his central nervous system became so advanced that he could not speak nor breathe and a tracheotomy had to be carried out to open an airway to his lungs. His skin began to peel off, and all this was in addition to the usual symptoms of acute thallium poisoning. Finally, on 19 November, after suffering for nearly 3 weeks of agony, he died.

Time had also run out for Graham Young.

The medical authorities were already worried about the Bovingdon bug and a team of doctors, headed by the local Medical Officer for Health, Dr Robert Hynd, had earlier visited the premises to try and discover the cause of the mysterious illness. A contaminated water supply was one theory, radioactivity was another, and a virus infection was a third possibility. Experiments with radioactive materials were being carried out on the nearby airfield and as radiation is known to cause hair loss this seemed a possible, if unlikely, cause, but it was the theory that the local newspaper thought was the most likely cause of the Bovingdon bug.

Two people were eventually to turn the spotlight on Young. First Diana Smart, who worked with him, observed that he was never affected by the bug, and she thought he might be a carrier of the virus and said as much to the Managing Director, Mr Foster. His assistant, Mr Philip Doggett, also informed him independently that Young showed an unhealthy interest in poisons. At the end of the third week in November, and the day after Fred Biggs died, the company decided that it was necessary to have a meeting of all who worked at Hadland's. The firm's doctor, Dr Arthur Anderson addressed them at a lunch time meeting in the work's canteen. He explained the three possible causes of the mysterious disease and how tests carried out at the works had found nothing to account for it.

He then threw the meeting open for general discussion, at which Young stood up and began to air his superior knowledge about thallium poisoning, saying that he thought this was the most likely cause. He talked about its symptoms and especially its classical symptom of hair loss. When he sat down, he had in fact solved the mystery, but sealed his own fate. After the meeting Dr Anderson sought him out and probed his general medical knowledge, which he found to be limited to toxicology.

The managers of the company were now convinced that Young was at the root of their troubles, but where had he got thallium from? There was none in the stores at the factory. On the following day they decided to call in the police, and they ran a check on Young via Scotland Yard's criminal records, which drew a blank, in other words there was no indication that he had previous convictions. They were pressed to search again, and then the awful truth emerged that Young was an ex-Broadmoor inmate convicted of poisoning his family.

Later that evening Young was arrested at his aunt's home in Sheerness where he was spending the weekend. The following day he made a full confession, even to the extent of suggesting that Jethro Batt be given Dimercaprol and potassium chloride as antidotes. His room at Maynard Road was searched where his collection of poisons was found together with his 'diary' which was a history of the October poisonings in the order in which he had carried them out, together with a progress report of his victims' progress. His poison hoard contained, among other things, about 3 g of thallium acetate and 32 g of antimony sodium tartrate.

Young was charged on Tuesday 23 November with the murder of Fred Biggs. Then followed intense police and forensic activity into the effects of, and detection of, thallium poisoning. A post-mortem examination of Fred Biggs was carried out by Professor Hugh Molesworth-Johnson, a senior lecturer in forensic medicine at St Thomas's Hospital Medical School, and he was able to find all the symptoms of thallium poisoning although not the element itself. Its presence was confirmed when samples were handed to Nigel Fuller of the Metropolitan Police Forensic Science Laboratory. Analysis of these showed 120 ppm of thallium in the gut; 20 ppm in the kidney; 5 ppm in the muscles and bone; and 10 ppm in the brain tissue. Perhaps the most dramatic feature of the enquiry was Fuller's analysis of Bob Egle's cremated remains which luckily had not been scattered, but taken to Gillingham in Norfolk, the town where he was born. The ashes weighed 1780 grams and contained 9 mg of thallium, corresponding to 5 ppm, the same as in the bones of Fred Biggs.

Just to make sure that such an amount of thallium did not occur naturally in the human frame, the forensic scientist, Nigel Fuller, who carried out this work, also analysed the remains of another cremation and detected no thallium whatsoever. The analysis of Bob Egle's ashes led to Young's being convicted of his murder. This made legal history as the first time a poisoner had been caught even though his victim had been consumed in the flames of a crematorium furnace, which would have destroyed any trace of an organic toxin.

Young claimed to have given the following doses of thallium to his victims:

Bob Egle 18 grains (in two doses), which amounts to 1,200 mg.

Fred Biggs 18 grains (in two doses), which amounts to 1,200 mg.

David Tilson 5–6 grains, equivalent to 325–390 mg.

Jethro Batt 4 grains, equivalent to 260 mg.

Young measured the doses visually against the amount of powder produced by a crushed headache tablet containing 5 grains (325 mg). We can see that these amounts of thallium are on the low side and would not explain the effects observed in their victims. Almost certainly, Young gave Tilson and Batt more than these doses which are less than those once prescribed by doctors to produce hair loss in patients. At times he wrote comments in his diary that indicated he felt some remorse for what he had done, and he wrote 'I feel rather ashamed in my action in harming J' meaning Jethro Batt.

Young appeared at the magistrate's court at Hemel Hempstead on 22 March, and his trial was scheduled for May. Young's insistence on pleading not guilty made it hard to engage a barrister to represent him and two postponements were necessary before Sir Arthur Irvine QC agreed to defend him. The trial was at last arranged for 19 July 1972. The venue was St Albans Crown Court, the judge was Mr Justice Eveleigh, and the prosecution was led by Mr John Leonard QC.

Young's second trial was no doubt his finest hour and he gave a remarkable performance. Convincing lies and point scoring repartee with the prosecution provided entertaining copy for the newspaper reporters and their readers. He said his 'diary' were notes for a novel he was hoping to write and that he had only made his confession to the police in order to put an end to the interview so he could get some rest.

Passages from his 'diary' were read out in court:

'Di (Diana Smart) irritated me yesterday so I packed her off home with an attack of sickness. I only gave her something to shake her up. I now regret that I didn't give her a larger dose, capable of laying her up for a few days.'

'F (Fred Biggs) is now seriously ill . . . In a way it seems a shame to condemn such a likeable man to such a horrible end. But I have made my decision and therefore he is doomed to premature decease . . . He is unconscious and it is likely he will decline in the next few days. It will be a merciful release for him as, if he should survive, he will be permanently impaired. It is better that he should die. It will remove one more casualty from the crowded field of battle.' Somehow F continued to cling on to life and a few days later the diary read: 'It is extremely annoying. F is surviving far too long for my peace of mind.'

The jury was not told about Young's previous conviction and sentencing of course, nor of his confession to having dispatched his stepmother with thallium many years ago. The evidence against him was overwhelming in any case, and the jury were only out for an hour and returned with a verdict of guilty with respect to the murders of Bob Egle and Fred Biggs. Before the judge passed sentence, Young's counsel addressed the court and spoke of his client's release from Broadmoor. He said that Young would prefer to go to prison rather than return there, and so he was sentenced to life imprisonment, and this was to be served at Parkhurst Prison on the Isle of Wight.

Young obtained a certain notoriety as an evil cult figure and his weird drawings were sold for £20 as a set of reproductions. Some were even given away free to diners in a fashionable London restaurant, Borsh'n'tears, in June 1975. Young died of a heart attack on 1 August 1990 at the age of 42, although some claim his death was suicide.

Graham Young's story was told in the film *A Young Poisoner's Handbook* released in 1995 in which Hugh O'Conor plays Young. The film is relatively light-hearted considering its subject matter and at times comes close to farce, but maybe that was only the way in which this story can be told, and indeed there were elements of farce in the real life plot. It is still debatable whether Young read *The Pale Horse* but he might well have done because it was a popular novel at the time he poisoned his stepmother with thallium. He denied ever reading the book but its black magic theme might well have appealed to him since he was also interested in the occult.

Perhaps there really is something mystical about thallium or it may be just a strange coincidence that those who have been fascinated by this poisonous metal have been equally fascinated by the occult, namely Crookes, Christie, and Young.

Other poisonous elements

I T is often said that the dose makes the poison, and indeed if the human
body is stressed with an excess of anything it will respond in a way that
may ultimately damage it to such an extent that it causes its own
destruction. We can even take in too much oxygen or water. Too much
oxygen damages the brain and has been known to kill premature babies and
deep sea divers, while too much water has killed those who were dying of
thirst and suddenly gulped it down. This disrupts the salt balance in their
blood which then stops the heart muscle. While these are extreme examples,
there are other less obvious elements that can be dangerous in excess, which
is what this chapter is mainly about. In it we will look at those elements that
are only moderately toxic and consequently have rarely been used in criminal cases. They are, in alphabetical order: barium, beryllium, cadmium,
chromium, copper, fluorine, nickel, potassium, selenium, sodium, tellurium,
and tin. Of course there are other deadly elements, such as chlorine gas, and
while this has been used to kill people in warfare – see Chapter 5 – it has
never to my knowledge been used to commit a murder. (On the other hand,
every day it *saves* countless human lives when it is used to chlorinate drinking water. Then it is guilty of killing nothing more than disease-causing
microbes.)

An element may be encountered as the pure substance, when it is rarely
toxic, as its insoluble compounds, which again are unlikely to be toxic even if
taken, or as its soluble compounds, which may well cause symptoms. The
importance of solubility is well demonstrated by the first of our metals:
barium.

Barium can make a deadly meal

Barium can stimulate metabolism to the extent that it will cause the heart to beat erratically (known as ventricular fibrillation), and its soluble salts are highly toxic. They paralyse the central nervous system at low doses and the heart at higher doses. The symptoms of barium poisoning are vomiting, colic, diarrhoea, tremors, and paralysis. Barium has occasionally killed patients who were given the wrong barium compound as the barium meal they swallowed prior to having an X-ray taken of their stomach or gut. As it progresses through the body, insoluble barium sulfate is highlighted because it absorbs X-rays whereas the surrounding tissue allows them to pass through. Any abnormal constrictions due to cancerous growths are thereby located.

Under normal circumstances a barium meal is made with barium sulfate, which is completely insoluble and so safe to ingest. Nor does it react with the hydrochloric acid of the stomach to form soluble barium chloride, which could be absorbed and which would be toxic. However, if the barium meal were made of barium carbonate, which looks just like the sulfate, then it could be fatal. This would react in the stomach to form soluble barium chloride and on rare occasions it has killed patients. The error would soon be realized as the patient given the wrong material started vomiting and experiencing diarrhoea. Sometimes death occurred within 10 minutes of taking a large dose although most who were poisoned this way survived for 24 hours or more. A fatal dose of barium carbonate is around a gram but a typical barium meal would be several grams of barium sulfate, and if the equivalent amount of barium carbonate were substituted then there would be little hope of saving the patient's life unless immediate action were taken.*

Barium has no biological role in humans although the average adult contains around 22 mg in their body because it is present in many things that we eat, such as carrots (which contain 13 ppm dry weight), onions (12 ppm), lettuce (9 ppm), beans (8 ppm) and cereal grains (6 ppm). Brazil nuts have been reported with as much as 10,000 ppm (1%) of barium. Even so it is no threat to our health.

There have been very few cases in which murders were committed with

---

* The antidote is sodium sulfate which forms insoluble barium sulfate, but this has to be administered very quickly.

barium salts. In 1994 at Mansfield, Texas, a 16-year-old schoolgirl Marie Robards murdered her father by feeding him barium acetate which she had stolen from the Mansfield High School chemistry laboratory. His death was diagnosed as due to heart failure and Marie would have got away with her crime had she not told a class mate of what she had done. She is reputed to have confessed to her friend and was apparently driven to doing this a result of appearing in the school's production of Shakespeare's *Hamlet*, which is all about the murder of a father and the evil that results. In 1996 Marie was brought to trial, convicted, and sentenced to 28 years in jail.

Beryllium: a gem of a metal but deadly

In 1955 the science writer Isaac Asimov wrote a prophetic short story called 'Sucker Bait' in which a space expedition goes to investigate a fertile planet where the original colony of settlers all suffered from a mysterious disease, which made breathing progressively difficult, and which killed them within a few years. The planet had abundant plant life and seemed ideal for human settlement. So what had happened? Their symptoms suggested a slow poison and yet tests revealed nothing, until it was finally discovered that the planet's soil contained high levels of beryllium, and that was what had killed them.

Thankfully, beryllium is a rare metal on Earth and our soils contain only about 2 ppm. Plants absorb very little because it has no biological role, but they do pick up some, with the result that the average person has a trace of beryllium in their body – around 35 micrograms – although this is too small to affect our health. Beryllium is related to magnesium, which is an essential element of human nutrition, and it can mimic this and displace it from certain key enzymes which then malfunction. Victims who are contaminated with excess beryllium suffer inflammation of the lungs, a condition known as berylliosis, or chronic beryllium disease, which leaves them breathless. It has long been recognized as an industrial disease among certain metal workers. Brief exposure to a lot of beryllium at one time, or exposure to a little over a long period, is likely to cause it.

Happily few cases of berylliosis are now reported, which is just as well because the condition cannot be cured, but its worst symptoms can be alleviated with steroids. Although the lungs are particularly sensitive, it is not because beryllium accumulates there. If beryllium dust is breathed in it is quickly absorbed into the blood stream and carried to other sites in the body, generally to the bone where it concentrates. The disease may take up to 5 years to manifest itself and kills about a third of those who are affected, while the rest are permanently disabled.

There was fear of a major beryllium disaster in 1990 when an explosion occurred at a Russian military plant near the Chinese border that was processing beryllium for nuclear warheads. The blast scattered a cloud of debris containing 4 tonnes of beryllium oxide dust over the town of Ust-Kamenogorsk, which has a population of 120,000. Subsequent monitoring of the inhabitants showed that about 10% of the people had been affected, as indicated by raised beryllium levels in their blood. To the relief of the Russian authorities these were only slightly raised above the normal background level.

There have been several deaths due to industrial exposure to beryllium and those most at risk were involved in making or disposing of fluorescent lamps. More than 400 workers in the USA went down with berylliosis. Other processes that exposed workers to beryllium were the manufacture of alloys, such as that with nickel which makes excellent springs, and that with copper, which is used to make spark-proof tools used in industries where there are risks of explosions. In the UK there were 30 deaths from berylliosis of the lungs in the second half of the last century. The use of beryllium was discontinued long ago although it is recognized that the disease may lie dormant for decades, and in one case a man who had worked as a metal machinist died of berylliosis survived 29 years after he had been heavily exposed to the element. He worked as a metal machinist.

There are no cases on record of people being deliberately poisoned with beryllium compounds, although there is a detective novel based on this, and that is *The Beryllium Murder*, which was published in 2000, and written by retired Berkeley physicist Camille Minichino. It is set in Berkeley and begins with a supposedly accidental death by beryllium poisoning, but of course there's more to this death than first appears.

Cadmium accumulates

Cadmium is an accumulative poison and by the age of 50 a person will have around 20 mg of it in their body, mainly in their liver. If the level of cadmium in this vital organ exceeds 200 ppm it prevents re-absorption of proteins, glucose, and amino acids, and damages the filtering system, leading to kidney failure. Cadmium is on the list of the UN Environmental Programme's top ten most hazardous pollutants.

Cadmium can never be totally excluded from our diet; it is present in foods like liver, shell-fish, and rice. Some plants have an ability to absorb cadmium, such as lettuce, spinach, cabbage, and turnip, while the mushroom *Amanita muscaria* can absorb cadmium in large amounts. Plants growing on contaminated land, such as that around old zinc mines, can have high levels of the metal, and food that is grown on such soil should not be eaten. Even when such land is turned over to sheep grazing they too have been shown to accumulate cadmium in their kidneys and livers.

The human daily intake of cadmium may be as low as 10 microgram or as high as 100 micrograms, but the average is probably less than 25 micrograms. The World Health Organization recommends that the maximum safe daily intake should not exceed 70 micrograms. The problem with cadmium is that this metal is chemically very similar to zinc which is needed for several enzymes in the body. Thankfully the human gut is able to keep out most of the cadmium that enters the stomach, but some slips past the defences and then it is trapped by the enzyme metallothionen. This is a protein that can bind to several cadmium atoms and transport them to the kidneys where in theory they are then ready to be excreted in the urine. Unfortunately cadmium binds so strongly to this enzyme that it tends to accumulate rather than be flushed away. The result is that a cadmium atom remains in the human body on average for around 30 years, and this is why there is so much concern about the effect of this metal on human health.

There have been several serious incidents of cadmium poisoning, and it has been known to kill within days. Inhaled cadmium oxide fumes are particularly dangerous, and this was how a team of construction workers was poisoned while working on the Severn Road Bridge in England in 1966. They used an oxyacetylene torch to remove steel bolts but were unaware that the bolts were galvanized with a thick layer of cadmium to prevent corrosion. The fumes that were given off poisoned them and the following day the men were all ill, experiencing breathing difficulties and coughing violently. One of them had to be taken to hospital, where he died a week later of acute cadmium poisoning. The others were also admitted for treatment but they survived.

Fuchu is a town of 45,000 inhabitants situated on the Ashida River in

West Honshu and about 200 miles north-west of Tokyo. It was there that a curious disease appeared in the mid-1950s which the locals called Itai-Itai (loosely translated as 'Ouch-Ouch' on account of the sound its sufferers made every time they moved). The cause was the rice that the inhabitants were eating which was heavily contaminated with cadmium and this originated from the spoil heaps of the Kamioka mine owned by the Mitsui Company. The dietary intake of cadmium was around 600 micrograms per person per day, and eventually around 5,000 people were affected.

There have been instances where the high level of cadmium in a corpse has led to the arrest of the person suspected of giving it to the victim. This is what happened to 46-year-old John Creamer, a painter and carpenter who lived in Pinellas Park, near St Petersburg on the west coast of Florida. He took his 37-year-old wife Jayne to Orlando on St Valentine's Day in 2002, where they had a celebratory dinner. She died a few hours later. Forensic analysis showed that the level of cadmium in her blood was 12 times higher than normal. Creamer was arrested in December 2002 following accusations by Jayne's relatives that he had poisoned his wife. Indeed her sister said that Jayne had e-mailed her on the day she died, saying she thought her husband had slipped something into her gin. Analysis of Mrs Creamer's blood had found a high level of alcohol along with cadmium and the anti-depressant drug Xanax that she was taking. After Creamer's arrest the police found containers at their home containing cadmium salts, although these may well have been legitimate paint pigments.*

In October 2003, the charges were dropped against Creamer and he was released from jail. And the reason? The previous month, the medical examiner, Dr Shashi Gore, who had originally ruled that Jayne had been poisoned with cadmium, had learned that cadmium blood levels could often give spuriously high readings. Gore had ordered more forensic tests to be carried out on Jayne's liver and kidneys, and those revealed normal levels of cadmium. Clearly there was no way that a successful prosecution could be brought.

An apparent 'outbreak' of cadmium poisoning appears to have afflicted Indiana County, Pennsylvania, following the death of 61-year-old Thomas Repine who appeared to die of a heart attack in 2002. Some members of his

---

* Cadmium yellow, which is cadmium sulfide (CdS), is one such pigment that was widely used for many years.

family found his death to be suspicious and as a result of their enquiries his body was exhumed and a high level of cadmium was found in his blood. Other seemingly suspicious deaths in the locality were brought to the attention of the coroner after tests for cadmium on corpses also uncovered elevated levels of cadmium, sometimes as high as 1,000 micrograms per litre of blood. (The US Environmental Protection Agency says the acceptable level is 5.) Whether a serial cadmium poisoner is at large remains to be seen, but it seems unlikely because there were no links between the various deaths, and environmental sources of the cadmium are now suspected.

Chromium: too much of a good thing can be very bad indeed

Chromium is an essential element for humans, although we require very little and the total amount in the average person does not generally exceed 2 milligrams. It can still be several times this amount with no adverse effect, provided it is not there as chromate, which is especially toxic.*

The daily intake of chromium varies according to the diet; and is likely to be in the range 15–100 micrograms, while the standard dose in multivitamin tablets is 25 micrograms. Chromium is an essential element because it is needed to help the body digest the energy molecule glucose. Lack of chromium causes mild diabetes, but cases of chromium deficiency are extremely rare and they are treated with a soluble chromium(III) salt such as chromium acetate. The foods with most chromium are oysters, calf's liver, egg yolk, peanuts, grape juice, black pepper, potatoes, and carrots.

Those working with chromium compounds are vulnerable to an industrial disease known as chrome ulcers, which was first reported in 1827 among workers in Glasgow, Scotland, and it became a regular feature in all the industries in which chromates or chromium salts were employed, such as chrome-plating, French polishing, calico printing, and leather tanning. The ulcers would suddenly appear but only after many months of exposure. They were about a centimetre in diameter, and exposed raw flesh which itched unbearably. In industries where workers were breathing chromium-laden dust there could be ulceration of the nasal cavity. The incidence of

---

* Chromate ion has the chemical formula CrO_4^{2-}.

lung cancer among those working with chromates was three times higher than that among the general population.

Chromate pigments are bright yellow and in the past have been extensively used to colour paints, plastics, rubber, ceramics, and floor coverings and they were based on lead chromate (known as chrome yellow) or barium chromate (known by various names such as lemon chrome and Steinbuhl yellow). These generally posed no threat to the users although the workers who produced them were susceptible to chrome ulcers.

Chromium is not seen as a major environmental pollutant although it has caused problems in rivers taking untreated industrial waste, especially that from tanneries. Soluble chromate in soils gradually turns into chromium(III) salts and because most of these are insoluble the chromium then becomes unavailable to plants. In this way the food chain is protected. There are localities where chromium has polluted the environment and may have affected people, both in terms of the health and their finances. The effect on health is generally perceived as an increased risk of getting cancer and as such is almost impossible to quantify. The financial damage is more quantifiable because it makes homes built on contaminated land impossible to sell.

The danger of polluted ground water from the dumping of chromate waste was the theme of the film *Erin Brockovich*, which was released in 2002. It concerned the fight to get compensation for the people of Hinkley, California, whose homes were blighted by it. Julia Roberts was the young campaigning single mother and Albert Finney played the part of the cynical old lawyer.

There are no cases on record of deliberate poisoning using chromium compounds either in real life or in murder mysteries. People have committed suicide by swallowing the chemical potassium dichromate or the pigment chrome yellow, of which a fatal dose would be around 5 grams.

Copper can kill

The enzyme *cyctochrome c oxidase* is required by cells to produce energy, while the enzyme *superoxide dismutase* is needed to protect them against free radicals. Both contain copper atoms which is why copper is an essential element for human life. The organs of the body which require most copper are the brain, liver, and muscles, and it is in these organs that surplus copper is stored until it is needed, although this must not be at the expense of their own functioning, in which case the copper begins to act as a poison, even to the extent of putting life in danger. Too little copper can be equally threatening.

The genetic illnesses known as Wilson's disease and Menke's disease are caused by the body's inability to utilize copper properly. In Wilson's disease the copper builds up in the brain to dangerous levels, while in Menke's disease there is a dire shortage of copper because the gene for making the copper-transporting protein is missing. The result is retarded growth, and an early death in infancy because Menke's disease is incurable. However, Wilson's disease is treatable with drugs that facility removal of excess copper from the body.

The average person has 70 milligrams of copper in their body and to maintain this they need to take in at least 1 mg of copper per day, while breast-feeding women need around 1.5 mg. It has been suggested that these levels are too low and that an intake of 2 mg per day would be more beneficial for everyone, and this seems to be supported by the effects that a low-copper diet had on volunteers. They were found to have increased levels of cholesterol, higher blood pressures, and a lack of energy.

It is not difficult to boost the intake of copper by choosing to eat copper-rich foods, and particularly meat, especially that of lamb, pork, and beef, where it is present as easily digested copper-protein. Among poultry, duck has the highest level, but most copper is to be found in sea-foods like oysters, crab, and lobster. The plant foods with most copper are almonds, walnuts, Brazil nuts, sunflower seeds, mushrooms, and bran. Eat lots of these foods and you could take in as much as 6 mg of copper a day. At one time food processing companies used copper salts to enhance the greenness of canned vegetables such as peas but regulations now prevent them from doing so.

Some researchers have suggested that we get *too* much copper and that this antagonizes the iron and zinc in our bodies because it can displace these metals from their active sites. This may be why it has an adverse effect on sperm production because these have a high level of zinc-containing enzymes. Those who have been exposed to too much copper have been adversely affected by it; none more so than workers on fruit farms and vineyards where copper sulfate is widely used as a pesticide spray in the form of Bordeaux mixture to combat fungal and bacterial diseases such as mildew, leaf spot, blight, and apple scab.

A few people have committed suicide by drinking copper sulfate solution and it has been estimated that as little as a gram can be fatal, although the irritant effect of this in the stomach automatically triggers off vomiting, making death unlikely. Only those who have taken a much larger dose have successfully killed themselves. Those who survive such attempts, however, are likely to suffer major damage to the stomach, intestines, kidneys, and brain. There have been accidental deaths due to copper, such as the young child which ate the sample of copper sulfate in a toy chemistry set, which is why this apparently safe chemical, with its beautiful blue crystals, is no longer included in these educational toys.

There are cases of suicidal and accidental deaths due to copper poisoning but would-be murderers are unlikely to choose copper as their poison, no doubt because the only readily available salt is copper sulfate with its easily detectable sky-blue colour and metallic taste. Even so, three girls, two aged 14 and the other 15, who were pupils at the H.J. Cody School in the small resort town of Sylvan Lake, Alberta, Canada, decided to poison one of their class-mates with it on 17 April 2003. They stole some copper sulfate from the school laboratory and stirred it into a Slush Puppy drink which they bought for the intended victim at a local convenience store. The blue-coloured drink consists of fruit juice and finely crushed ice and was the perfect way of disguising the poison. Unfortunately the drink was passed around and seven girls sampled it, including the two who had added the copper sulfate, although they only took small sips. Within a few hours the other girls started to feel ill, and were vomiting, shivering, complaining of intense headaches, with dry and burning mouths. They were taken to the Sylvan Lake Medical Center for treatment. Happily, none of them died.

The three girls were charged with attempted murder when they came to court in August 2003, but they pleaded guilty to the lesser charge of adminis-tering a noxious substance with the intent to endanger life, along with the lesser charges of theft and criminal negligence. On Wednesday 12 November they were sentenced to 60 days' detention in a youth jail, after which they were to spend a month in close supervision in the community, and were then on probation for 18 months. As is the case with juvenile crimes, none of the girls was identified by name.

Fluoride can be fatal

The element fluorine is a highly toxic and reactive gas, used in industry in the making of all kinds of chemicals, of which the best known is Teflon. *Fluoride*, on the other hand is relatively unreactive and is how the element occurs in nature, being the negatively charged fluorine atom F⁻. That it had a role to play in living things came to light as long ago as 1802, when it was detected in ivory, bones, and teeth. By the mid-19th century it had been found to be present in blood, seawater, eggs, urine, saliva, and hair. Although it appeared to be in all living things, this did not *prove* that it was an essential element. That had to wait until tests showed that if it was excluded from the diet of laboratory animals they failed to grow properly, were anaemic, and infertile. Today it is regarded as essential for humans, but only in tiny doses.

The amount of fluoride in the average person is between 3 and 6 grams, which delivered as a single dose could be fatal and, as we shall see in one case of murder, a teaspoonful of sodium fluoride killed within hours. The reason the fluoride we have in our body is perfectly safe is that it is mainly immobilized in our teeth and bones, where it is tightly bound to the calcium phosphate of which they are made. The product of this reaction is fluoroapatite, a much harder material, and as tooth enamel it can better resist decay by dental caries, which is why fluoride is added to water supplies and toothpastes. Were too much fluoride allowed to roam free in our body then it would pose a serious threat because the fluoride ion has a powerful effect on enzymes, effectively blocking their activity.

The average person takes in between 0.3 and 3 mg of fluoride a day in their diet. Fluoridated tap water may provide much of this, but most people get their fluoride from foods such as chicken, pork, eggs, cheese, and tea, a cup of which provides 0.4 mg. Fish are particularly rich in this element because they come from an environment, seawater, which contains 1 ppm of fluoride. Mackerel has 27 ppm (fresh weight).

There is a fine dividing line between the right amount of fluoride and too much fluoride. It was known for centuries that cattle which grazed where volcanic dust had settled would become ill and lame. Now we know it was due to their high fluoride intake, and research in Iceland in 1970 showed that grass affected in this way could contain as much as 0.4% fluoride (dry weight). Humans as well as animals can be affected by too much fluoride and suffer from fluorosis, the first signs of which are mottled teeth. Later, there may be a hardening of the bones, which can lead to a deformed skeleton. In certain parts of India, such as the Punjab, the condition is endemic, especially where villagers drink water from wells with levels of fluoride up to 15 ppm. About 25 million Indians suffer a mild form of fluorosis, and many thousands showing skeletal deformities.

Death due to fluoride poisoning is rare and when it has occurred it has generally been accidental, as was the case at a hospital in the USA in 1943 where patients were served scrambled eggs to which sodium fluoride had been added instead of sodium chloride (common salt): 163 were taken ill and 47 died. The sodium fluoride had been purchased as a pesticide, but wherever sodium fluoride was sold as rat, ant, and cockroach poison, there was always the possibility it would be used to kill unwanted humans as well. Products such as Ant-Bane and Bee Branch Roach Killer were on sale in the USA, and these consisted of around 80% sodium fluoride. The sudden death of Mrs Mamie Furr in Bogalusa, Louisiana, one January day in 1949 was due to such a product.

On the morning of the 25th she had gone to work as usual to a local box factory and at lunchtime she returned home where she was invited to have a coffee by her next door neighbour, Mrs Cola Leming. The two women were seen by a neighbour to be drinking and talking together on the top step of the back porch of the Leming home. The conversation might well have become less than friendly because Cola had previously run off with Mamie's husband Will the previous October and they had lived together as man and wife in New Orleans. Although they had returned to Bogalusa, and to their respective spouses, Cola still had hopes of one day marrying Will and had persuaded him to begin divorce proceedings.

Mamie began to feel decidedly ill within minutes of returning to her home, and she started vomiting and was soon vomiting up blood. She called another neighbour and asked her go fetch her husband. An ambulance was called and Mamie was rushed to hospital but she died within an hour of being admitted. Meanwhile Mrs Leming was seen by a neighbour as she sat on a rocking chair on her back porch muttering to herself: 'My God, what have I done?'

What she had done was stir some sodium based insecticide into Mamie's coffee. Forensic analysis of the contents of Mamie stomach, vomit, and a table cloth taken from Mrs Leming's kitchen showed clear evidence of sodium fluoride. The local coroner had little difficulty in arriving at a verdict of death from fluoride poisoning. Cola Leming was arrested and brought to trial in March 1950. Her defence lawyer tried to make out that Mamie had several times threatened suicide but this did not carry much weight with the jury. Indeed they learned that soon after her return on New Year's Day 1949 she had again vowed to have Will Furr as her partner, no matter what the cost. That cost was to be found guilty of murder and although she was not sentenced to death she was sentenced to spend the rest of her natural life at hard labour in the Louisiana State Penitentiary.

Nickel can be nasty

Nickel can poison whole communities and environments. No place on earth suffers from nickel pollution like the industrial town of Monchegorsk in the Murmansk region of northwest Russia where its mining and refining sustains the town's economy. Its inhabitants have a particularly high incidence of respiratory and lung diseases caused by breathing the polluted pall of nickel-laden smoke that hangs over the town. The local environment is blighted with grassless fields and leafless trees poisoned by this element.

Toxic though it is, nickel appears to be an essential element for humans, but we may only need as little as 5 micrograms a day. Why nickel is required, and why our bodies retains about 15 mg of this metal, is still not clearly understood, although in some species it is related to growth. The average daily intake is estimated to be around 150 micrograms. Canned baked beans provides a meal rich in this element because one of the enzymes of the bean from which this delicacy is made is the *Jack bean urease* enzyme, containing 12 nickel atoms per molecule. Another rich source of nickel is tea.

It may be essential, but nickel causes problems when in contact with the human body. These problems are of three kinds depending on the form in which we encounter it: touching the metal or its alloys; ingesting one of its soluble salts; or breathing its dust or nickel carbonyl vapour. Actual contact with the metal, or an alloy such as stainless steel, can cause contact dermatitis which manifests itself as 'nickel itch'. People who are susceptible to this are advised to avoid wearing things made of stainless steel such as watches, garment fasteners, spectacle frames, and ear rings. Previous generations of women who wore stockings held up with suspenders became allergic to their metal fastenings, which were invariably made of stainless steel and millions suffered this way, primarily because it transpires that 10% of women are sensitive to nickel. (Only one man in a hundred is affected this way.)

The other threats from nickel are to those who work with the metal or solutions of its salts in industry. Epidemiological studies show increased risk of lung and nasal cancer among nickel refinery workers and even among those who left the industry many years ago. The reason is thought to be due to nickel atoms substituting for zinc and magnesium atoms in the key enzyme *DNA polymerase* which is involved in replicating our individual DNA. The nickel ion is slightly different and so affects the behaviour of this enzyme resulting in the formation of a rogue sequence of DNA and a cancerous cell.

Once it enters the body, nickel binds itself to albumin and is transported via the blood to certain organs where it tends to accumulate, such as the kidneys, the liver, and the lungs. It is excreted primarily via the urine. Fatal poisoning by nickel compounds is extremely rare although a 2-year-old child who ate 15 g of nickel sulfate crystals died within 4 hours. A group of people on kidney dialysis machines suffered extreme toxic symptoms when a nickel salt was accidentally added to the solution they were being dialysed with. They survived. Nickel salts may be deadly but they are not on sale to the public and as far as can be ascertained, no one has ever used them as homicidal agents.

The deadliest form of nickel is nickel carbonyl, a compound that caused several deaths among workers at a nickel refinery in Wales before its dangers were appreciated. The compound came to light in 1888 when the industrialist Ludwig Mond and his assistant, Carl Langer, investigated the problem of leaky valves through which carbon monoxide gas (CO) was passing. They discovered that the gas was reacting with the nickel from which the valves were made and forming volatile nickel carbonyl, $Ni(CO)_4$, a liquid which boils at 43°C, and which has a musty, sooty odour. From this they developed a process, the Mond Nickel Process, for making very pure nickel but, when it was put into production, there were sudden and unexplained deaths among those operating the plant.

A few breaths of nickel carbonyl vapour will immediately cause a painful throat and tight chest. The person affected soon develops a headache, feels sick, and becomes dizzy. If exposure is brief then these symptoms will clear up after a few days. This apparent recovery may still happen if the person has been heavily exposed to the vapour, but then more serious symptoms begin to appear a week or so later with the lungs being especially badly affected. In such cases the risk of dying is high.

One of the most remarkable cases of nickel carbonyl poisoning happened in 1957 and was of a 25-year-old man who was accidentally sprayed with it and quickly became breathless and started to turn blue. First aid was administered in the form of pure oxygen gas and he was given the antidote diethyldithio-carbamate (DDC) and rushed to hospital where his urine showed one of the highest levels of nickel ever recorded: 2 ppm. Because he was given immediate treatment he eventually survived and made a full recovery. In other cases where the level of nickel in the urine has been much less the person has died and one man with 0.5 ppm succumbed despite being given DDC.

Potassium, the essential deadly poison

In the body, potassium has many functions, the more important of which are operating nerve impulses and contracting muscles. Potassium is present as the positively charged ion K^+ and it concentrates inside cells, which is where 95% of the body's potassium is located, unlike sodium and calcium, which are more abundant outside cells. A cell membrane has millions of tiny channels through which potassium ions flow and every one is capable of transferring hundreds of potassium ions per second in and out of the cell. This activity is responsible for the transmission of nerve impulses from the brain, because it moves like a wave along the nerve fibre just as if it were an electric current.

Some channels only permit potassium to pass through, and the toxin of the black mamba kills its victim by specifically blocking them. The same result can be achieved by injecting a concentrated solution of potassium chloride into the blood stream which prevents the movement of potassium out of the cell because there is already too much potassium on the outside pushing to get in. All body functions are affected, but none more dramatic-ally than the heart muscle which stops beating. Murders have been commit-ted with potassium chloride, and doctors and nurses have been known to end the lives of terminally ill patients by giving them injections of potassium chloride solution. Such injections are used in the USA to execute criminals convicted of murder.

Deadly though it can be, potassium is still a major dietary requirement, lack of which causes muscle weakness. Indeed it is not generally appreciated that the need for potassium salts in the diet is much greater than for sodium salts. The recommended daily intake is 3.5 g, double the 1.5 g a day recommended for sodium. Vegetarians take in a lot more potassium than non-vegetarians because potassium is abundant in all plant foods. We must have a regular supply of dietary potassium because we have no mechanism for storing it in the body, yet few people are affected by a deficiency of this metal because almost all we eat contains potassium. Some foods are particularly rich in it, such as raisins, almonds, peanuts, and bananas; one banana will provide a quarter of our daily requirement. Other common foods with lots of potassium are potatoes, bacon, bran, mushrooms, chocolate, and fruit juices.

Salt substitutes are 60% potassium chloride and 40% sodium chloride, and using this in cooking and flavouring is not life-threatening but life-saving if it reduces the amount of salt in the diet of those suffering from heart disease. There are rare cases of excess ingestion by humans proving fatal, and a person who ate half an ounce of this salt (14 g) died as a result, although the normal amount of potassium chloride needed to cause a serious toxic response is more like 20 g.

Potassium chloride was the chosen murder weapon of one notorious serial killer whose victims were babies and young children. She was Beverley Allitt who was employed as a children's nurse at the Grantham and Kesteven General Hospital in Lincolnshire, England. Over a period of 10 weeks she injected 10 children with potassium chloride. Thankfully she only succeeded in killing four of them.

Although Allitt trained to be a nurse, she had repeatedly failed her nursing exams, but in February 1991 she was taken on temporarily at the local hospital because they were short staffed. Her first victim was a 7-week-old baby, Liam Taylor, who was admitted to the children's ward on 21 February suffering from congestion of the lungs. His parents were reassured by Allitt that their baby was in good hands, but when they returned to visit him a few hours later they were informed that he had been rushed into emergency care, although he was now recovering. The parents asked if they could stay the night at the hospital and were shown to a bedroom specially assigned for this purpose, happy in the knowledge that Allitt had elected to work a night shift specially to be near Liam in case he needed help quickly, which around midnight he did. Allitt suddenly signalled for help because Liam's heart had stopped beating. Although doctors and nurses struggled to revive the baby, it was in vain; he was dead.

On 5 March 1991, an 11-year-old boy, Timothy Hardwick, who suffered from cerebral palsy, was admitted to the hospital after a severe epileptic fit and was put under Allitt's care. She was particularly solicitous in attending to the boy, but soon after she was left alone with him his heart stopped beating. She summoned help immediately, but despite the efforts of a paediatric specialist who rushed to the boy's aid, he died. An autopsy was carried out, but found no obvious cause of death, and so his death was recorded as due to epilepsy.

A few days later, on 10 March, a 1-year-old girl Kayley Desmond was admitted to the children's ward suffering from lung congestion. Allitt was assigned to her care and although the little girl appeared to be getting better she suddenly went into cardiac arrest. The crash team at the hospital managed to revive her and she was then rushed to the larger hospital at Nottingham where she recovered fully. Doctors there noticed that she had a puncture below her armpit, and this came to their attention because there was a bubble of air under the skin. We now know that Allitt had injected potassium chloride solution from a partly filled needle and before she had removed all the air in the syringe. (It was little wonder that she had failed her nursing examinations.)

Frustrated at her failure to kill Kayley, Allitt changed her *modus operandi* and decided to try insulin as her weapon of attack. On 20 March, 5-month-old Paul Crampton came to the hospital with severe bronchitis. He suddenly went into a coma but the doctors revived him and noted that his blood sugar level had fallen dangerously low, this being the effect that insulin can have. Despite the efforts of the medical staff, Paul had two further attacks and so he too was sent to Nottingham for more specialist care – and survived.

The next day Allitt returned to using potassium chloride and her victim was 5-year-old Bradley Gibson who suddenly suffered a heart attack, but the doctors were able to save him. When he had another heart attack later that night he too was despatched to Nottingham and he too recovered. It was the same with her next victim, 2-year-old Yik Hung Cha, who had fallen from a window and fractured his skull. He suffered in exactly the same way as Bradley, and he too was saved by being rushed to Nottingham. Allitt's next victims were not so lucky.

Katie and Becky Phillips were twins who had been born prematurely and had been looked after at the hospital before being sent home in March 1991.

On 1 April Becky was readmitted to the children's ward suffering from gastroenteritis. Allitt was responsible for her care but the baby went into convulsions that evening, which a doctor diagnosed as being caused by her illness. Her parents stayed by her bedside watching over her but she died in the night. An autopsy revealed no obvious cause of death. Then her twin sister Katie was admitted to the hospital and in the course of the next two days she had two heart attacks, during which her parents were impressed with the strenuous efforts Allitt made to save their daughter's life. However, she was transferred to Nottingham where they discovered that five of her ribs were cracked (probably as attempts had been made to re-start her heart) and that her brain was damaged through lack of oxygen. However, Katie's parents were so grateful that her life had been saved, and thinking this was due to the prompt action of Allitt, they asked the nurse to be the baby's godmother, and this she did when the child was christened a few days later.

Meanwhile other children in Allitt's care continued to be struck down with unexpected complications, but were saved thanks to the efforts of other nurses and doctors. At first the hospital staff suspected that a virus was the cause and they had the children's ward decontaminated, but it was only when that had no effect that they began to suspect that human malevolence was behind these mysterious attacks and that pointed to the only person who was always around when they happened: Allitt. Their suspicions were soon to be confirmed.

Claire Peck was in hospital suffering from an asthmatic attack that was so severe that doctors had inserted a tube into throat to help her breathe. The little girl suddenly had a heart attack and Allitt was the only nurse in the ward at the time. The emergency team rushed to help the young girl and revived her, but soon after they left the child, again under Allitt's sole care, she immediately had another attack. This time they could not revive her and Claire Peck died. An autopsy now included analysis of her blood and this showed an unnaturally high level of potassium. The hospital authorities called in the police, and they arrested Allitt.

Allitt came to court in March 1993 and after a trial lasting nearly two months she was found guilty and given 13 life sentences, the highest such penalty handed down to a woman. Why did she do it? The answer would appear to be that she was suffering from a syndrome called Munchausen by Proxy. This rare mental state was first diagnosed in 1977 and it manifests itself in people who maliciously attack those who are in their care.

Selenium is smelly

Selenium is another example of an element that is essential yet toxic in tiny amounts. From the days of its discovery in 1817 it was known to be an element to avoid, and one of the first to suffer from it was its discoverer, the Swedish chemist Jöns Jacob Berzelius. He was alerted to its most noticeable side effect when his housekeeper complained about his appallingly bad breath, accusing him of eating raw garlic. Although he did not realize it, Berzelius was emitting a gas we now know as dimethyl selenium, and which is one of worst smelling gases ever discovered. Nor was it his only encounter with this obnoxious element. On another occasion he was overcome by breathing hydrogen selenide gas (formula H_2Se) which he had prepared without realizing that this gas is deadly. He was ill for 2 weeks.

Prolonged exposure to selenium, as occurred in some industries, led to anaemia, loss of weight, dermatitis – and social isolation. Selenium was treated as a pariah element, to be avoided if possible. And then in 1975 it was shown to be something we cannot avoid, it was proved to be an essential element for humans. Yogesh Awasthi, based at Galveston in Texas, found that it was part of the antioxidant enzyme, *glutathione peroxidase*, which eliminates peroxides before they can form dangerous free-radicals. Selenium was there to *protect* the body. In 1991, Professor Dietrich Behne at the Hahn-Meitner Institute in Berlin found selenium in another enzyme, *deiodinase*, which promotes hormone production in the thyroid gland. Every one of our trillions of living cells contains more than a million atoms of this element, and we have about 14 mg of selenium in our body. The recommended maximum daily intake is 450 micrograms; above this we risk selenium poisoning, the most obvious symptom of which is extremely foul breath and body odour, as we seek to rid ourselves of an unwanted excess.

The daily intake of selenium depends on the foods we eat. The average person takes in about 65 micrograms per day which is enough to prevent selenium deficiency, although it is less than the recommended intake of 75 micrograms for men. Selenium levels are highest in hair, kidneys, and testicles, where it is needed to protect sperm. Most people get their selenium from breakfast cereals and bread, especially wholemeal bread, two slices of which will provide up to 30 micrograms depending on the soil of the farm from which it came. Foods particularly rich in selenium are Brazil nuts, molasses (black treacle), tuna, cod, salmon, liver, kidney, peanuts, and bran.

There is no evidence that selenium has ever been used as a homicidal agent, nor is it ever likely to be because its tell-tale odour on the breath of the victim would immediately alert those around them to what they had been given.

Sodium under suspicion

There are lots of sodium compounds but the one that accounts for most of this element in our lives is salt (NaCl) in which the sodium is present in its most stable state, the positively charged sodium ion Na^+. It is not technically a poison but it is a cause of concern because doctors believe that too much salt puts unnecessary stress on the human frame and especially in those with a predisposition to high blood pressure and heart disease, resulting in a situation where the body is less able to remove unwanted salt and so it compensates by retaining more water. This causes increased pressure on the arteries, resulting in high blood pressure with all the risks to health that this implies. Nevertheless, it is necessary to take in a regular supply of sodium because it is continually lost from the blood stream as it is filtered out by the kidneys.

The amount of sodium a person consumes each day varies from individual to individual and from culture to culture; some people get as little as 2 g per day, some as much as 20 g. Foods which have a lot of salt are tuna, sardines, eggs, liver, butter, cheese, and pickles. Vegetables generally have very little. Although a lot of sodium is recycled within the body, a lot is lost via urine (which contains around 350 ppm), the faeces, and the sweat glands.

Blood tastes salty because it contains a lot of salt, 0.35% in fact. Our skeleton contains a lot also (1%), as do most tissues of the body, which is why the total body burden of this element is 100 g. Sodium in the body is mainly in the fluid outside cells, the situation being just the opposite to that which pertains with potassium. Blood needs a lot of sodium to regulate osmotic pressure and blood pressure, as well as to help solubilize proteins and organic acids.

Salt is a blessing in tropical countries, where it has saved millions of lives. Diarrhoea, and the resulting dehydration it causes, is reported to kill millions of children a year. The simple answer to this potentially fatal illness is to drink a solution of glucose and salt and the United Nations children's organization, UNICEF, distributes millions of sachets for making up such solutions.

While such use of salt may save lives, it also has the power to harm, if not to kill. Too much salt taken in a single dose will provoke the toxic response of vomiting. This would suggest that it would be impossible to murder somebody by feeding them salt, but it was just how 39-year-old Susan Hamilton of Edinburgh, Scotland, continually poisoned her daughter. Hamilton appeared before a jury on 6 June 2003 and, after a 3 week trial, she was found guilty of 'assault and the endangerment of life' and sentenced to 4 years in jail. She committed the crime 3 years earlier, on 10 March 2000, when she had injected concentrated salt solutions into the feeding tube that doctors had inserted into her daughter's stomach.

Her daughter, who could not be named for legal reasons, was born in 1991, and had muscle problems that made it hard for her to swallow and the doctors to whom she was taken thought she had a wasting disease. When she was 4, it was decided to feed the girl via a nasogastric tube which is inserted up one nostril and then goes down into the stomach, but this too had appeared not to be successful and eventually she had had a percutaneous endoscopic gastrostomy tube inserted directly into the stomach. All this may have been unnecessary in view of what her mother was doing, but at the time it seemed the right course of treatment.

Time and time again Hamilton reported that her daughter was very ill and she was admitted to Edinburgh's Royal Hospital and each time her examination revealed very high levels of sodium in her blood. Because there are medical reasons why sodium levels can be higher than normal, this did not at first raise suspicions, but eventually the doctors treating the child came to the conclusion that she was being deliberately poisoned. Neither the special foods that the girl was being given, nor the medication that she was taking would raise her blood sodium levels to the extent being observed. The police were called in and when they went round to the Hamilton's home they discovered a syringe which contained drops of liquid. This was taken away for forensic analysis and the liquid identified as salty water.

Hamilton's last attempt to poison her daughter had almost killed the child, but in any case it had left her permanently damaged. The hospital records revealed that she must have been repeatedly injecting solutions of salt into her daughter's feeding tube and on 17 separate occasions she had made the child so ill that hospitalization was necessary. Indeed the poor girl was thought at one stage to be developing leukaemia and a 'Dream Come

True' trip had been organized to take her to Euro Disney in the belief that she was terminally ill.

Hamilton was administering the equivalent of 2 teaspoonfuls of salt to the girl to make her ill but on the final occasion of March 2000 she had given her so much that the child suffered a stroke which then left her permanently brain damaged. The explanation for the mother's behaviour was Munchausen by Proxy syndrome.

Tellurium and its tell-tale aroma

Tellurium is chemically similar to selenium but unlike selenium it tends to accumulate in the body. Selenium produces really bad breath and body odour, but tellurium is much worse. The two elements are chemically very similar but, unlike selenium, there is no biological role for tellurium, and yet we have around 0.7 mg in our body. More than this, and those around us would soon be driven away by the stench we were emitting.

The daily intake of tellurium for the average person is judged to be around 0.6 mg, and it is part of our diet, coming as it does from plants. These can take in tellurium from soil and can have levels as high as 6 ppm, with onion and garlic having the most. The tellurium that we ingest is absorbed into the blood stream and excreted in the urine, although around 10 micrograms is disposed of by converting it to the volatile dimethyl tellurium, which it can expel via the lungs or the sweat glands.

Those who worked with tellurium often developed what was called 'tellurium breath' even when there was as little as 10 micrograms of the element per cubic metre of air. Those who come into contact with tellurium are advised to take extra vitamin C which reduces considerably the smell of tellurium on the breath. In 1884, tests were carried out on volunteers who were given 0.5 microgram of tellurium oxide, TeO_2, by mouth. In an hour it was detectable on their breath and the smell did not disappear for 30 hours. Some men, given doses of 15 mg, were still found to have tellurium breath eight months later!

Tellurium has caused occasional deaths. As little as 2 g of sodium tellurite (Na_2TeO_3) can be fatal and this was discovered accidentally in 1946 when three soldiers were given it from a mislabelled bottle as part of their medical treatment. Two of them died within 6 hours. Acute tellurium poisoning results in vomiting, inflammation of the gut, internal bleeding, and respiratory failure. Chronic poisoning produces garlic breath, tiredness, and indigestion.

It seems unlikely that anyone would ever have been deliberately poisoned with a tellurium compound in view of the tell-tale signs that would immediately arouse suspicion.

Tin is safe till you make it organic

There is evidence that the human body needs tin, and it is essential to some creatures. Rats fed on a tin-free diet failed to grow properly, but recovered when a tin supplement was given, suggesting it has some important role. Whether we need it or not is still unproven, but we have around 30 mg of tin in our body, and that comes from the food we eat. Plants growing in uncontaminated soil can have up to 30 ppm of tin; while plants growing on contaminated soil can have very high levels and sugar beets grown adjacent to a chemical plant had 0.1% (dry weight) while vegetation near a tin smelter reached levels of 0.2%.

The average person takes in around 0.3 mg of tin a day, of which 0.2 mg comes naturally from food. Less than 3% of the tin in our food is absorbed by our body, and that which is absorbed is mainly excreted in the urine but some tends to collect in the skeleton and liver, which is why over time the amount slowly builds up, although not to anything like a dangerous level. Tin has always been part of the human diet, but it greatly increased when canned food was introduced in the 1800s. The lacquering of the inside of cans has considerably reduced the amount of tin that leaches into the food they contain. The USA has imposed limits of 300 ppm for tin in canned food; the UK limit is 200 ppm.

Inorganic tin compounds are generally regarded as non-toxic. However, organotin compounds are toxic, especially when tin atoms have three organic groups attached. Such organotin compounds are able to penetrate biological membranes and once inside a cell they upset various metabolic processes with fatal results. Trimethyl tin, and especially triethyl tin, are particularly toxic to humans, but larger organic groups are much less toxic which is why some, such as tributyl tin (TBT), have been widely used as pesticides, including putting it on clothes that are likely to become sweaty, such as T-shirts, to counteract the bacteria that breed on them and emit unpleasant odour molecules.

In France in the 1960s and early 1970s, diethyl tin di-iodide was pre-scribed as the drug Stalinon by doctors for the treatment of staphylococcal skin infections. Diethyl tin has only two organic groups attached and was thought to be safe, but the treatment led to several deaths, which were thought to be due to contamination by triethyl tin iodide.

Tin became an environmental problem when TBT was added to marine paints which were introduced in the 1960s, proving to be immensely popular since it prevented the drag caused by marine growths on hulls. It reduced the time spent in dry-dock, to the extent that some ships needed repainting only once every 5 years. Although TBT saved energy and other resources, to an estimated $7 billion per year, by the 1980s it was clear that species such as oysters, marine snails, and dog-whelks, which flourish in coastal waters, were undergoing strange sexual changes or becoming infertile. The cause was TBT and as little as 1 nanogram per litre of seawater (1 ppb) was capable of causing the observed changes. The use of TBT has now been outlawed in most countries.

As far as records show, there has been no case in which a tin compound has been deliberately used as a homicidal agent.

Appendix

—◄+►—

Table A.1 Essential elements in the body of an average 70 kg (155 pound) person

Elements	Present as	Total amount in body
Oxygen	Mainly as water[a] and present everywhere	43 kg
Carbon	Everything but water	12 kg
Hydrogen	Mainly as water[a] and present everywhere	6.3 kg
Nitrogen	Protein, DNA, etc.	2 kg
Calcium	Bone,[b] teeth, cell messenger	1.1 kg
Phosphorus	Bone,[b] teeth, DNA, ATP	750 g
Potassium	Electrolyte, mainly inside cells	225 g
Sulfur	Amino acids, particularly in hair and skin	150 g
Chloride	Electrolyte balance	100 g
Sodium	Electrolyte, mainly outside cells	90 g
Magnesium	Metabolic electrolyte	35 g
Silicon	Connective tissue	30 g
Iron	Haemoglobin	4,200 mg
Fluoride	Bones and teeth	2,600 mg
Zinc	Enzyme component	2,400 mg
Copper	Enzyme cofactor	90 mg
Iodine	Thyroid hormones	14 mg
Tin	Not known	14 mg
Selenium	Enzyme, antioxidant	14 mg
Manganese	Enzyme component	14 mg
Nickel	Enzyme component	7 mg
Molybdenum	Enzyme cofactor	7 mg
Vanadium	Lipid metabolism	7 mg
Chromium	Glucose tolerance factor	2 mg
Cobalt	Part of vitamin B_{12}	1.5 mg

[a] Water accounts for around 60% of the body's weight.
[b] Bone accounts for around 13% of the body's weight.

Glossary

———◆———

Words in italics refer to other entries in the glossary.

Acute illness is severe and quickly comes to a crisis as compared to a *chronic* illness which persists for a long time and is less threatening.

Alternative names for chemicals are listed under the individual elements *antimony, arsenic, lead*, and *mercury*.

Analytical samples for assessing the amounts of chemical elements that are present must be brought completely into solution. This can be done in several ways, for example by heating a carefully weighed amount of the tissue with concentrated nitric acid at 140°C, after which sulfuric and perchloric acids are added and the heating taken to much higher temperatures (up to 300°C) to ensure complete oxidation of all organic material. Soil samples are generally dissolved using aqua regia, and for particularly intractable materials hydrofluoric acid is used. Microwave heating is also widely employed. The final solution should be perfectly clear showing that there is no undissolved material. It is then ready for analysing by methods such as *atomic absorption spectroscopy* or *inductively coupled plasma*, the latter in conjunction with *mass spectrometry*.

Antidotes are listed under their most common names, such as *BAL, Thimerosal, Versenate*, while those specific to an individual toxin are listed under *arsenic antidotes, antimony antidotes, lead antidotes, mercury antidotes*, and *thallium antidote*.

Antimony is chemical element number 51, with atomic weight 122, chemical symbol Sb, and is a member of group 15 of the periodic table. Antimony can exist in two forms: the metallic form is bright, silvery, hard, and brittle, melts at 631°C, and has a density of 6.7 kg per litre; the non-metallic form is a grey powder. Antimony can exist in two oxidation states: antimony(III) and antimony(V), the former being the more stable.

    Antimony compounds have had various names over the centuries and these are as follows:

Common, historical, or medicinal names	Chemical names	Chemical formula
Regulus of antimony	Elemental antimony	Sb
Stibine	Antimony hydride	SbH_3
Tartar emetic, *Kalii Stibyli Tartras, Brechweinstien*	Antimony potassium tartrate, potassium tartaroantimonite(III)	$K(SbO)C_4H_4O_6.\frac{1}{2}H_2O$
Antimony sulfide, *stimmi, golden sulphuret, Kermes mineral, spiessglanz,* stibnite	Antimony trisulfide, antimony(III) sulfide	Sb_2S_3
Antimony oxide	Antimony(III) oxide	Sb_2O_3
Butter of antimony	Antimony(III) chloride	$SbCl_3$
Antimonyl chloride, powder of Algaroth	Antimony oxide chloride	$SbOCl$

Antimony analysis used to be carried out by the Marsh test (see page 150) but the results were so similar to those of arsenic that this method of analysis had difficulty distinguishing the two elements. As with arsenic, the first necessity is to bring all the antimony into solution from the material under investigation – see *analytical analysis*. When all the tissue has been dissolved, the solution can be subjected to various tests. If antimony in oxidation state V is present it needs to be reduced to oxidation state III, and this can be done by adding potassium iodide solution. (The iodide is converted to dark brown iodine, which is then reduced back to colourless iodide with ascorbic acid.)

Antimony analysis used to involve converting the antimony into the gas stibine (SbH_3) and identifying and measuring that. In the 1960s, a better method of analysis that could detect microgram amounts was introduced and this required its conversion in solution to deep blue potassium tetraiodoantimonite, which has a wavelength of 330 nm that could be used to measure the amount of antimony present by measuring its intensity. It is possible to measure smaller amounts of antimony by *atomic absorption spectroscopy*. Today it is possible, using *inductively coupled plasma* techniques to measure even smaller amounts, at the nanogram level (parts per billion), which is sometimes needed for environmental and biological research into antimony, and sometimes even in criminal investigations when it can help identity the sources of minute traces of lead to which it has been added.

Antimony antidotes are *chelating agents* of which the most used has been *BAL*, also known as *Dimercaprol*. This should be administered in 200 mg lots every 6 hours for four days and then twice daily thereafter. If the antidote is not immediately available then first aid treatment should be given to the victim in the form of lots of strong tea, the tannin of which forms a complex molecule with the antimony and thereby delays its absorption. When the person has been poisoned by inhaling the antimony gas stibine (SbH_3) then a blood transfusion needs to be given without delay.

Antimony potassium tartrate has the chemical formula $K_2[Sb(O_2CCH(OH)$ $CH(OH)CO_2)_2Sb]$ and it consists of two antimony atoms linked by two tartrate ions, with each antimony bonded to various oxygens in the molecule.

Arsenic is chemical element number 33, with atomic weight 75, chemical symbol As, and is a member of group 15 of the periodic table. Arsenic can exist in two forms: grey arsenic which is a metal with a density of 5.8 kg per litre; and yellow arsenic which has a density of 2 kg per litre. Metallic arsenic is brittle and tarnishes easily, and on heating it does not melt but it sublimes at a temperature of 616°C; it reacts with oxygen to form white arsenic trioxide (As_2O_3).

Arsenic compounds have had various names over the centuries and these are as follows:

Common and medical names	Chemical names	Chemical formula
Arsenic, white arsenic, Trisenox	Arsenic trioxide, arsenic(III) oxide	As_2O_3
Realgar, red arsenic	Tetraarsenic tetrasulfide	As_4S_4
Scheele's green	Copper arsenite	$CuHAsO_3$
Fowler's solution	Potassium arsenite	KH_2AsO_3 and K_2HAsO_3
Arsenious acid	Not a chemically recognized entity and it exists only in solution	
Orpiment	Diarsenic trisulfide	As_2S_3
Lewisite	Dichloro(2-chlorovinyl)arsine	$ClCH{=}CHAsCl_2$

Arsenic antidotes are chelating agents of which *BAL*, also know as *Dimercaprol*, is the best. It should be injected at the rate of 150 gm every 4 hours, and which will normally save the life of anyone who has knowingly been exposed to a large dose of arsenic. Another antidote is *Versenate*.

Atomic absorption spectroscopy (AAS) detects the amount of metal in an *analytical sample* by vaporizing it in a hot flame, or with a laser, and measuring the radiation that is absorbed by its atoms, which both identifies them and permits the amount present to be calculated. The method is quick and has the ability to measure quantities at the nanogram level (parts per billion).

BAL is short for British Anti-Lewisite, whose chemical name is 2,3-dimercapto-propan-1-ol, and whose common name is Dimercaprol. Its chemical formula is $HSCH_2CH(SH)CH_2OH$. BAL was developed to treat soldiers who might be exposed to arsenical warfare agents such as Lewisite and proved such an effective antidote that it became the standard medical treatment for arsenic poisoning in any form. In the 1940s, researchers in the USA began testing BAL on animals to see if it was an antidote to lead, and when this proved successful they used in on human volunteers, and again it is part of the treatment for lead poisoning. Today all hospital pharmacies hold BAL as the drug of choice for dealing with the toxic elements *arsenic, antimony, lead*, and *mercury*, and for all heavy metal poisoning. BAL became the cure for Wilson's disease in which there is an accumulation of copper in the brain and liver that leads to complete disability.

Calcium-EDTA – see *chelating agents*

Chelating agents are chemicals that bind to a metal via two or more of their atoms and these grip the metal atom rather like the claws of a crab. (The word 'chelate' is derived from the Greek word for a crab's claw.) The ones used in medical treatment to remove metals from the body are *BAL, calcium-EDTA*, Dithizone, Dithiocarb, and DMPS. They rely on their chelating ability to latch onto metal atoms in the blood, or to prise them out of enzymes, and transport them to the kidneys where they can be excreted.

Calcium-EDTA, also known as *Versenate*, is a mixed sodium and calcium salt of the molecule ethylenediaminetetraacetic acid (*EDTA*).

Dithizone has the chemical formula $C_6H_5N=N.CS.NH.HNC_6H_5$ (where C_6H_5 is a phenyl group) and is known to chelate heavy metals particularly well on account of its sulfur atom, but it may cause side effects when used as an antidote.

Dithiocarb is sodium diethyldithiocarbamate $Et_2N.CS.SNa$, which is also good at chelating heavy metals on account of its sulfur atoms.

DMPS is short for 2,3-dimercapto-1-propanesulfonic acid. This is chemically similar to BAL and has the chemical formula is $HSCH_2CH(SH)CH_2SO_3H$. Its sodium salt is also used and it has the name Unithiol.

Chronic illness One that persists for a long time and is less likely to be life-threatening than an *acute* illness which is severe and represents a crisis in the health of a patient.

Dimercaprol is the common name for *BAL*.

DMPS – see *chelating agents.*

EDTA is an abbreviation for ethylenediaminetetraacetic acid which has the formula

$(HO_2C)_2NCH_2CH_2N(CO_2H)_2$. Sodium and calcium salts of this complex acid make excellent *chelating agents*. See also *Versenate*.

EPA is short for the US Environmental Protection Agency.

FDA is short for the US Food and Drugs Administration.

Grain was the smallest weight previously used by apothecaries before the metric system of weights became standard. A grain was a 480th part of a troy ounce, of which there were 12 to the imperial pound. A grain is equivalent to 65 milligram.

Inductively coupled plasma (ICP) is an analytical technique using high frequency energy to heat argon gas in which the sample to be analysed has been sprayed as an aerosol. The frequencies can be as high as 90 megahertz and the generators have power outputs of up to 10 kilowatts. As a result the argon gas can reach temperatures of almost 10,000°C which is enough to break down all molecules into their individual atoms and excite their electrons. As these electrons drop back to their normal energy levels they emit light of a characteristic wavelength and the intensity of this light is measured to reveal the amount of that element in the sample. Tiny amounts of an element can be measured by ICP in conjunction with optical emission spectrometry (ICP-OES) or with *mass spectrometry* (ICP-MS).

Lead is chemical element number 82, with atomic weight 207, chemical symbol Pb, and is a member of group 14 of the periodic table. Lead melts at 334°C and has a density of 11.4 kg per litre. It is a soft metal and its compounds can exist in two oxidation states: lead(II) and lead(IV), of which the former is the more stable. Most lead compounds are insoluble. Lead has four natural isotopes: lead-204, lead-206, lead-207, and lead-208 of which all but the first are the end products of the radioactive decay of elements of higher atomic number such as uranium and thorium. By measuring the amount of lead in rocks that contain uranium it is possible to estimate their age. Also the ratio of isotopes in a sample of lead, and especially the lead-206/lead-207 ratio, varies according to the source of the lead and this enables the deposit from which it was mined to be identified.

Lead compounds have had various names over the centuries and these are as follows:

Common name	Chemical names	Chemical formula
Chrome yellow	Lead(II) chromate	$PbCrO_4$
Galena, lead glance	Lead(II) sulfide	PbS
Litharge, yellow lead oxide	Lead(II) oxide	PbO
Red lead, minium	Lead tetroxide	Pb_3O_4
Sugar of lead, *sapa*	Lead(II) acetate	$Pb(CH_3CO_2)_2$
TEL	Tetraethyl lead	$Pb(CH_2CH_3)_4$
White lead, basic lead carbonate	Lead(II) carbonate hydroxide	$2PbCO_3.Pb(OH)_2$

Lead poisoning interferes with the body's ability to make the haem molecule which is a key part of the haemoglobin of red blood cells which transports oxygen. Lead causes a build-up of aminolevulinic acid (ALA) from which haem is made, and it is the excess of this in the body which produces the many symptoms of lead poisoning.

Lead antidotes are the *chelating agents* calcium-EDTA, BAL, and DMPS. Some of them, like calcium-EDTA, produce a rapid drop in blood lead levels but this may rise again if treatment is stopped as additional lead is released from the tissues. If too much antidote is given then the skeleton 'de-leads' too rapidly, and this in itself can cause severe lead poisoning.

Mass spectroscopy (MS) is an analytical technique in which a beam of molecules is ionized, thereby breaking the molecule into various fragments. The beam of particles is then passed through a strong electric and a magnetic field which separates them according to their mass and charge. When these ions are detected they can be identified by their mass to charge ratio and this can reveal which molecule was present in the beam. This technique is particularly useful in differentiating different isotopes of the same element, and is often used in conjunction with *inductively coupled plasma*.

Methyl mercury is the term used to describe compounds in which there is a methyl group (CH_3) chemically bonded to a mercury atom, as in methyl mercury chloride ($H_3C-Hg-Cl$) or dimethyl mercury ($H_3C-Hg-CH_3$). Methyl mercury compounds are particularly dangerous because they can pass across the blood-brain barrier that protects the brain.

Mercury is chemical element number 80, with atomic weight 200.5, chemical symbol Hg, and is a member of group 12 of the periodic table. It is that rare example of a metal that is liquid at room temperature; its freezing point is *minus* 39°C and its boiling point is 357°C. It is unreactive towards acids.

Mercury compounds have had various names over the centuries and these are as follows:

Common name	Chemical name	Chemical formula
Calomel	Mercury(I) chloride	Hg_2Cl_2
Cinnabar, vermilion	Mercury(II) sulfide	HgS
Corrosive sublimate	Mercury(II) chloride	$HgCl_2$
Red precipitate	Mercury(II) oxide, red form	HgO
Yellow precipitate	Mercury(II) oxide, yellow form	HgO

A table of mercurial compounds used in medicine is given in Chapter 2.

Mercury analysis requires the mercury to be in the soluble ionic form, when its presence can be shown by the addition of test reagents such as diphenylcarbazone, which gives a blue colour. An older test used to be the addition of a few drops of potassium iodide solution which would precipitate the mercury ions as mercury iodide, which is bright yellow, and which redissolved if more potassium iodide was added. Modern methods can measure the exact amounts of mercury present in a given sample by using *atomic absorption spectroscopy* or *inductivity coupled plasma spectroscopy*, although if there are large amounts of mercury present it may still be easier to user the older technique of precipitating it as insoluble mercury sulfide and weighing the precipitate.

A new visual test for detecting mercury in water, and at concentrations as low as 0.5 ppm, was developed in 2004 by a group at Imperial College London, headed by James Durrant. The sensor contains a ruthenium dye that is anchored to titanium dioxide particles. When it comes in contact with mercury ions, it changes colour from red to orange, and it can even detect mercury in the presence of similar ions such as copper and cadmium, which interfere with the older methods of analysis.

Mercury antidotes are *chelating agents* of which BAL has been most used. If someone is known to have taken a large dose of a mercury compound then BAL will be useful in tying it up but it has to be given within 3 hours. An injection of 300 mg of BAL at the start of treatment, followed by 150 mg at 6-hourly intervals, will generally save life, and if treatment is rapid then the patient will be free of the mercury within 48 hours. In cases where the mercury has had time to be absorbed into the body removal is slower and the symptoms of mercury poisoning will be evident for a long time.

The antidote is not capable of saving the victim in some instances when the damage to the kidneys is too severe. The fight to save the life of such a person requires the use of an artificial kidney and this, together with a low protein diet, will generally succeed in saving the patient. Even before BAL was available, about 70% of people survived acute mercury poisoning and with modern treatment this is more likely to offer a chance of survival in excess of 95% and only those for whom there has been a long delay before the start of treatment are at risk of dying.

Nuclear Magnetic Resonance spectroscopy (NMR) is used to identify the locations of atoms in molecules by the radio frequency which they absorb when exposed to a powerful magnetic field. This causes the nuclei of atoms such as hydrogen and carbon-13, to flip their spin from one direction to the other. The energy required to do this is affected by the electrons surrounding the nucleus and these are affected by the chemical bonds of the molecule.

Neutron activation analysis (NAA) is a sensitive technique of analysis that can even measure the amount of certain metal elements in a single strand of hair. The drawback is that it requires access to a nuclear reactor. When atoms are exposed to the neutrons inside such a reactor they can absorb them and be converted to specific, short-lived, isotopes which then decay emitting characteristic radiating such as gamma rays. Measuring these identifies the atoms that are present and this method is sensitive enough to detect nanograms (billionths of a gram) and even picograms (trillionths of a gram) of material.

Organo mercury and organo lead compounds are those in which there is a direct carbon-to-metal chemical bond. The term *organo* derives from the word organic, which in chemistry refers to the chemistry of carbon and its compounds. The simplest organic group is the methyl group (CH_3) in which three of the four bonds that carbon can form are to hydrogen atoms, the fourth bond being the one to the metal atom. Thus there are methyl mercury compounds with the $H3C-Hg$ component and methyl lead compounds with the H_3C-Pb component. More complex organo groups have extended carbon components such as the ethyl group which has two carbons (CH_2CH_3), for example tetraethyl lead is $Pb(CH_2CH_3)_4$. Another group that can be attached is the phenyl group which is derived from benzene and has the formula C_6H_5 and is a six-membered ring with a hydrogen on each atom except the one attaching itself to the metal.

ppb is short for parts per billion and is one microgram in a kilogram, or one microgram in a litre. A microgram is a millionth of a gram, and while this may seem a small quantity it represents many trillions of atoms.

ppm is short for parts per million, and is one milligram in a kilogram, or one milligram in a litre. A milligram is a thousandth of a gram.

Prussian blue is the common name for potassium ferric ferrocyanide whose chemical formula is $KFe^{III}[Fe^{II}(CN)_6]$. Prussian blue has been known for centuries and was the blue of blue ink.

Tartaric acid has the chemical formula $CO_2H.CH(OH).CH(OH).CO_2H$ and it can exist in three forms known as meso-, d-, and l-tartaric acid.

Tartar emetic is the common name for *antimony potassium tartrate*.

Thallium is chemical element number 81, with atomic weight 204, chemical symbol Tl, and is a member of group 13 of the periodic table. It is a soft, silvery-white metal which melts at 304°C and has a density of 11.9 kg per litre, so it is slightly heavier than lead. It is a reactive metal, tarnishing in moist air, and is readily attacked by acids. Thallium can exist in two oxidation states: thallium(I) and thallium(III). In the lower state as the ion Tl^+ it resembles potassium. The higher oxidation state is present in seawater, but that in rocks and soils is in the lower oxidation state. A test on the urine of a person who is suspected of suffering from thallium poisoning can be done by adding a drop of an alcoholic solution of dithiocarbazone which produces a cherry red colour if the test is positive.

Thallium antidote is *Prussian blue* or $KFe^{III}[Fe^{II}(CN)_6]$ and it consists of a three dimensional network of molecular cages. In alternate cages reside a potassium ion and this can be exchanged for thallium ions to give the more stable compound $TlFe^{III}[Fe^{II}(CN)_6]$, which is thereby removed from the body.

Thimerosal is the mercury compound sometimes used as an antimicrobial agent to protect sensitive materials such as vaccines. It is a derivative of benzene with the chemical formula $CH_3CH_2HgSC_6H_4CO_2Na$. This has the salt group (CO_2Na) and mercury group (CH_3CH_2HgS) on adjacent carbon atoms of the benzene ring. Because the mercury atom is attached to an ethyl group, rather than a methyl group, thimerosal does not have the toxicity generally associated with methyl mercury compounds.

Versenate is the common name for the disodium salt of *EDTA* and it was introduced in the 1950s as a chelating agent for heavy metal poisoning. When it was first tried on six patients suffering from lead poisoning it was found to be an excellent antidote. Indeed one child who was in a coma responded so well that she was sitting up feeding herself, and talking again, within two days. Nowadays the preferred form of giving EDTA is as the calcium disodium salt of EDTA known simply as calcium-EDTA.

X-Ray fluorescence spectroscopy (XRF) is an analytical technique that uses high-energy radiation to knock an electron from the innermost orbit of an atom. This electron is then replaced by an electron from the next orbit and as it does so it emits an X-ray which is characteristic of atoms of that element and of no other. In the case of Napoleon's wallpaper, discussed in Chapter 6, the radiation was supplied by an isotope of promethium (Pm-147).

Bibliography

General

Ball, P., *Bright Earth: the Invention of Colour*, Viking, London, 2001.

Bowen, H.J.M., *Environmental Chemistry of the Elements*, Academic Press, London, 1979.

Butler, I., *Murderer's England*, Robert Hale, London, 1973.

Camps, F.E. (ed.), *Gradwohl's Legal Medicine*, 2nd edn, John Wright & Son Ltd., Bristol, 1968.

Cooper, P., *Poisoning by Drugs and Chemicals, Plants and Animals*, 3rd edn, Alchemist Publications, London, 1974.

Cox, P.A., *The Elements: Their Origin, Abundance and Distribution*, Oxford University Press, Oxford, 1989.

Drummond, J.C. and Wilbraham, A., *The Englishman's Food*, Pimlico, London, 1994.

Duffus, J.H. and Worth, H.G.J. (eds), *Fundamental Toxicology for Chemists*, The Royal Society of Chemistry, Cambridge, 1996.

Emsley, J., *The Elements*, 3rd edn, Oxford University Press, Oxford, 1995.

Emsley, J., *Nature's Building Blocks*, Oxford University Press, Oxford, 2001.

Evans, C., *The Casebook of Forensic Detection*, John Wiley & Sons Inc., New York, 1996.

Feldman, P.H., *Jack the Ripper: the Final Chapter*, Virgin Books, London, 2002.

Fergusson, J.E., *The Heavy Elements*, Pergamon, Oxford, 1990.

Finlay, V., *Colour*, Hodder and Stoughton, London, 2002.

Glaister, J., *The Power of Poison*, Christopher Johnson, London, 1954.

Hunter, D., *Diseases of Occupations*, 5th edn, Hodder and Stoughton, London, 1976.

Jacobs, M.B., *The Analytical Chemistry of Industrial Poisons, Hazards, and Solvents*, 2nd edn, Interscience, New York, 1949.

Kaye, B.H., *Science and the Detective*, VCH, Weinheim, 1995.

Kelleher, M. and Kelleher, C.L., *Murder Most Rare: the Female Serial Killer*, Dell Publishing, New York, 1998.

Kind, S., *The Sceptical Witness*, Hodology Ltd, Forensic Science Society, Harrogate, 1999.

Lenihan, J., *The Crumbs of Creation*, Adam Hilger, Bristol, 1988.

Martindale: The Extra Pharmacopoeia, 27th edn, The Pharmaceutical Press, London, 1977.

McLaughlin, T., *The Coward's Weapon*, Robert Hale, London, 1980.

Mann, J., *Murder, Magic and Medicine*, revised edn, Oxford University Press, Oxford, 2000.

Montgomery Hyde, H., *Crime Has its Heroes*, Constable, London, 1976.

Ottoboni, M.A., *The Dose Makes the Poison*, 2nd edn, Van Nostrand Reinhold, New York, 1991.

Polson, C.J. and Tattersall, R.N., *Clinical Toxicology*, EUP, London, 1965.

Rentoul, E. and Smith, H., *Glaister's Medical Jurisprudence and Toxicology*, 13th edn, Churchill, Edinburgh, 1973,

Root-Bernstein, R. and Root-Bernstein, M., *Honey, Mud, Maggots, and Other Medical Marvels*, Macmillan, London, 1997.

Roscoe, H.E. and Schorlemmer, C., *Treatize on Chemistry*, Macmillan & Co., London, 1913.

Rowland, R., *Poisoner in the Dock*, Arco, London, 1960.

Simpson, K., (ed.), *Taylor's Principles and Practice of Medical Jurisprudence*, Vol. II, 12th edn, Churchill, London, 1965.

Stevens, S.D. and Klarner, A., *Deadly Doses: a Writer's Guide to Poisons*, Writer's Digest Books, Cincinnati, Ohio, 1900.

Stolman, A. (ed.) and Stewart, C.P., 'The absorption, distribution, and excretion of poisons' in *Progress in Chemical Toxicology*, vol. 2, p. 141, 1965.

Stone, T. and Darlington, G., *Pills, Potions and Poisons*, Oxford University Press, Oxford, 2000.

Sunshine, I. (ed.), *Handbook of Analytical Toxicology*, Chemical Rubber Co., Cleveland, Ohio, 1969.

Thompson, C.J.S., *Poisons and Poisoners*, Harold Shaylor, London, 1931.

Thorwald, J., *Proof of Poison*, Thames & Hudson, London, 1966.

Timbrell, J., *Introduction to Toxicology*, Taylor & Francis, London, 1989.

Waldron, W.A., 'Health Standards for Heavy Metals', *Chemistry in Britain*, p. 354, 1975.

Weatherall, M., *In Search of a Cure*, Oxford University Press, Oxford, 1990.

Wilson, C. and Pitman, P., *Encyclopaedia of Murder*, Arthur Barker, London, 1961.

Witthaus, R.A., *Manual of Toxicology*, William Wood, New York, 1911.

Wootton, A.C., *Chronicles of Pharmacy*, Milford House, Boston, 1910 (republished 1971).

Alchemy

Clegg, B., *The First Scientist: a Life of Roger Bacon*, Constable, London, 2003.

Cobb, C., *Magick, Mayhem, and Mavericks*, Prometheus Books, Amherst NY, 2002.

Fara, P., *Newton: the Making of a Genius*, Picador, London, 2002.

Greenberg, A., *A Chemical Mystery Tour: Picturing Chemistry from Alchemy to Modern Molecular Science*, Wiley-Interscience, New York, 2000.

Greenberg, A., *The Art of Chemistry: Myths, Medicines and Materials*, Wiley-Interscience, New York, 2003.

Mackay, C., *Extraordinary Popular Delusions and the Madness of Crowds*, Richard Bentley Publishers, London, 1841. (Reprinted by MetroBooks New York 2002.) This book has a 158-page chapter on the alchemists.

Marshall, P., *The Philosopher's Stone: a Quest for the Secrets of Alchemy*, Macmillan, London, 2001.

Morris, R., *The Last Sorcerers*, The Joseph Henry Press, Washington DC, 2003.

Multhauf, R.P., *The Origins of Chemistry*, Oldbourne, London, 1966.

Schwarcz, J., *The Genie in the Bottle*, W.H. Freeman, New York, 2002.

Szydlo, Z., *Water Which Does Not Wet Hands: the Alchemy of Michael Sendivogius*, Polish Academy of Sciences, Warsaw, 1994.

Mercury

Banic, C. *et al.*, 'Vertical distribution of gaseous elemental mercury in Canada' in *Journal of Geophysical Research*, vol. 108, p. 4264, May 2003.

Barrett, S., 'The mercury amalgam scam; how anti-amalgamists swindle people' at http://www.quackwatch.org/01QuackeryRelatedTopics/mercury.html

Caley, E.R., 'Mercury and its compounds in ancient times' in *Chemical Education*, vol. 5, p. 419, 1928.

Cook. J., *Dr Simon Forman: a Most Notorious Physician*, Chatto & Windus, London, 2001.

Devereux, W.B., *Lives and Letters of the Devereux, Earls of Essex*, Volume II, John Murray, London, 1853.

Freemantle, M., 'Chemistry for water' in *Chemical & Engineering News*, 19 July 2004.

Goldwater, L.J., 'Mercury in the Environment' in *Scientific American*, p. 224, May 1971.

Goldwater, L.J., *Mercury: a History of Quicksilver*, York Press, Baltimore, 1972.

Goodman, J. (ed.), *The Christmas Murders*, Allison & Busby, London, 1986.

Holmes, F., *The Sickly Stuarts*, Sutton Publishing, Stroud, Glos., 2003.

Irwin, M., *That Great Lucifer: a Portrait of Sir Walter Ralegh*, Chatto and Windus, London, 1960. (Somewhat inaccurate.)

McElwee, W., *The Wisest Fool in Christendom*, Faber and Faber, London, 1958.

McElwee, W., *The Murder of Sir Thomas Overbury*, Faber & Faber, London, 1952.

Mitra, S., *Mercury in the Ecosystem*, Trans Tech Publications, Switzerland, 1986.

Rimbault, E.F. (ed.), *The Miscellaneous Works in Prose and Verse of Sir Thomas Overbury, Kt.*, Reeves and Turner, London, 1890. (Also contains notes and a biographical account.)

Rowse, A.L., *Simon Forman*, Weidenfeld and Nicolson, London, 1974.

Rowse, A.L., *The Elizabethan Renaissance: the Life of the Society*, Macmillan, London, 1971.

Smith, W.E. and Smith, A.M., *Minamata*, Chatto & Windus, London, 1975.

Somerset, A., *Unnatural Murder: Poison at the Court of James I*, Weidenfeld & Nicolson, London, 1997.

White, B., *Cast of Ravens: the Strange Case of Sir Thomas Overbury*, John Murray, London, 1965.

Arsenic

Beales, M., *The Hay Poisoner: Herbert Rowse Armstrong*, Robert Hale, London, 1997.

Bentley, R. and Chasteen, T.G., 'Microbial methylation of metalloids: arsenic, antimony, and bismuth' in *Microbiology and Molecular Biology Reviews*, vol. 66, p. 270, 2002.

Bentley, R. and Chasteen, T.G., 'Arsenic curiosa and humanity' in *Chemical Educator*, vol. 7, p. 51, 2002.

Christie, T.L., *Etched in Arsenic*, Harrap, London, 1969.

Gerber, S.M. and Saferstein, R. (eds), *More Chemistry and Crime*, American Chemical Society, Washington DC, 1997.

Gunther, R.T. (ed.), *The Greek Herbal of Dioscorides*, translated by John Goodyear, Oxford University Press, Oxford, 1934.

Heppenstall, R., *Reflections on the Newgate Calendar*, W.H. Allen, London, 1975.

Irving, H.B., *Trial of Mrs Maybrick*, Notable British Trials Series, William Hodge & Co., Edinburgh, 1930.

Islam, F. S. *et al.*, 'Role of metal-reducing bacterial in arsenic release from Bengal delta sediments' in *Nature*, vol. 430, p. 68, 2004.

McConnell, V.A., *Arsenic Under the Elms*, Praeger, Westport, Connecticut, 1999.

Meharg, A., *Venemous Earth*, Macmillan, London, 2005.

Nriagu, J., *Arsenic in the Environment: Human Health and Ecosystems*, John Wiley & Sons Inc., New York, 1994.

Norman, N.C. (ed.), *Chemistry of Arsenic, Antimony and Bismuth*, Thomson Science, London, 1998.

Odel, R., *Exhumation of a Murder*, Harrap, London, 1975.

Przygoda, G., Feldmann, J., and Cullen, W.R., 'The arsenic eaters of Styria: a different picture of people who were chronically exposed to arsenic' in *Applied Organometallic Chemistry*, vol. 15, pp. 457–462, 2001.

Vallee, B.L., Ulmer, D.D. and Wacher, W.E.C., 'Arsenic Toxicology and Biochemistry' in *Archives of Industrial Health*, vol. 58, p. 132, 1960.

Whittington-Egan, R., *The Riddle of Birdhurst Rise*, Harrap, London, 1975.

Antimony

Adam, H.L., *The Trial of George Chapman*, Notable British Trials Series, William Hodge, Edinburgh and London, 1930.

McCormick, D., *The Identity of Jack the Ripper*, 2nd edn, revised, John Long, London, 1970.

Wilson, W., *A Casebook of Murder*, Leslie Frewin, London, 1969.

Farson, D., *Jack the Ripper*, Michael Joseph Ltd., London, 1972.

Jones, E. and Lloyd, J., *The Ripper File*, Arthur Barker, London, 1975.

McCallum, R.I., *Antimony in Medical History*, Pentland Press, Durham, England, 1999.

Roughead, W., *Trial of Dr Pritchard*, Notable British Trials Series, Edinburgh and London, 1925.

Shotyk W. *et al.*, 'Anthropogenic impacts on the biochemistry and cycling of antimony', in A. Sigel, H. Sigel, and R.K.O. Sigel (eds), *Biogeochemistry, Availability, and Transport of Metals in the Environment*, vol. 44, p. 177, Marcel Dekker, New York, 2004.

Shotyk, W. *et al.*, 'Antimony in recent, ombrotrophic peat from Switzerland and Scotland', in *Global BioGeochemical Cycles*, vol. 18, Art. No. GB1017, January 2004.

Sylvia Countess of Limerick CBE, chairman, *Expert Group to Investigate Cot Death Theories: Toxic Gas Hypothesis*, Final Report May 1998, Department of Health, London.

Lead

Baker, G., 'An inquiry concerning the cause of endemial colic of Devonshire' in *Medical Transactions of the Royal College of Physicians*, p. 175, 1772.

Beattie, O. and Geiger, J., *Frozen in Time: Unlocking the Secrets of the Franklin Expedition*, E.P. Dutton, New York, 1988.

Beattie, O., Baadsgaard, H. and Krahn, P., 'Did solder kill Franklin's men?' in *Nature*, vol. 343, p. 319, 1990.

Boulakia, J.D.C., 'Lead in the Roman world', in *American Journal of Archaeology*, vol. 76, p. 139, 1972.

Chisholm Jr., J.J., 'Lead Poisoning' in *Scientific American*, p. 15, February 1971.

Dagg, J.H., Goldberg, A., Lochhead, A. and Smith, J.A., 'The relationship of lead poisoning to acute intermittent porphyria' in *Quarterly Journal of Medicine*, vol. 34, p. 163, 1965.

Gilfillan, S.C., 'Lead poisoning and the fall of Rome', *Journal of Occupational Medicine*, vol. 7, p. 53, 1965.

Griffin, T.B. and Knelson, J.H. (eds), *Lead*, Georg Thieme, Stuttgart, 1975.

Hammond, P.B., 'Lead poisoning; an old problem with new dimensions' in F.R. Blood (ed.), *Essays in Toxicology*, vol. 1, p. 115, Academic Press, New York, 1969.

Hernberg, S., 'Lead poisoning in a historical perspective', *American Journal of Industrial Medicine*, vol. 38, p. 244, 2000.

Macalpine, I. and Hunter, R., *George III and the Mad Business*, Alan Lane, London, 1969.

Martin, R., *Beethoven's Hair*, Bloomsbury, London, 2001.

Nriagu, J.O., *Lead and Lead Poisoning in Antiquity*, Wiley & Sons Ltd., New York, 1983.

Patterson, C.C., 'Lead in the environment' in *Connecticut Medicine*, vol. 35, p. 347, 1971.

Waldron, H.A. and Stofen, D., *Sub-clinical Lead Poisoning*, Academic Press, London, 1974.

Warren, C., *Brush with Death: a Social History of Lead Poisoning*, The Johns Hopkins University Press, Baltimore, Maryland, 2000.

Weiss, D., Shotyk, W. and Kempf, O., 'Archives of atmospheric lead pollution' in *Naturwissenschaften*, vol. 86, p. 262, 1999.

A website that discusses all aspects of leaded gasoline can be found at http://www.ChemCases.com/tel which is produced by the Kennesaw State University, Georgia, USA.

Thallium

Cavanagh, J.B., 'What have we learnt from Graham Frederick Young? Reflections on the mechanism of thallium neurotoxicity' in *Neuropathology and Applied Neurobiology*, vol. 17, p. 3, 1991.

Christie, A., *The Pale Horse*, Collins, London, 1952.

Deeson, E., 'Commonsense and Sir William Crookes' in *New Scientist*, p. 922, 1974.

Holden, A., *The St. Albans Poisoner: the Life and Crimes of Graham Young*, Hodder & Stoughton, London, 1974.

Lee, A.G., *The Chemistry of Thallium*, Elsevier, Barking, Essex, 1971.

Marsh, N., *Final Curtain*, Collins, Toronto, 1948.

Matthews, T.G. and Dubowitz, V., 'Diagnostic mousetrap' in *British Journal of Hospital Medicine*, p. 607, June 1977.

Paul, P., *Murder Under the Microscope*, ch. 21, Macdonald, London, 1990.

Prick, J.J.G., Sillevis-Smitt, W.G., and Muller, L., *Thallium Poisoning*, Elsevier, Amsterdam, 1955.

Sunderman, F.W., 'Diethyldithiocarbamate therapy of thallotoxicosis' in *American Journal of Medical Science*, vol. 253, p. 209, 1967.

Van der Merwe, C.F., 'The treatment of thallium poisoning by Prussian blue' in *South African Medical Journal*, vol. 46, p. 960, 1972.

Young, W., *Obsessive Poisoner: Graham Young*, Robert Hale, London, 1973.

Other Poisonous Elements

Asimov, I., 'Sucker Bait' in *The Martian Way*, Grafton Books, London, 1965.

Baldwin, D.R. and Marshall, W.J., 'Heavy metal poisoning and its laboratory investigation', *Annals of Clinical Biochemistry*, vol. 36, pp. 267–300, 1999.

Brown, S.S. and Kodama, Y. (eds), *Toxicology of Metals*, Ellis Horwood, Chichester, England, 1987.

Cooper, P., *Poisoning by Drugs and Chemicals, Plants and Animals*, 3rd edn, Alchemist Publications, London, 1974.

Hunter, D., *Diseases of Occupations*, 5th edn, Hodder and Stoughton, London, 1976.

Minichino, G., *The Beryllium Murder*, William Morrow & Company, New York, 2000.

Ottoboni, M.A., *The Dose Makes the Poison*, 2nd edn, Van Nostrand Reinhold, New York, 1991.

Simpson, K. (ed.), *Taylor's Principles and Practice of Medical Jurisprudence*, vol. II, 12th edn, Churchill, London, 1965.

Witthaus, R.A., *Manual of Toxicology*, William Wood, New York, 1911.

Glossary

Bennett, H. (ed.), *Concise Chemical and Technical Dictionary*, 3rd edn, Edward Arnold, New York, 1974.

Budavari, S. (ed.), *The Merck Index*, 13th edn, Merck & Co. Inc., Rahway NJ, 2001.

Greenwood, N.N. and Earnshaw, A., *Chemistry of the Elements*, 2nd edn, Butterworth Heinemann, Oxford, 1997.

Hawley, G.G., *The Condensed Chemical Dictionary*, Van Nostrand Reinhold, New York, 1981.

Pearce, J. (ed.), *Gradner's Chemical Synonyms and Trade Names*, 9th edn, Gower Technical Press, Aldershot (UK), 1987.

Sharp, W.A. (ed.), *The Penguin Dictionary of Chemistry*, 3rd edn, Penguin, London, 2003.

Index